# AN INTRODUCTION TO
# TREES
## FOR SOUTH EASTERN AUSTRALIA

# AN INTRODUCTION TO

# TREES

## FOR SOUTH EASTERN

## AUSTRALIA

## K. J. SIMPFENDORFER

M.ScFor., Dip. For. (Vic.), M.I.F.A., F.R.H.S.

INKATA PRESS

Inkata Press Proprietary Limited
Melbourne
First published 1975
Reprinted 1981
(c) K. J. Simpfendorfer
National Library of Australia Card Number and
ISBN 0 909605 03 3
Library of Congress Catalog Card Number 74–79011

Designed by Arthur Stokes
Type set in Monotype Baskerville by
Dudley E. King Pty. Ltd.
Melbourne
Printed in Hong Kong

# CONTENTS

# FOREWORD

Most, if not all, people at some time or another need and appreciate the beauty and peace of mind engendered by trees. Even the early settlers in Australia faced with the effort of clearing land for crops and pastures still appreciated their value. They planted trees around their homes and farms for protection from the weather and to improve the beauty of their surroundings. Much has been learnt from these plantings. While many species proved unsuitable there are also many fine, old trees in parts of south-eastern Australia which originate from this period.

Since World War II there has been a great upsurge in tree planting as more and more people have come to appreciate the utilitarian and aesthetic value of trees in the landscape. Perhaps one of the most interesting aspects has been the rapid increase in the number of Australian native species which have been introduced into cultivation. However, it would be a mistake to conclude that the use of them is a recent development; many species of eucalypts, acacias, native pines, etc. were propagated in the oldest of the Forests Commission nurseries when it was established at Macedon in 1872. Over the past one hundred years new species have been continually introduced and tested under field conditions. Some were immediately successful and were widely accepted within a few years, others required the development of special nursery techniques, while still others were unsuitable for local conditions and were discontinued.

In this publication, descriptions are given of some of the more satisfactory species, both native and introduced. Most of them have been well proved in practice, but to keep the reader up to date the more promising of recent introductions have also been included. It takes many years to assess the potential of a new introduction. Some species fail in the nursery, some are difficult to transplant, while others grow well for a few years and then deteriorate or prove to be unsatisfactory. Many new species, both native and introduced, are tried in the Forests Commission nurseries each year, but only time determines which are successful under cultivation and so are worth propagating.

The information is based on the experience of Forests Commission staff over the past one hundred years and that of the author over the past thirty years. Mr Simpfendorfer has used his detailed knowledge of tree selection and culture to great advantage in this publication. I am sure the reader will gain much of value, and will find that it provides encouragement to many to increase planting as a contribution to the beauty of trees in the landscape.

Dr F. R. Moulds, *Chairman*,
Forests Commission, Victoria.

# Acknowledgements

The author wishes to acknowledge, with thanks, the assistance of the following in the preparation of this publication:

The staff of the National Herbarium, Melbourne for correcting the botanical names, common names, and countries of origin.

Many members of the staff of the Forests Commission, Victoria, for information and assistance readily given.

The Forests Commission, Victoria, and the Public Service Board of Victoria for giving permission for this book to be published as a private venture.

# INTRODUCTION

Primitive Man's greatest discoveries are often stated to be the domestication of fire and the invention of the wheel. An equally great, if not greater, event was the realization that seed was the genesis of a new plant. Once this was understood the change in civilization, from a nomadic tribal culture to the development of settled communities and city states, became possible.

How or when this change occurred is not known. Possibly an observant individual noted that some plants, when damaged or blown down, may regenerate from branchlets embedded in the soil and that these could be moved to another site to grow into a replica of the original. This would have been a relatively easy step, but the discovery that the hard grains left after a flower had withered and died could produce a new plant probably required a good deal of original and thoughtful observation.

It is apparent that this discovery was made in many parts of the world since some form of agriculture has existed for a long time in most continents. By the time of the older civilizations several food plants were under cultivation and the role of the peasant farmer had become established. Plants for decorative purposes, such as those in the Hanging Gardens of Babylon, were being cultivated but the actual number of recognized species was small, even in Greek and Roman times. Most of these were named only because of their use for food, medicinal purposes or witchcraft, and, as it appears from history, for poisoning emperors, political opponents and the next in line of promotion. Very few works of reference on plants from these periods have survived, but of the few, that of the Roman herbalist Dioscorides (first century A.D.) is the most noteworthy.

During the Middle Ages little progress was made. Some plants such as cereal crops, grape vines, a few trees (cedar and yew), some vegetables (notably turnips), and medicinal herbs were the only ones which were widely known. What knowledge there was beyond these was largely retained in monasteries, many of which specialized in a wide range of plants particularly aromatic and perfumed plants. At the end of the Middle Ages and beginning of the Renaissance probably no more than one thousand were identifiable by name.

In 1597 the Englishman John Gerard published his *Herball*. This was one of the first new works to be published for about 1500 years. In it he described the common plants of England but by modern standards his descriptions are not very detailed. Take, for example, this description of the oak:

The common Oke groweth to a great tree; the trunke or body wherof is covered over with a thicke rough barke full of chops or rifts; the armes or boughes are likewise great, dispersing themselves farre abroad: the leaves are bluntly indented about the edges, smooth, and of a shining greene colour, whereon is often found a most sweet dew and somewhat clammie, and also a fungous excrescence, which wee call Oke Apples. The fruit is long, covered with a browne, hard, and tough pilling, set in a rough scaly cup or husk: there is often found upon the body of the tree, and also upon the branches, a certaine kind of long white mosse hanging downe from the same: and sometimes another wooddie plant, which we call Misseltoe, being either an excrescence or outgrowing from the tree it selfe, or of the doung (as it is reported) of a bird that hath eaten a certaine berry.

The Oke doth scarcely refuse any ground; for it groweth in a dry and barren soile, yet doth it prosper better in a fruitful ground; it groweth upon hills and mountaines, and likewise in vallies: it commeth up every where in all parts of England, but it is not so common in other of the South and hot regions.

The Oke doth cast his leaves for the most part about the end of Autumne: some keepe their leaves on, but dry all Winter long, untill they be thrust off by the new Spring.

Acornes if they be eaten are hardly concocted, they yeeld no nourishment to mans body, but that which is grosse, raw, and cold. Swine are fatted herewith, and by feeding thereon have their flesh hard and sound.

Gerard's *Herball* appears to be more a descriptive catalogue of plants with medicinal value since some which are known (from other references) to have existed at the time, such as Scots pine and sycamore, are not listed. On the other hand two introductions from the New World, 'Tabaca' (tobacco) and 'Mays' (maize), are included.

The great upsurge in exploration which commenced in the late fifteenth century led to the discovery of vast numbers of new plants. North America had many plants with parallels in Europe such as pines, firs, spruces, elms, oaks, poplars, willows. A similar, although not so close, relationship extended to the Far East temperate countries—but beyond this all was new. Not only new species but new genera and families had to be described; very few of the Old World names were applicable. All of the tropical regions, South America and Africa abounded with new species but the greatest diversity and variation from the known pattern was found in the Australasian region.

Some of the ferns, grasses and herbs found in Australia by the early botanists could be referred to 'old' genera, but for the flowering plants and the few conifers new genera were needed for most species. The peak of the naming has now passed although new species are still being located and described, not only from the more remote parts of the continent but sometimes from the long settled areas where some observant person has noticed something different.

When the first settlers arrived, clearing the bush was necessary for survival. The pioneers tackled the problem with great energy and found the limits of suitable agricultural country by years of trial and error, and hard work. How many undescribed but typically Australian species disappeared during this phase of settlement is unknown, but doubtless there were some. By about one hundred years ago most of the good natural country had been selected and, with the decline in gold mining, jobs and farms had to be found for unemployed miners. Land settlement schemes were the order of the day and some grandiose projects were undertaken. The large-scale alienation and clearing of land belonging to the Crown stirred the conscience of some people and the first active conservationists began to press their claims. In the prevailing climate of public opinion they were not too popular, but these dedicated people did manage to have some valuable areas reserved even though these reserves were much smaller and fewer than was necessary.

It is cold comfort to know now, one hundred years later, that they were often right; many of the land settlement schemes from the 1870s onwards have not been successful. There have also been some excellent successes, particularly where irrigation water could be provided, but the area and number of failures is still large. It is easy to see success; the failures are reclaimed by Nature and too easily forgotten. One only has to look at old survey plans to see the large amount of country which was surveyed for settlement but which has since reverted to forest.

Not everyone, nor even every settler struggling to clear his block, was concerned solely with getting rid of trees. From the first days of settlement trees have been planted for aesthetic purposes, because of sentimental feelings for 'home', or as shelter. It is due to their successes and failures that much of the knowledge of species suitable for planting has been accumulated and which can be used with profit today.

Since World War II there has been a big increase in planting. Before, people tended to stick with the tried and proven types but now they are more interested in trying a wide range of different species. Probably more new types, mostly Australian, have been propagated in the last thirty years or so, than in the previous one hundred years. Although many may not be conscious of it, these people have been practising conservation in a sound manner since tree planting is essentially conservation and is the basis of conservation. People's interests in conservation vary; they may be interested in a particular animal, bird, plant, soil, water or scenery, but plants are the basis of all conservation objectives. Given the right habitat, animals and birds will thrive; with adequate tree cover, soil and water are safe while scenery loses much of its appeal if there are no trees.

Trees may be planted for conservation ideals, for profit, for amelioration of the local climate, for aesthetic purposes, or simply for the satisfaction of seeing things grow. Whatever may be your reason, if it involves planting that is the essential justification and it is hoped that you will find something of interest and value in this publication.

# EXPLANATIONS

This publication has been prepared primarily for individuals and organizations undertaking plantings on a fairly large scale. It has been written for the assistance of landowners planning farm forests, windbreaks, shade, etc.; for road authorities in roadside plantings; for municipalities in the planning of parks, playgrounds and other public amenities; for community organizations establishing golf courses and similar recreational facilities; for school planting projects; and for many others.

However, this does not mean that it cannot be used by the town or city dweller. The information given is just as applicable to the planting of a single tree as to thousands; it is only a matter of scale. Whether you are embarking on a large project or just adding a tree or two to a suburban garden, the information is equally applicable and, it is hoped, equally interesting and useful.

It has been written with a minimum of technical terms. Everyday words have been used in the everyday sense and, while to the scientifically disposed some loss of precision in meaning may have resulted, the advantages of plain words is believed to more than justify the lack of phraseology and terminology understood only by those with technical training.

Part I is essentially background which aims to give an understanding of trees and their requirements. It is rather general, so the reader should be aware that while exceptions are inevitable, under average circumstances it is unlikely that they will be significant. The pattern followed in each section has been to give the theoretical basis and to relate this to the practical aspect, while providing reliable answers to some of the questions most frequently asked by interested people. Not every question may be answered but the information given is that which has been shown to be of most interest over some thirty years experience in forest extension work.

Part II comprises descriptions of trees suitable for planting in south-eastern Australia. Virtually every plant has a functional or amenity value under certain circumstances. In the selection of the species most suitable for amenity plantings in country districts and towns, the following guidelines have been used:

1 They must be satisfactory for some form of amenity use such as farm forests, windbreaks, shade, parks, roadside, etc.

2 Under reasonable conditions, they should grow to at least 3 metres in the medium to high rainfall districts and 2 metres in the lower rainfall areas.

3 They must be relatively easy to establish in quantity in the field.
Plants which are essentially garden shrubs have not been included.

## HOW TO USE THIS BOOK

At the end of the book, tables of characteristics (pages 278–312) and of uses (pages 313–347) are given. Should you wish, for example, to select trees for a particular rainfall, soil type, etc., reference to the tables of characteristics (pages 278–312) will show the suitable species and reference to the selected species in Part II—Species Descriptions (pages 125–261) will give further details. Similarly, if you want trees for a windbreak, park, shade, roadside planting, etc., you will find them tabulated in the tables of uses (pages 313–347).

# PART I

# BACKGROUND

# OF THEORY

# EVOLUTION

Once upon a time there was no planet Earth. How it came to be formed is beyond the purpose of this book but the event is estimated to have taken place some 4500 to 6000 million years ago. Our interest begins about 4500 to 5000 million years ago when a solid Earth had been formed. At this stage its atmosphere is believed to have comprised hydrogen, nitrogen, methane, ammonia, water vapour, and possibly a little carbon dioxide. Except for that bound up in water vapour, oxygen in quantity was conspicuously absent. Most of the atmosphere probably originated within the Earth, escaping to the surface by fissures, volcanoes, etc., and due to the physical turbulence of the early history of the Earth, there was an abundance of assorted gases.

The atmosphere was not as we know it now. Ultra-violet light from the Sun split the water into hydrogen and oxygen; the hydrogen probably escaped into space and the oxygen reacted with methane, ammonia, etc. to form more elaborate substances. In time, this physico-chemical process gave a sea rich in organic compounds. At some time and under appropriate conditions the necessary molecules accidentally came together and amino acids, the essentials of protein, were formed. But how nucleic acid, the other essential of life, was first formed has not yet been explained.

The greatest event in evolution, the beginning of life, then occurred. The problems of understanding life has always puzzled philosophers and has led them to deduce the vitalistic theory, embodying the spirit of life concept, contrasted with the mechanistic approach which regards life as probably arising from the accidental aggregation of the right molecules. Whichever is correct, life emerged in a very simple and primitive form. Numerous forms probably occurred but one which was better fitted to contend with the environment must have become dominant to the point of excluding all others since every living thing today—animal or plant—has the same basic manner of cell functioning.

The only food for these simple organisms was the organic matter formed in the sea under the influence of ultra-violet light. As the original elements and simple molecules were elaborated into more complex organic compounds and as there was no form of life to undertake decay, and so recycle the elements, it is probable that food supplies decreased and possibly even became limiting. The breakdown of compounds for food released carbon dioxide and at this stage some forms acquired the ability to manufacture food from the steadily increasing supply of carbon dioxide and other elements available by using the energy of sunlight. This may be regarded as the second major step in the evolution of life. It involved a change from a 'consumer' type of society, and therefore one with a limited existence, to a 'producing' or 'manufacturing' type, that is one with a continuing potential.

The 'manufacturing' process is believed to have involved the breakdown of carbon dioxide into carbon and oxygen. Most of the oxygen was probably released into the atmosphere but some, together with the carbon, reacted with the methane, ammonia and hydrocarbons to form more complex compounds. By these processes the atmosphere was gradually 'purified' and eventually converted to one of predominantly oxygen and nitrogen similar to that of today. Little more is known about the stages of this process but it prepared the way for the first plants.

These early plants were probably very simple photosynthetic organisms. It is not known for how long they existed as, being soft-bodied, they left very little fossil record. The oldest known specimen, a single bacterium found in South Africa in 1969, is estimated to be about 3500 million years old. It is probable that life was well established in the Archaeozoic period, ending about 2500 million years ago, since limestone deposits which are mainly calcium carbonate are known from this age; that is, enough oxygen had been released by the living organisms to saturate the sea with carbonates and cause the deposition of limestone (calcium carbonate). (Carbonates comprise carbon and oxygen.) A sea saturated with calcium carbonate also opened the way for the development of more rigid plant and animal structures as calcium is necessary for the formation of cell walls, bones and shells. Probably many organisms evolved during the next period, the Proterozoic period, as at the beginning of the Cambrian period which followed it (about 600 million years ago) a large number of hard-bodied fossils of surprising diversity and complexity suddenly appear in the geological record. Presumably the abundant supply of calcium carbonate in the sea made the transition from soft-bodied to hard-bodied possible. Many were plants but some were animals and, since animals are consumers of oxygen, their development must have followed rather than preceded that of plants.

Evolution is related to the geological time scale and the following periods have been used in describing the development of succeeding stages:

|  |  |  | *Duration of period (millions of years ago)* |
|---|---|---|---|
| Archaeozoic |  |  | more than 2500 |
| Proterozoic |  |  | 2500–600 |
| Palaeozoic | Cambrian |  | 600–500 |
|  | Ordovician |  | 500–440 |
|  | Silurian |  | 440–400 |
|  | Devonian |  | 400–350 |
|  | Carboniferous |  | 350–270 |
|  | Permian |  | 270–225 |
| Mesozoic | Triassic |  | 225–180 |
|  | Jurassic |  | 180–135 |
|  | Cretaceous |  | 135–70 |
| Cainozoic | Tertiary | Palaeocene | 70–60 |
|  |  | Eocene | 60–40 |
|  |  | Oligocene | 40–25 |
|  |  | Miocene | 25–7 |
|  |  | Pliocene | 7–2 |
|  | Quaternary | Pleistocene | 2–0·01 |
|  |  | Recent (Holocene) | 0·01 to present |

The third major evolutionary step was the emergence during the Cambrian period of the first plants containing chlorophyll. Chlorophyll is the substance which enables a plant to use the energy of sunlight, carbon dioxide and other nutrients to synthesize sugars, and release oxygen in the process. Concurrently the first fungi evolved but, as they lacked chlorophyll, they could only obtain their food by breaking down more elaborate substances, releasing carbon dioxide in the process for reuse by plants. Thus the cycle of growth, decay and reuse, the foundation of a continuing form of life, was established.

No spectacular changes seem to have occurred during the following Ordovician and Silurian periods but the early Devonian (some authorities say late Silurian) period saw the fourth major step, the appearance of the first land plants. At this stage there was no soil as we know it today. There were no terrestrial algae, lichens, etc. to break down the rocks; the best materials available would have been pockets of fine materials produced by chemical or mechanical weathering. On the other hand its mineral content and so nutrient supply would have been very good. Possibly small areas of eroded material deposited in the sea were uplifted and provided an anchorage for plants.

The first plants probably lived in the tidal zone since two significant changes were necessary for a life on land:

(a) *The development of a root system.* True sea plants, for example seaweeds, do not have a root system but anchor themselves to rocks by means of a structure known as a holdfast and absorb water and nutrients from the surrounding water through their 'foliage'.

(b) *The development of rigidity.* In the sea some support is obtained from the water but on land a thickening of the cell wall and the development of special tissues to enable the plant to remain upright were required.

Some of the oldest known land plants, *Baragwanathia* and *Yarravia*, were first described in 1935 from specimens found on Mt Pleasant, north east of Melbourne, in a cutting on the old road from Thornton to Alexandra. The sediments were formerly regarded as Silurian but are now more often referred to as Lower Devonian. Land plants are believed to have originated in the Southern Hemisphere, probably in the Australian region, and migrated northwards, the migration reaching a peak during the Jurassic period.

Land plants were well established by the late Devonian period. Horsetails, club-mosses and large fern-like plants were the dominant ground flora while the rather curious *Cordaitales*, the extinct ancestor of the present day conifers, first appeared. Upright and rigid stems were developed although some of the horsetails were jointed and flexible. True forests probably did not exist as distribution appears to have been patchy with the tallest plant only reaching a height of about 6 metres. They did not have true leaves; probably they were like the present-day *Psilotum nudum*, Whisk Fern, a very primitive leafless plant from Florida, about 30 to 50 centimetres in height, which has its chlorophyll in the stems and branchlets.

True leaves first appeared in the late Devonian to early Carboniferous periods on the *Filicophyta* or the progenitors of today's ferns. Appearance of true leaves was accompanied in the Carboniferous period by an increase in plant height and in the luxuriance of the foliage. Club-mosses and horsetails reached 25 metres and fern-like plants were 15 metres tall. Their remains form the large Carboniferous coal deposits of the Northern Hemisphere. Despite their abundance and dominance in this era, relatively few species survive today; most of the survivors are small plants only a few centimetres tall although a few species may reach 1 metre.

The other major event of significance in the Carboniferous period was the appearance of the first seed plants. Up to this stage plants had alternation of generations, a three phase form of reproduction. The plant (sporophyte) produced spores which grew a prothallus carrying male and female organs (gametophyte), with the resulting fertilized cell giving rise to another plant. (The same reproductive process is present today in ferns, mosses, and liverworts.) Seed plants have a two stage reproduction in which the parent plant produces seeds which give rise directly to another plant, thereby eliminating the prothallus stage. Seeds have an advantage over spores in that they contain the embryo plant together with a reserve food supply which aids germination and initial growth. The seed plants rapidly became the dominant flora. The first seed plants were the *Gymnospermae* represented today by the conifers and cycads. (The term 'gymnosperm' means naked seed since the seed is produced exposed on a cone scale or similar structure.)

For the next 120 to 130 million years, during the Permian, Triassic and Jurassic periods there was no great leap forward in evolution. It seems to have been a period of consolidation and development rather than the introduction of new concepts. The conifer-like species increased in size reaching 50 metres in height. Probably the most notable event was the disappearance, for no apparent reason, of the formerly widespread and abundant fern-like *Glossopteris* species and their replacement by true ferns.

The Cretaceous period beginning about 135 million years ago was the most significant in the development of modern flora. Except for the conifers and ferns, practically all the trees and shrubs grown today originated or evolved from this period. It is notable for four main features. Firstly, the *Angiospermae*, meaning vessel seeds (in which the seed is produced in an enclosed chamber or ovary), were evolved. Secondly, the flowering plants appeared in the early Cretaceous. Thirdly, there was an almost immediate and spectacular increase in the number and proliferation of species. Finally, as far as distribution and evolution of flora in Australia is concerned, the Great Flood occurred in mid-Cretaceous (see pages 13–14).

The 70 million years since the close of the Cretaceous period has been the era of the angiosperms. No major step in evolution, such as the appearance of chlorophyll or seed, occurred but the angiosperms consolidated their position increasing in number and diversity of species. Most, but by no means all, of the modern genera seem to have evolved only in the last 20 million years or so. Accompanying this there was a trend towards simpler and more efficient structures as they adapted to life around them, for example, a decrease in the number of stamens as the efficiency of insect pollination improved.

A summary of the main steps in plant evolution is given in Table I (7). In brief, evolution took place in stages; as a new adaption proved superior to the old it flourished for a time only to be replaced in turn by a further improvement. It was more a step by step process rather than a steady and consistent rate of change.

Table I. Evolutionary Stages of Plants

| First appearance | | Main characteristics | Present day representatives | Approx. number of species today |
|---|---|---|---|---|
| Era | Millions of years ago | | | |
| Archaeozoic | origin to 2500 | First living organism | | |
| Proterozoic | 2500–600 | Development of organisms capable of using the energy of sunlight and of producing oxygen, i.e. the first plants | Algae-like bacteria | 5000 |
| Cambrian | 600–500 | First plants containing chlorophyll and capable of photosynthesis as it is understood today | Red algae ⎫<br>Brown algae ⎬<br>Green algae ⎭ | 75 000 |
| | | | Mosses ⎫<br>Liverworts ⎭ | 20 000 |
| | | | Fungi | 70 000 |
| Early Devonian | 400 | Plants first established on land | | |
| Devonian | 400–350 | Upright plants with more or less rigid stems. Roots develop—with differentiation into stems and roots (*Psilophyta*) | Club-mosses | 250 |
| Late Devonian | 350 | True leaves appear (*Filicophyta*) | Ferns | 10 000 |
| Carboniferous | 350–270 | Seed plants appear (*Gymnospermae*) | Conifers ⎫<br>Cycads ⎭ | 500 |
| Cretaceous | 135–70 | Flowering plants emerge (*Angiospermae*)<br>(a) monocotyledons | Grasses ⎫<br>Palms ⎬<br>Lilies<br>Orchids ⎭ | 50 000 |
| | | (b) dicotyledons | Other flowering plants ⎫⎬⎭ | 200 000 |

## DERIVATION OF MODERN GENERA

Plants preceded animals by hundreds of millions of years and, like people, each individual has a different genetic make-up. Obviously the one most suitable for the conditions will stand the best chance of survival, and pass its favourable characteristics to its offspring. Survival of the fittest has been the basis of evolution since the beginning of life on Earth. Many water plants had developed from the first simple primeval cell by the time the move to land took place. Most modern trees are derived from only two groups of the numerous species which appeared on land in the Devonian period—the conifers from the *Cordaitales*, and the flowering trees probably from the ancestors of the present-day order *Ranales*. The former has long been extinct while the latter, very widespread in ancient times, is now confined to relatively few species.

A diagram of the appearances of modern genera is given in Figure 1. The appearance period given is the epoch in which the fossil remains can be related to a particular genera. There are often older fossils which look like a modern species and may be related to it, but identification with some degree of certainty is not possible. Moreover the period shown is based on the known

Figure 1. First appearance of some modern genera.

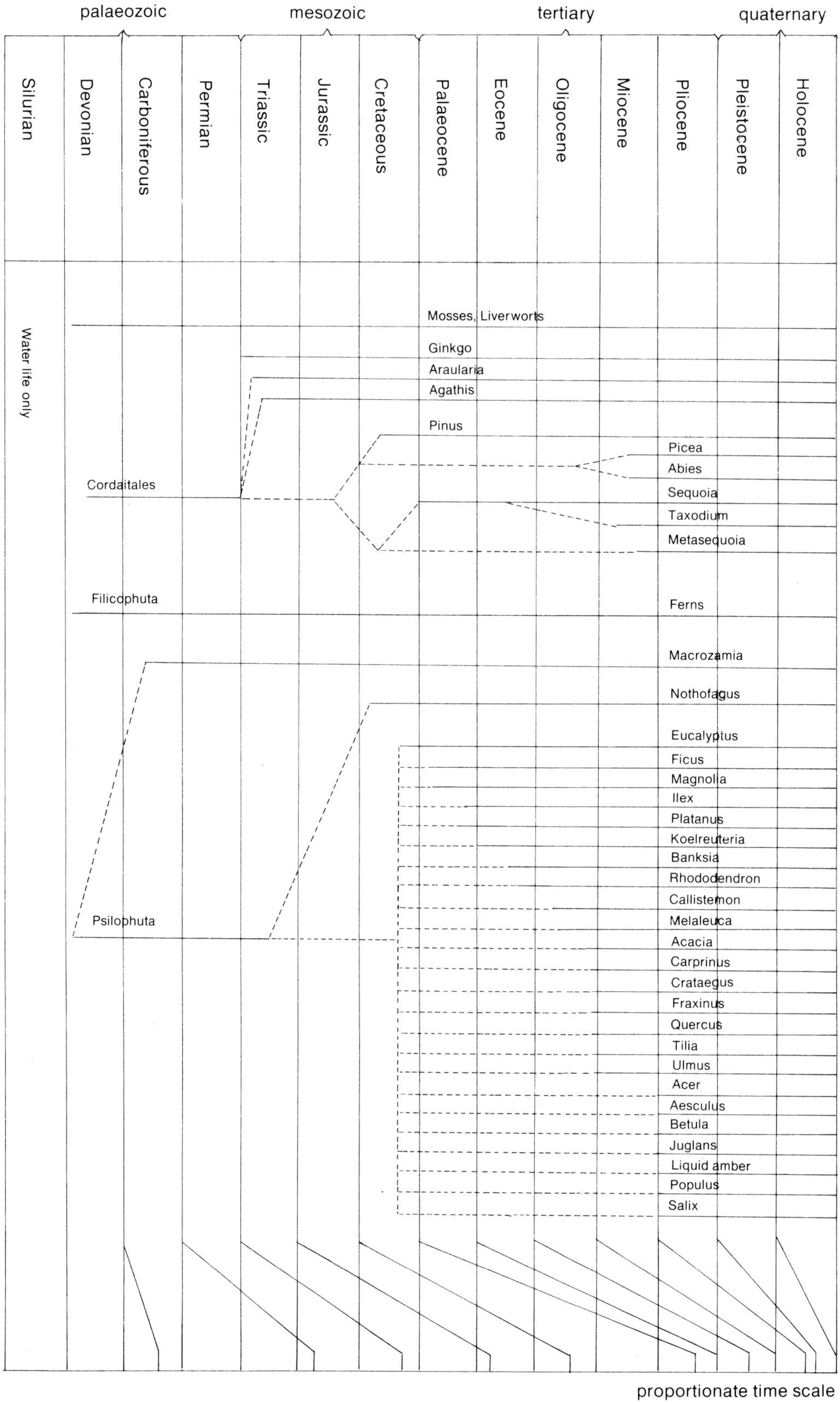

palaeozoic    mesozoic    tertiary    quaternary

Silurian | Devonian | Carboniferous | Permian | Triassic | Jurassic | Cretaceous | Palaeocene | Eocene | Oligocene | Miocene | Pliocene | Pleistocene | Holocene

Water life only

Mosses, Liverworts

Ginkgo

Araularia

Agathis

Pinus

Picea

Abies

Sequoia

Taxodium

Metasequoia

Cordaitales

Filicophuta

Ferns

Macrozamia

Nothofagus

Eucalyptus

Ficus

Magnolia

Ilex

Platanus

Koelreuteria

Banksia

Rhododendron

Callistemon

Melaleuca

Acacia

Carprinus

Crataegus

Fraxinus

Quercus

Tilia

Ulmus

Acer

Aesculus

Betula

Juglans

Liquid amber

Populus

Salix

Psilophuta

proportionate time scale

8

oldest specimen; in the future still older ones may be found, showing the origin to be at an earlier period.

It is important to appreciate that evolution is not a straight linear progression. Several branches may have arisen from the original ancestral type, followed by repeated sub-branching during geologic time. Each small branch may have continued its own development and evolution to the present day or may have died out and disappeared. Consequently forms existing today may have comparatively little in common with their progenitor or with each other even though they came from the same ancestral source.

There is a considerable element of reasonable supposition, rather than proof, in deriving an evolutionary line or genealogy (the science of phylogeny). A few probable connections are shown in Figure 1. The conifer record is fairly clear. They preceded the flowering plants by about 150 million years, and have survived with relatively little change for some 250 million years, so a more continuous record is available. Their length of existence and comparative abundance make them one of the most successful groups evolved to date.

Although flowering plants are of much more recent origin, the record of succession is vague and incomplete by comparison. Unlike the harder cones of conifers, the flowers and fruits are usually soft so their preservation as fossils is not so good. Much of the information available is based on the examination of pollen found in swamps, peat, etc. and relating it to the pollen of modern species. Even though this has been a profitable line of investigation, the line of development of the flowering plants is not as clear as may be hoped. It can be surmised that plants classified into the same family today, for example,—*Eucalyptus*, *Callistemon*, and *Melaleuca* are all in Family *Myrtaceae*—are branches from a common ancestor, and it is probably generally correct, but the element of doubt remains. Evolution can work both ways: (a) plants with a common ancestor can diverge markedly (divergent evolution) because, for example, the environment changes over a long period or (b) species with different ancestors may converge (convergent evolution) because they are in an environment suitable to both of them. Therefore present-day similarities do not necessarily indicate a common ancestor, nor do differences necessarily mean a different origin.

## DISTRIBUTION THEORIES

Very few plants are found in all countries even though climate, soil, etc., may be similar and suitable; each area or region has its own characteristic flora. On the other hand, if land plants originated from common ancestral stock, how did they manage to spread to all continents? Yet even among the flora of areas separated by oceans there are numerous examples of similarities. For example, the genera *Nothofagus* (southern beeches) and *Araucaria* (Southern Hemisphere conifers) are found in Australia, New Zealand and South America. The family *Proteaceae* which is abundant in South Africa is also well represented in Australia and, to a lesser extent, in New Zealand and South America.

In the past world of plants there are also many puzzling cases. The small *Glossopteris* 'fern', abundant in the Carboniferous and Permian periods, appears almost simultaneously in the fossil records of Australia, New Zealand, South

Africa, India, South America and Antarctica. Moreover coal seams are formed from plants growing in temperate to subtropical climates, yet large coal deposits are found in Antarctica.

When one considers the possible number of variations which could arise in a group of plants, it is difficult to believe that the same variation occurred simultaneously in several different land masses and that a similar evolutionary pattern followed, for example, *Nothofagus* could evolve in three different land masses separated for a long time by ocean.

Many theories have been advanced to explain these and many other apparent inconsistencies. Of these, the more durable ones are:

*Land Bridges*   This postulates the existence of land bridges to enable migration of plants, for example, between Australia and South America. Certainly land bridges have existed and functioned in many regions, but there are other regions where geological history does not support this theory.

*Permanence of Continents*   This theory visualizes that the area of the continents and oceans has always been the same. As a continent subsided beneath the sea a compensating movement saw the emergence of a land mass which enabled plants to migrate until they reached the areas where they occur today. Again plant distribution and geological history only partly support this argument.

*Pendulum Theory*   This postulates that the Earth, as well as having the north-south axis around which it rotates, also has an east-west axis passing through Ecuador and Sumatra around which it has oscillated several times during its history; each oscillation would have taken tens of millions of years. Hence the Arctic and Antarctic would, at some time, have been in the tropics and the forests which formed the coal deposits could have flourished. Although it explains some cases of plant distribution, too many are known which are inconsistent with it.

*Polar Origin*   The Polar Origin theory postulates that only two centres of plant origin occurred, one at each of the poles, from which dispersion took place radially towards the equator via land existing at the time. Detailed research of the last few decades has shown that there are several centres of origin within more temperate zones, so this theory is not completely tenable.

*Continental Drift*   This theory postulates that originally there was only one major land mass and one ocean. Fractures or rift lines developed and various parts of the land mass separated and drifted apart to form new continents. This theory is now generally accepted as the most plausible explanation and as such is worth looking at in greater detail.

The Theory of Continental Drift was first proposed in detail by Wegener in 1915. It fell into disrepute between the 1920s and the early 1960s but investigations in plant distribution and the earth sciences, particularly the theory of plate tectonics, have led to its general acceptance.

On looking at a map of the world it is apparent that some of the continents fit together in a general way, for example, South America fits into the west of Africa and North America, with a slight adjustment, against Europe. The Theory of Continental Drift postulates that in the past the continents were joined

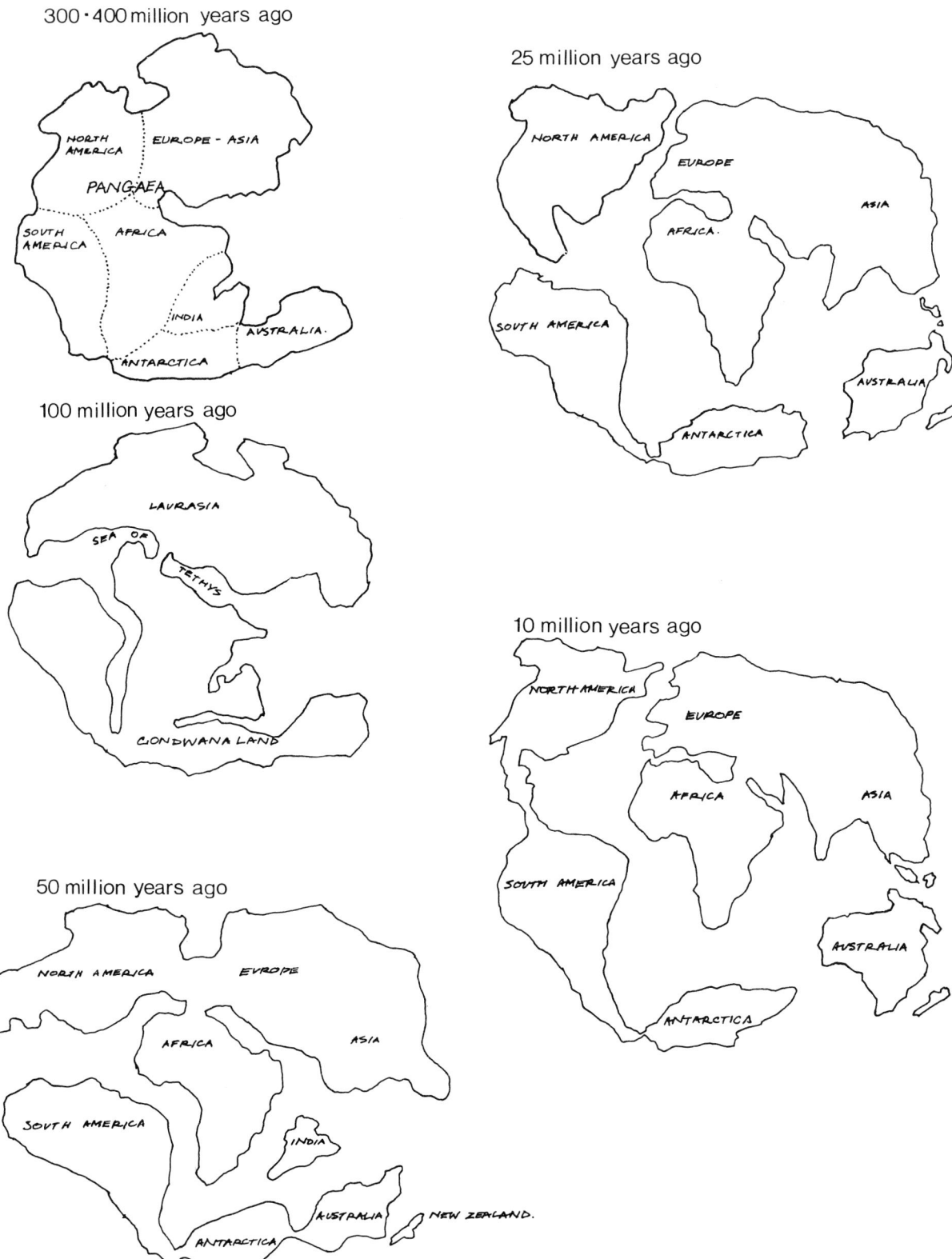

Figure 2. Simplified illustration of Continental drift. Compiled from various sources basically Mollweide Projection.

into one land mass and have drifted apart, but the lack of any geological explanation of how this could occur reduced the plausibility of the theory until the principles of plate tectonics were established in the mid 1960s.

Plate tectonics suggest that the Earth's crust is not solid and continuous but comprises rigid and distinct plates, about 40 to 80 kilometres thick, floating on the Earth's molten interior. About ten major and ten minor plates are recognized. Earthquakes and volcanic activity occur along the margins as plates slide past or over each other at the geologically rapid speed of 1 to 6 centimetres per year. Upwelling of lava at the plate margins tends to force plates apart which implies that the Earth may be expanding rather than shrinking as traditionally thought. (Equipment to check this was left on the Moon by Apollo 15.)

In the late Palaeozoic (300 to 400 million years ago), all the present continents formed one land mass, Pangaea ('all lands'). An east-west fracture developed forming two super-continents Laurasia (Europe, Asia, and North America) and Gondwanaland (South America, Africa, India, and Australia) with the Sea of Tethys in between, and only linked in the present Middle East region (Figure 2).

In the Northern Hemisphere (Laurasia), North America separated from Europe, while the Eurasian land mass rotated about 30° clockwise to bring Japan from a polar latitude to its present location.

In the Southern Hemisphere changes were much more extensive. The first separation was in the mid-Cretaceous (110 million years ago) when South America split off the west side of Africa. Shortly afterwards, about 20 million years later, separation occurred between Africa and the rest of Gondwanaland, leaving the latter as an L-shaped pre-continent joining South America, Antarctica (via Cape Horn) and Australia. About 10 million years later (80 million years ago) New Zealand split off on the eastern side and drifted eastwards. The early separation of Africa and New Zealand from the general land mass in the early- to mid-Cretaceous period, when flowering plants first appeared, has given them a longer period of isolation leading to a more distinctive flora being evolved in each country.

Approximately 65 million years ago India split off the east coast of Africa and drifted north east eventually colliding with the Asian continent about 30 million years ago. Continuing pressure led finally to the crumpling up of the Himalayas, mainly in the last 2 million years. In its journey across the tropics the previous typical southern flora was presumably lost in the hotter environment and replaced by Asian forms when land contact was again made. This gives an explanation of why the earlier fossil flora of India is typically Southern Hemisphere but its present-day flora is distinctively Asian.

Australia separated in the Eocene (45 million years ago) and has drifted northerly about 15° of latitude while Antarctica continued its southerly drift also for about 15° latitude. Antarctica and South America remained joined via Cape Horn until less than one million years ago.

At the time Australia, New Zealand, South America and Antarctica were in contact; the pre-continent was in the temperate zone (about 45° to 50° latitude). This caused the Antarctic coal deposits to be formed and the ancestral *Nothofagus*, which had appeared by mid-Cretaceous, to migrate to each country before they drifted apart. Continental Drift also explains why the fossil

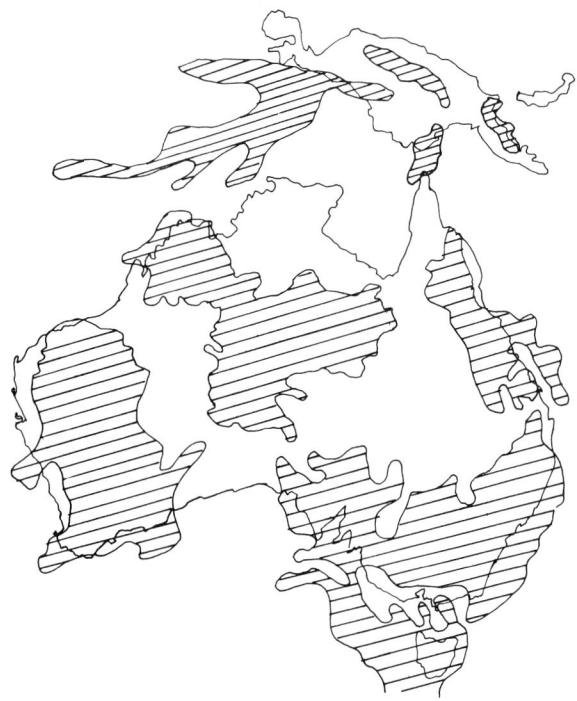

Figure 3. Probable land area (hatched) during Mid-Cretaceous times (after Laseron and Brunnschweiler).

fern *Glossopteris*, for example, has a world-wide distribution; it first appeared in the Carboniferous-Permian period when there was only one continent.

However, it needs to be remembered that concurrently the normal geological activities of subsidence, emergence, folding, faulting, vulcanism, ice ages, etc. would have been superimposed on continental drift. These have all had their effect and they become more significant the nearer we approach to modern times.

Although it has nothing to do with trees, it is interesting to speculate on the likely course of events over the next 50 to 100 million years. The evidence suggests that Africa will split lengthwise along the Great Rift Valley, the eastern part will drift into the Indian Ocean and the western part continue its northerly drift finally eliminating the Mediterranean Sea. South America will continue moving northward, and eventually fill the Gulf of Mexico. India is now drifting eastward, and Australia will continue its northerly drift wedging Indonesia against Asia and finally becoming a part of south-eastern Asia. The combined eastern thrust of India and northern thrust of Australia will cause major mountainous areas to arise in south-eastern Asia and completely change its topography.

*The Great Flood* Unlike continental drift which affected plant distribution on a world-wide scale, the Great Flood of the Cretaceous was purely Australasian in extent. But it had a decisive effect on the evolution and distribution of native plants as we know them today.

More than half of the present continent of Australia was covered by sea. The land was reduced to four large islands with a big inlet covering most of southern Victoria and the adjoining parts of Bass Strait. A fifth island was located north west of Darwin while nearly all of New Guinea was under water. (Figure 3.)

This happened in mid-Cretaceous times (about 110 million years ago) and lasted for 10 to 20 million years. By this time the great proliferation of angiosperms (flowering plants) in the early Cretaceous (110 to 135 million years ago) had begun, so the Flood effectively separated the then existing flora into four main island groups. With the rapid development characteristic of the whole period, each island group tended to evolve in its own direction. While there has been some intermingling of species since then, even today each island area has its own characteristic flora. By reference to Figure 4, the development of the four main flora associations (viz. south west, south east, north east, and central north west) can be appreciated.

A long period of geological stability followed in which the climate of 'Australia', including the inland, was uniformly warm and moist over the whole continent.

During the Miocene the present Nullarbor Plain and lower Murray regions were submerged which split the southern Australian flora into three groups (viz. south west, Adelaide area, and south east). Limestone deposits of great depth were deposited and on re-emergence the flora which had developed on acid soils did not readily invade the adjoining limestone areas. Subsequently some species did adapt but the discontinuities are still recognizable.

## LAST MILLION YEARS

Except for the formation of the Kosciusko plateau and the Eastern Highlands about 5 to 7 million years ago during the Pliocene, conditions were stable until the beginning of the Pleistocene. Geologically there were not many changes but the last million years was a period of violent fluctuations in climate. It apparently varied several times between warm and cool, wet and dry, and on more than one occasion was warm and moist over most of the continent. During the adverse periods, the inland sand dune systems were formed. These dunes, with the periodic changes in climate and coastal flooding between the Ice Ages, eventually broke up the continuous distribution of the vegetation. Isolated remnants and pockets, such as Palm Valley in central Australia which is 800 kilometres from the nearest occurrence of the same species, originated in this period.

*Ice Ages* Four Ice Ages are generally recognized; the first about 500 000 years ago and the last as recently as 10 000 years ago. The last Ice Age was mainly in the Northern Hemisphere where northern USA, Canada, and northern Europe-Asia were covered by ice to a depth of about 3 kilometres. In Australia, ice was very limited, being confined to Tasmania and 800 to 1000 square kilometres in the Kosciusko region. At its maximum it was only about 30 metres in depth.

In the Northern Hemisphere it had a devastating effect on plants and their distribution. Plants in Asia and North America were able to migrate southwards to warmer climates but in Europe their migration was halted by the Mediterranean Sea. Only those able to tolerate the cooler climate survived, hence the number of species found in Europe is small compared with North America and Asia.

In Australia the effect of the ice itself was negligible, but the formation of ice sheets over the land locked up such enormous quantities of water that the sea dropped to about 85 metres below its present level. Many of the shallow seas around Australia became dry land; Tasmania, New Guinea and many coastal islands were joined to the mainland while the Gulf of Carpentaria and Arafura Sea were above sea level. To the north west much of the sea from Malaysia through Indonesia to New Guinea became dry land with the deeper straits narrowed in width and extent. Plant and animal migration was made easier and, even though it happened only 10 000 years ago, intermingling of plants has clearly occurred.

Melting of the ice at the end of the last Ice Age raised the sea level forming Bass and Torres Straits as well as numerous coastal islands. The evidence suggests that melting of ice sheets is still continuing and, if the polar ice caps do disappear, the sea level would rise by another 30 to 40 metres which would have a considerable effect on most coastal towns and cities.

*Arid Period*   The last great event affecting the present distribution of vegetation over southern Australia was the Arid Period of only 5000 years ago. For some unknown reason the climate suddenly became drier which led to widespread arid conditions and probably (as some evidence shows) warmer conditions.

This was not just a severe drought. Vegetation succumbed over large parts of southern Australia while the sandy soils of areas such as south-eastern South Australia, the Wimmera and Mallee regions became mobile. Many new sand dunes were formed and old ones (formed during the Pleistocene period) were rejuvenated by the above-normal winds which seem to have accompanied the onset of aridity.

How long the Arid Period lasted is a matter of speculation. It was certainly a number of years, possibly even several centuries. Plants could only survive in the cooler, moister pockets where a range of habitats existed. The best example is the Grampians of western Victoria, while other isolated mountainous regions such as Mt Buffalo in Victoria, the Flinders Ranges in South Australia and the Stirling Range in Western Australia were also significant plant refuges. It is probable that many of the unusual plants now found in these localities were much more widespread before the Arid Period. How many species became extinct is not known but it is a fair guess that there could have been quite a number.

The end of the Arid Period was not sharply defined but rainfall gradually increased to present-day levels. Species with the most favourable means of seed dispersion were the first to re-colonize the bare land and largely formed the vegetation associations present when the first white settlers arrived. Plants with a less efficient means of dispersal did not spread so far or fast and still tend to be concentrated in the refuge areas of the Arid Period.

## WALLACE'S LINE

In 1876, Wallace divided the world into six major biogeographical regions and noted that the boundary between the Malayan and Australasian realms was unusually distinct. In fact it was so sharp that he was able to describe it as starting to the south east of the Philippines, passing between Borneo and Celebes, and southwards between Lombok and Bali, two islands which are only 24 kilometres apart. This boundary has become known as Wallace's Line and is widely accepted, although it is now realized that there is some intermingling of species on either side.

The presence of such a sharp boundary is interesting in a region where the migration of plants would have been assisted by land bridges, particularly during the last Ice Age. It can be partly explained by geological history. If we accept some blurring of species on each side of the Line, its location corresponds approximately with an ocean trench 1800 to 3600 metres deep running just east of the Celebes and southwards to the east of Timor. It is a narrow strip which has not been above the sea since Cretaceous times (about 100 million years ago) and has acted as a barrier to migrations, although not an insuperable one. A strip of water is not such a barrier to plant and animal migration as is often thought. For example, the 1883 eruption of Krakatoa completely destroyed all life on the island yet within fifty years it was revegetated with a young forest, and 1200 forms of animal and plant life were present, even though the nearest land was 40 kilometres away.

## PRESENT-DAY FLORA

The outcome of all this activity is that present-day Australian flora comprises three main floristic elements.

*Indo-Malaysian* This is typical northern New South Wales and Queensland rain forest. Large numbers of species occur; up to one hundred tree species per hectare including conifers (*Araucaria*), vines (lianes) and epiphytes. Buttressed tree bases and large leaves are usual. It exists in discontinuous blocks from Darwin across the north, reaching its peak development in the Queensland rain forests and extending down the east coast to eastern Victoria and Wilsons Promontory.

*Antarctic* Typical cool high rainfall, mountain forests which are characterized by *Nothofagus* species. Species are very few per hectare, understorey species are sparse, and vines are absent. Epiphytes, mosses, and ferns occur on the tree trunks. It reaches its peak development in the myrtle beech forests of Tasmania but extends to Victoria and up the east coast to Queensland and New Guinea. It is also common in New Zealand, Chile and southern Argentine.

*Australian* Eucalypt species predominate. Few species occur per hectare, understorey varies from herbs to heavy undergrowth and except in high-rainfall mountain areas, formation is rather open. Very broadly this group can be sub-divided into east and west, i.e. that occurring in the eastern states and that in the west. Minor variations occur such as the native pine, casuarina, etc. areas but all of these still belong to the typical Australian element.

# TAXONOMY

## HISTORY

The beginnings of plant classification, usually referred to as systematics or taxonomy, are lost in antiquity. Primitive tribes recognized the value of certain plants for weapons, medicinal, or ceremonial purposes and gave them names. How they acquired these names would be a very interesting study in social development and behaviour as they are a reflection of the way the plant affected the development and well-being of the tribe.

Names occur in the ancient writings from Assyria, Egypt and China as well as in the Bible but the earliest descriptive records of any consequence are survivals from the Greek and Roman eras. The first classifications were on the basis of size, i.e. trees, shrubs, undergrowth, grasses; the concept of natural relationships between plants did not arise until much later. Most of the works were by medical practitioners and listed mainly plants of medicinal interest. The number of species recognized was very small; Theophrastus (370–287 B.C.) listed only 500 species but Pliny the Elder (A.D. 23–79) had increased the named species to nearly 1000 by including timber trees, ornamentals and so on. Many of the scientific plant names used today are actually the old Latin common names for example, *Betula* (birch), *Acer* (maple), *Quercus* (oak), and *Ulmus* (elm).

Little progress on classification was made during the Middle Ages when the herbalists were about the only people who were interested in plants. The herbals of the Greek Dioscorides (first century) remained the standard work of reference until 1597 when John Gerard published his *Herball*. During these sixteen centuries, the cultivation of unusual plants was almost entirely confined to monasteries where fruits, vegetables, and medicinal and perfumed herbs were the main interest. However, as with most arts and sciences, renewed interest was taken during the Renaissance. The most significant aspect was the gradual change from grouping plants as trees, shrubs, etc., to a more systematic basis in which similar plants were grouped together. Binomial names (that is a two-word name, the first indicating the genus and the second the species) were first used, though not consistently, by Bauhin in 1620. By 1694 plant names had increased to the 10 146 published by Tournefort, although there were many species duplicated under different names.

Important steps towards evolving a logical system had been made by the early 1700s but it was the Swedish naturalist Linnaeus (1707–1778) who produced the first comprehensive and systematic listing of plants. In 1753 he published his *Species Plantarium* in which he described 7300 plants. It was the first work both to use the binomial system consistently and to use a systematic basis for classification.

Previously it was usual for the name both to identify and to describe the plant. This was satisfactory when only two or three plants were involved in a group but in some of the larger groups ten to twelve words were necessary to 'name' a plant. This may have been acceptable in a more leisurely age but by the beginning of the eighteenth century it had become cumbersome and unwieldly. Linnaeus simplified the nomenclature by using the name for identification only, deleting all descriptive material.

His system of classification was based on the sexual characteristics of the flowers. He recognized twenty-four main classes of stamen arrangement (including one for plants with no proper flowers), and four classes for stamen and pistil arrangement when these parts were not in the same flower. Every plant could immediately be placed in a group, even though it sometimes meant that plants similar in general characteristics were placed in different groups and vice versa. The system was widely accepted although subject to criticism from some quarters. Most of the critics were constructive but some, such as the Russian Siegesbech, described the use of a sexual basis as 'lewd', 'loathsome harlotry', and deplored the teaching of 'so licentious a system' to the young. An Englishman Browne regarded it as a 'vulgar error'.

Linnaeus' assistants and students visited Russia, Siberia, Spain, North and South America, West and South Africa, India, China, Indonesia, New Zealand and the South Pacific area but not Australia. They brought back thousands of new species which were named and described by Linnaeus. Today many of these botanical explorers are commemorated in plant names; Kalm (*Kalmia*), Thunberg (*Thunbergia*), Sparrman (*Sparmannia*), Osbeck (*Osbeckia*), and Alstromer (*Alstromeria*). Linnaeus himself is commemorated in the genus *Linnaea*, which contains only one species *L. borealis*.

Linnaeus realized that his system was an artificial one and recognized that as knowledge increased a system based on natural relationships would be evolved. Evolution, discussed since the days of the Greek philosophers, was to be a significant factor and as a greater understanding of it developed it led to a more natural arrangement. This happened much more quickly than Linnaeus would have expected as by the early 1800s natural systems were in general use and his system was largely superceded. Even so he remains one of the greatest naturalists, because he not only classified all the known plants but also all the animal kingdom including fishes and insects.

The binomial system remains the basis of all plant and animal classifications. Most of Linnaeus' names have been retained but his primary classification into twenty-eight main plant groups has been replaced by more natural systems. The main emphasis during the past 150 years has been on classifying into larger groups such as families, orders, etc., clarifying the evolutionary and natural relationships between them, and in naming the enormous number of species discovered since the late 1700s.

In 1789 de Jussieu introduced the terms 'monocotyledons' and 'dicotyledons' to describe the two primary divisions of flowering plants. Bentham and Hooker in *Genera Plantarium* (1862–1883) grouped plants into 202 families. Engler and Prantl in *Die naturlichen Pflanzenfamilien* (1909) used evolutionary characteristics to a much greater extent than their predecessors and theirs is the system now in general use in Australia. In this system the monocotyledons culminate in the highly specialized orchids and the dicotyledons in the equally highly evolved composites (dandelions, daisies and dahlias, etc.). Significant modifications and contributions were made by de Candolle (1778–1841), Robert Brown (1773–1858), Eichler (1839–1887) and Engler (1844–1930). Bessy's (1845–1915) system based on principles or 'dicta' which are used in determining 'primitiveness' has found some support, especially in USA. More recent systems of classification are those of Tippo (1942), Pilger and Melchior (1954), and Bold (1970). Constant revision is still in progress as new geological evidence and new species are being discovered particularly in regions which are largely unexplored botanically, such as Central Africa, New Guinea and the south-western Pacific.

In determining the evolutionary sequence of flowering plants, certain criteria have been generally accepted as indicating primitive or more modern characteristics. Of these, the main ones are:

- Petals. Presence or absence of petals, and, if present, whether free or united. Separate petals are the most primitive; united petals or no petals are more recent.
- Separate parts are more primitive than united ones.
- Position of the stamens. Stamens emerging from below the ovary are the oldest; stamens above the ovary the more recent.
- Number of parts. Many parts indicate an old plant as evolution has tended towards fewer parts.
- Reduction of sepals to scales, hairs, etc. is a recent development.
- A simple vascular system denotes an older plant.
- The presence of secretory tissues suggests more recent evolution.
- Dry fruits are frequently more ancient than succulent types.
- Woody tissue is usually older than fibrous tissue.

These are only general indicators; no single one is conclusive. Each criterion must be considered in association with the others and with the overall morphology of the plant and therefore there is room for subjective assessment. It is only to be expected that different systems of classification have been proposed by different authorities depending on the relative importance given by them to individual characters. For example, some botanists consider that the flowering plants are derived from the *Gnetales*, a small group of specialized conifers. This implies that since conifer cones are wind-pollinated, the earliest flowers were also wind-pollinated and had no need of petals, that is they were similar to the catkin plants of today, (willows, etc.). Willow-like plants have in fact been found in the earliest fossil remains of flowering plants. On the other hand, there is also fossil evidence to show that the flowering plants may have been derived from the cycads which gave rise to the Order *Ranales* (the magnolia/buttercup order). The other flowering plants may have been derived from this order by a reduction and simplification of structures. The latter view seems to be favoured by most modern botanists.

There are other viewpoints but whichever view is correct it is obvious that it must affect the deductions used to determine the course of evolution and, since the major classifications are largely based on associations of similar plants, the arrangement of the different major groups. Classification is therefore still in a state of flux but as it concerns the main groupings, rather than the individual plants, it has little affect on everyday usage. In the next section, Classification, a broad outline is given of a classification in fairly wide use (basically Engler and Prantl's), but the reader should not overlook the fact that other systems do exist.

## CLASSIFICATION

Living organisms are divided primarily into the Animal and Plant Kingdoms. The essential differences of interest to us are that animals can move, need to break down elaborated material for energy, and release carbon dioxide; while plants do not move, build up food materials from the elements or elementary compounds, and release oxygen. These differences are obvious until the lower organisms, viruses and bacteria are considered where the distinctions are not so clear. However, in most natural systems viruses and bacteria are classified in the Plant Kingdom.

The major divisions of the Plant Kingdom are:
(a) Viruses
(b) Bacteria
(c) Thallophyta
(d) Bryophyta
(e) Pteridophyta
(f) Spermatophyta

### (a) *Viruses*

Viruses are the simplest form of life known and are unlike any other organism. They can be crystallized like salts and do not feed, grow, breathe, or move but when conditions are favourable they can reproduce themselves, i.e. a real case of 'animal, vegetable or mineral'. Size varies from 10 to 200 millimicrons and reproduction is asexual. They are the cause of some important plant diseases such as 'yellows' and 'mosaics' in tomato and tobacco but otherwise are not of great interest in the Plant Kingdom.

### (b) *Bacteria*

True bacteria are simple unicellular organisms varying in size from 0·0003 to 0·025 millimetres. Some are immobile, others can move by the to and fro whip-like action of their flagella. Their metabolic rate is very high and under favourable conditions they can reproduce remarkably rapidly. Reproduction is entirely asexual. Some are aerobic requiring air for survival, the rest are anaerobic and cannot live in the presence of oxygen.

(i) Saprophytic bacteria live on dead plant and animal tissue and break it down into other compounds by fermentation. Use is made of this in curing tobacco, retting flax, and in septic tanks.

(ii) Pathogenic bacteria live on live animal and plant tissue producing powerful toxins as by-products. They are primarily of interest to the Animal Kingdom but amongst plants they cause pear blight, cucumber wilt and a staining of willows.

(iii) Element or autotrophic bacteria live on inorganic materials and are of vital importance to plant growth. The nitrogen fixing bacteria (*Rhizobium* species) live in nodules on the roots of *Leguminosae,* etc. and convert free nitrogen from the air into more elaborate compounds. The nitrifying bacteria break down organic materials; *Nitrosomonas* change ammonia into nitrites and *Nitrobacter* convert nitrites to nitrates—nitrates and ammonia are the only two forms in which plants can use nitrogen. Some forms, notably *Azotobacter* and *Clostridium* can also convert nitrogen but do not need to live in symbiosis with a plant.

## (c) *Thallophyta*

The *Thallophyta* comprise (i) the algae which contain chlorophyll, (ii) the fungi which do not contain chlorophyll, and (iii) the lichens which are a mixture of algae and fungi. These are characterized by a sexual three-stage life cycle or 'alternation of generations' compared with the two-stage life cycle of higher plants. In brief, one generation produces a spore which on germination produces a thallus plant bearing the male and female organs. Gametes produced by these organs fuse to give a fertilized cell from which a new plant arises. (The plant bearing the spores is referred to as the sporophyte and the one with the male and female organs as the gametophyte.)

(i) Algae live mainly in water, except for a few land forms which occur only in moist situations. They are the simplest oxygen producing plants. About 15 000 species are known and are grouped mainly on the basis of colour.

Diatoms are brown unicellular plants, often grouped into colonies. They are the most abundant of the plants occurring in marine plankton.

Blue-green algae are unicellular, less than 0·01 millimetres long and are the simplest and lowliest of the green plants. Some have the ability to fix nitrogen, like some bacteria, and convert it into compounds usable by higher plants.

Green algae comprise by far the largest group. They are very variable, ranging from the green scum on the bark of trees and ponds to seaweeds almost as large as the brown algae. In the simpler forms the individual plant is unicellular but they often form colonies, grouped as filaments or plates, and cause the seasonal 'blooms' sometimes seen on water.

Brown algae are found almost exclusively in the sea as seaweeds. They can grow to large plants, for example, *Sargassum,* a free floating species of the Sargasso Sea, can grow to 100 metres in length. Around Australia the largest is bull kelp (*Sarcophycus potatorum*) which grows to about 30 metres in length. Most seaweeds are attached to rock by means of a specialized disc-like structure known as a 'holdfast'.

Red algae are the most colourful marine seaweeds but as they are found well below the tidal range they are not seen as often as the brown seaweeds. They are usually small, the largest growing to about 1 metre, and are the most highly developed form of algae. Red algae are the source of carrageen and are rich in iodine.

It is of interest to note that many of the seaweeds around the Australian coastlines are as typically Australian as the land flora.

(ii) Fungi are a very diverse group. About 70 000 species are known. Since they do not contain chlorophyll, they can only live by breaking down other organic substances and are the principal agents of decay. They range from unicellular plants to complex organisms with a mass of hyphae forming a mycelium. Hyphae join together to form an organized structure which produces the spores, for example, a mushroom is only the 'fruit' of the fungus, the 'plant' is the largely unseen mycelium.

Slime moulds (*Myxomycetes*) are 'unicellular' plants but are unique in that they have no cell wall. The 'cell' contents or protoplasm flows over the surface of the host. Hence they are plants which can move.

Algae-like fungi (*Phycomycetes*) are simple plants, typically arranged in filaments. Typical of this group are plant blights, downy mildews, and some moulds found on manure, fruit, and bread.

Sack fungi (*Ascomycetes*) are a large group of land plants. Over 20 000 species are known. They are much more highly developed than the previous two groups and bear their spores in sack-like structures. Typical representatives are the blue, green, and yellow moulds on leather, foodstuffs, etc., powdery mildews, yeasts and the cup-like fungi.

Basidia fungi (*Basidiomycetes*) are probably the best known. Spores are borne on the end of a stalk in a specialized structure or basidia. Included in this group are smuts, rusts, puff-balls, coral fungi, bracket fungi, toadstools and mushrooms.

(iii) Lichens are a remarkable group of land plants in that they comprise a fungal mycelium amongst which algae cells are interspersed. They are widespread and will survive in places too exposed or cold for other plants. Reindeer moss is a lichen (*Cladonia rangiferina*) and is one of the few of direct economic importance. One of the rock lichens is regarded as a delicacy in China and Japan. Three groups are recognized:

Crustose lichens are pressed closely to the surfaces on which they are growing (rock, bark, etc.).

Foliar lichens are more upright (suggesting foliage) and 2 to 3 centimetres high.

Fruticose lichens are much branched, often in an erect or hanging position (suggesting fruit).

## (d) *Bryophyta*

*Bryophyta* are essentially the mosses and liverworts. Some 23 000 species are known. They contain chlorophyll and are much more complex in structure than the *Thallophyta*, particularly the sex organs where fertilization occurs within the female organ. On the other hand the absence of a true root system or a highly organized vascular system separates them from higher plants. Alternation of generations is the dominant form of reproduction. The main groups are:

(i) True mosses; most people are familiar with the appearance of mosses. They are usually found growing in forested and moist situations. Some 14 500 species have been named but there are probably many more still awaiting identification.

(ii) *Sphagnum* mosses are world-wide in their distribution and are found in peaty and swampy areas, particularly at higher elevations or above the tree-line. There is only one genus, *Sphagnum*, containing about 300 species.

(iii) Liverworts are not so well known. They are small prostrate plants with plate-like, strap-like, or leafy plant bodies which are found growing on rocks, soil, etc. They rarely exceed a few centimetres in height. Some 8000 species are known.

(iv) Hornworts are not well known. They are simple plate-like plants with a special form of reproduction in a horn-like structure. They rarely exceed 2 centimetres in height and some 325 species are known.

### (e) *Pteridophyta*

This order and the *Spermatophyta* (below) comprise the vascular plants. The pteridophytes were among the first of the larger land plants. To support themselves in the less buoyant conditions of air (compared with water) they needed to develop structures, in the form of a vascular system, to give the required rigidity. As well as giving strength some of the vascular tissues serve also as conducting tissues.

Pteridophytes have alternation of generations but differ from the *Thallophyta* and *Bryophyta* as the conspicuous part of the plant bears the spores (the sporophyte) which gives rise to the small prothallus (gametophyte) bearing the sex organs. In the *Thallophyta* and *Bryophyta* the prothallus (or gametophyte) is the largest part of the organism in the cycle. There are four main groups in this order but the first three are small and can for convenience be regarded as one.

(i) 'Ancient' plants. These were very abundant in the late Palaeozoic era and are the main components of coalfields of that age. In evolutionary terms they are probably on the way to becoming extinct and are represented today mainly by the genera *Leucopodium* (creeping 'ground pines') containing about one hundred species, *Selaginella* (moss-like plants) containing about 500 species, and *Equisetum* (horestails) with about twenty-five species.

(ii) Ferns are the most abundant representatives comprising about 9000 species. They usually have an underground stem or a root-stock at ground level; in a few species such as tree ferns a stem is produced above ground. Bracken fern, with world-wide distribution, has an underground stem (rhizome) from which true roots arise, the above-ground portion being the leaf and its stalk. In tree ferns, the trunk is actually a mass of fibrous roots and leaf-stalks and quite unlike the usual tree trunk. The taller tree ferns may reach 15 to 20 metres, and include about 400 species of mainly tropical and subtropical plants. Ferns grow in moist situations as at the prothallus stage the male sperm requires a film of free water through which it swims to reach the female cell. Once past this stage ferns can be successfully transplanted to drier situations. The prothallus rarely exceeds 1 centimetre in diameter and the development, fertilization, and initiation of a new fern plant takes only a few months.

### (f) *Spermatophyta*

*Spermatophyta* comprise the true seed plants and are the largest group in the Plant Kingdom. More than 200 000 species have been described. As there are large areas of the world which are still not fully known botanically, it is probable that the number will be increased as new species are found. The number of 200 000 refers to species only and not to varieties. Thousands of varieties of some plants are known, for example, roses, rhododendrons, and many other garden plants. Most of these originated from very few species, in most cases only about five to ten wild types.

Alternation of generations is replaced as a means of reproduction and the parent/seed cycle is established. In the *Thallophyta* the male and female gametes fuse outside the organs, in the *Bryophyta* they fuse in the female organ, and in the

*Pteridophyta* some growth takes place in the female organ and develops into a new plant at that site. In the *Spermatophyta* the position is advanced one stage further as the seed partially develops as an embryo, is provided with some reserve food supply and a protective coat, and is then shed to establish a new plant at a new site.

There are two main subdivisions (i) the conifers or *Gymnospermae*, and (ii) the flowering plants or *Angiospermae*.

(i) Conifers or gymnosperms are geologically a very old group. About 500 species are known and they contain all the commercial softwood species. Their unifying feature is that they bear the seed on a scale and not contained in an ovary as in the flowering plants.

'Ancient' conifers. Included in this are species which are very old in evolutionary terms and which were previously much more widespread than they are today. They are probably on the way to extinction. The main ones are the cycads (about one hundred species), including the *Zamia* 'palms' and *Ginkgo* (one species).

'True' conifers total about 350 species and include all the well-known soft-woods—pines, spruces, firs, cedars, larches, hemlocks, redwoods, cypresses, araucarias, callitris, junipers, yews, podocarps, etc. Although also an old group in evolutionary terms and restricted in numbers of species, their abundance and widespread occurrence suggests they will survive for a long time to come.

*Gnetales* are a small group of mainly desert plants in which the female cone is reduced in size and simple vessels appear in the wood.

(ii) Flowering plants or angiosperms are by far the largest plant group. More than 200 000 species are known. Their common feature is that the seed is

| MONOCOTYLEDONS | DICOTYLEDONS |
|---|---|
| Grasses (including all the cereals), rushes, sedges, bulbs (lilies, daffodils, jonquils, tulips, etc.), bamboos, palms and orchids. | All other flowering plants. |
| Root system usually fibrous or bulbous. | Root system with a strong primary root, often woody. |
| Trunk without definite cambium. | Trunk with cambium. |
| Perennial species without annual rings. | Annual rings in perennial species. |
| Leaves usually long and narrow with parallel venation; stalk often not developed. | Leaves usually broader in shape with a net-like system of veins; stalk usually apparent. |
| Flower parts (petals, etc.) usually in threes or multiples of three. | Flower parts usually in fours or fives or small multiples of fours and fives or in larger numbers. |

formed in a closed chamber or ovary compared with the naked position on a scale as in the conifers.

Two main sub-divisions of the flowering plants are recognized; monocotyledons and dicotyledons. The cotyledons are the seed leaves within the seed which are the first to appear on germination; the first group has only one seed leaf and the second, two. This might appear to be a very minor feature for the major sub-division of a group as large as the angiosperms but it is a reliable and consistent one. Moreover, there are a number of other similarities within each sub-division which gives a more unified concept to the groups.

## SUBDIVISION OF PHYTA

The discussion of the main divisions in the Plant Kingdom from now on is generally confined to the *Spermatophyta*, as this group includes all the trees which are described later. Within the phytum or Division *Spermatophyta*, there are numerous classes, orders and families in descending order to genera and species. It is not within the scope of this publication to define the characteristics of each group, but a representative classification is given for a common Australian species, coastal wattle (*Acacia longifolia* var. *sophorae*) as an illustration.

|                         |                    |
| ----------------------- | ------------------ |
| Kingdom                 | *Plantae*          |
| Phytum or Division       | *Spermatophyta*    |
| Sub-division or Branch  | *Angiospermae*     |
| Class                   | *Dicotyledoneae*   |
| Sub-class               | *Polypetalae*      |
| Order                   | *Rosales*          |
| Family                  | *Leguminosae*      |
| Sub-family              | *Mimosoideae*      |
| Genus                   | *Acacia*           |
| Species                 | *longifolia*       |
| Variety                 | *sophorae*         |

A complete classification like this is usually only seen in taxonomic text books; in general usage only the genus, species and variety are used as this is adequate for general identification.

## NOMENCLATURE

Each plant known to science has its Latin scientific or botanical name and this name is accepted in all countries. Nomenclature is governed by a set of rules, the International Code of Botanical Nomenclature, drawn up and revised by successive botanical congresses. They are accepted internationally and all plants are now named in accordance with these rules. Common names, on the other hand, can be given by anybody but there are various organizations endeavouring to arrive at standard common names.

Common names, while very serviceable in a restricted locality, often cause confusion when used elsewhere; for example, the eucalypt blue gum in Victoria is usually the species *E. st johnii*, in Tasmania *E. globulus*, in New South Wales *E. saligna* and South Australia *E. leucoxylon*. Confusion like this led to the need, early in botanical history, for an international 'language'. To avoid injuring national pride a 'neutral' one was necessary so the scholastic language of Latin was accepted.

When a new plant is discovered, it is the privilege of the botanist describing it to give it a name of his own choosing. In selecting a name for either the genus or species, it is usual for it to be either (a) descriptive, for example, *Eucalyptus* is derived from *eu* meaning well and *kalypto* meaning to cover (Greek) which refers to the 'well-covered' manner in which the operculum, or cap, covers the flower in bud, or (b) commemorating a noted botanist, such as *Banksia* after Sir Joseph Banks the naturalist on Cook's voyages, and *Hakea* after the German botanist Hake, or (c) using the original Latin name, such as *Acer* for maples. Species names are usually derived from the first two sources as Latin names rarely referred to species, only to what is now regarded as the genus. The name, together with a complete botanical description, is then published in a botanical journal with a wide circulation and becomes the accepted and permanent name. Unfortunately a number of plants have been inadequately described or the descriptions have been published in a minor journal and a single plant has often been described at least twice under different names. In such cases the name published first takes priority and is accepted, for example, red gum was known for many years as *Eucalyptus rostrata* but an earlier description gave it the name of *E. camaldulensis*, so this is now the correct name. Often the second, but incorrect, name is printed in brackets and in different type after the correct name as the synonym to aid identification.

Name changes like these may be annoying but they are also necessary and need to be accepted. It is only with better knowledge, communication, and more caution on the part of the botanists, that duplication is avoided today. Once the records of the past 150 to 200 years have been tidied up, more stability should result.

The binomial system is used internationally irrespective of the system used to group plants in larger assemblages. The binomial system, as the name implies, gives a two-word name to each distinct plant. The first word is the genus (plural, genera) and includes similar plants, for examples, all eucalypts (gums, stringy-barks, boxes, peppermints, ironbarks, mahoganies, mallees, etc.) belong to the genus *Eucalyptus*. Similarly all wattles belong to the genus *Acacia*.

The second word is the species (plural, also species) or specific name which distinguishes it from other plants in the genus, for example, *Eucalyptus leucoxylon*. It usually refers to some distinctive characteristic of the species, in this case *leucoxylon* for the white wood (*leucos* white, *xylon* wood). Sometimes distinguished persons are commemorated, for example, *Eucalyptus muellerana* commemorating Baron von Mueller, an outstanding nineteenth-century botanist in Australia.

The species is regarded as the basic biological concept but so far no one has been able to define it satisfactorily. It is largely a matter of individual interpretation but in general terms a species can be considered as a group of individuals so alike that they appear to have common parents. This may sound vague but it works fairly well; everybody who has ever recognized a plant has unconsciously summed up the distinguishing characteristics of that species and so has 'described' a species.

A further distinction is the variety and applies when a species can vary in flowers, foliage, shape, etc. It is usually descriptive and is more common with ornamentals where a particular feature is of special value, for example, the white-flowered *E. leucoxylon* has a red-flowered form known as var. *rosea*.

Hybrids between two species are common, especially as foresters and horticulturists undertake artificial cross-breeding in search of improved strains. These are usually written as *Populus* x *generosa* meaning a cross between two *Populus* species and has been given the species name *generosa*. Often, particularly with natural hybrids, one or both parents are not known. However the case above is an exception in that it is known to be a cross between *P. trichocarpa* (male) and *P. deltoides* var. *angulata* (female).

Among ornamental plants, particularly garden plants, 'cultivars' are common. Cultivar is a word derived from 'cultivated variety'. These are plants which have been propagated vegetatively (cuttings, etc.), usually from a single plant which has some desirable characteristics. The cultivar name may be in English or in Latin and is often preceded by 'c.v.' which indicates a cultivar, for example, *Populus simonii* c.v. '*fastigiata*'. With the more ornamental garden plants the species is often left out, for example *Rhododendron arboreum* c.v. '*Alarm*' is commonly written just as *Rhododendron* '*Alarm*'.

In technical publications, the name is usually followed by some letters which represent an abbreviation of the author's name (the botanist who first described the species), for example, *Eucalyptus regnans* F. Muell. indicates that Baron Ferdinand von Mueller published the first description of this species. Listing the author's name is a further means of reducing the possibility of confusion with another species but this practice however is not usually followed in non-scientific publications.

The principles to follow in spelling sometimes cause confusion. In the botanical name a capital letter is only used for the genus as in *Eucalyptus muellerana* with a small letter for the species, even though it may be derived from a proper noun, such as Baron von Mueller in this case. (This style was approved by the International Code in 1961.) The 'i' or 'ii' ending of the specific name is often incorrectly used; the International Code of Botanical Nomenclature rules that 'The specific epithet, when adjectival in form and not used as a substantive, agrees grammatically with the generic name'. While this may be clear to the reader who is a Latin scholar, for the rest of us it means that in practice the ending should be 'ii' except when preceded by a vowel or 'er', for example, *Eucalyptus steedmanii*, *E. stoatei*, and *E. seiberi* respectively.

The spelling of common names, particularly Australian natives, is inconsistent. Incorrect forms like Lemon Scented Gum, Western Australian Red-Flowering Gum, Round-leaf Box, Narrow-leaved Peppermint, and Open-fruited Mallee are frequently seen. In current usage the endings 'ed' and 'ing' are generally not used while capital letters are used only with proper nouns or in headings. In compound words hyphens have been retained to aid pronunciation in a few cases, such as 'she-oak', otherwise the form without a hyphen such as 'bottlebrush' is preferred. Applying these principles then the correct versions of the common names given above would be respectively 'lemon scent gum', 'Western Australia red flower gum', 'round leaf box', 'narrow leaf peppermint', and 'open fruit mallee'. This is the style which should preferably be used and is now being generally accepted.

# ANATOMY

Trees are such a commonplace part of the scenery that few people are aware of the complex and diverse nature of their structure and growth. Yet for anyone planting a tree even an elementary knowledge of these matters would enable him to carry out the project with a more intelligent understanding of the processes involved.

## CELL

### (a) *Structure*

A plant, like all other living organisms, consists of an ordered accumulation of cells. They are the fundamental units of structure in plants and animals and the similarity in cell organization in both groups points to a common ancestry in evolution. It was the development of the microscope in the early 1600s that enabled Robert Hooke to observe the cellular structure of living material in 1665. It was Robert Brown who first examined the internal structure of the cell in 1831. In 1838 and 1839 Schleiden, a botanist and Schwann, a zoologist, published their hypotheses that the cell is the basic organizational unit of all plants and animals; a concept which is now universally accepted.

The ordinary cells in plants are box-like structures, more or less cubic in shape, but brick-like and tube-like cells are also common, particularly in specialized tissues. Size is usually about 0·01 to 0·5 millimetres in diameter and 0·5 to 1·0 millimetres in length but some, such as vessels, may be larger in diameter and up to several centimetres in length. The young cell wall is thin and flexible but becomes thick and rigid with age. As the cell grows, successive layers are laid down on the cell wall by the protoplasm in the cell so the youngest layers are on the inside. Long-chain cellulose molecules form the wall framework which, in woody species, becomes impregnated with lignin giving them their characteristic rigidity. Adjoining cells are cemented together with pectin, the setting ingredient of fruit jellies.

The inside of the cell is occupied by the protoplasm and the vacuole. When young, the cell is completely filled with protoplasm, but as it enlarges in growth, the protoplasm is spread around the walls and the centre is occupied by the vacuole. The vacuole is filled with cell sap, a solution of sugars and organic acids. Water diffuses in and out of the vacuole and helps to keep the cell turgid.

The protoplasm is the only truly living part of the cell. It is involved in or controls respiration, photosynthesis, absorption of water, reproduction, and most of the other vital plant processes. It is constantly moving around the cell—

known as 'protoplasmic streaming'. The fluid part is thick and gelatinous, like egg-white, and contains numerous granules and globules of various sizes. The most important of these is the nucleus. All the other constituents (solid and liquid) collectively form the cytoplasm.

The nucleus is a dark, fairly dense body. It is confined by its own membrane which contains the nuclear sap and within which is a network of darker strands. This network contains the two nucleic acids DNA (deoxyribonucleic acid) and RNA (ribonucleic acid). At cell division, they aggregate into the chromosomes (see below) which are the repositories of the genetic material transferring the plant's characteristics from generation to generation.

The remainder of the protoplasm or cytoplasm contains various functional granules, the largest and most important of which are the chloroplasts. These contain the chlorophyll molecules essential for photosynthesis. Chloroplasts reproduce by splitting independently of the cell division. Other granules of importance are the mitochondria or energy providers and the chromoplasts containing the colouring materials which give colour to the plant. (Figure 4)

(b) *Division*

The growth of a plant depends on the production and enlargement of new cells. Division is not just a matter of cutting a cell in two, the actual contents must be divided into two equal halves to share the genetic material. This is accomplished by a complex process known as mitosis.

Once cell division is initiated, the network in the nucleus thickens and aggregates into discrete bodies or chromosomes. Chromosomes vary widely in shape, size and number among species, but within each species of plant or animal the number, shape and size is constant. On division each chromosome splits lengthwise so that each gene, etc. is split into two identical pieces. The two halves separate and the halves migrate to opposite ends of the cell to form new nuclei. A cell wall now forms between the nuclei and so two new cells are created.

The whole process takes from half an hour to two hours depending on the species. It is continuing all the time and normally does not speed up very much during the growing season; the faster growth of spring is not due to faster cell division but rather to the individual cell growing to a larger size. Normally it takes about one and a half to two days from division for a cell to enlarge to its full size.

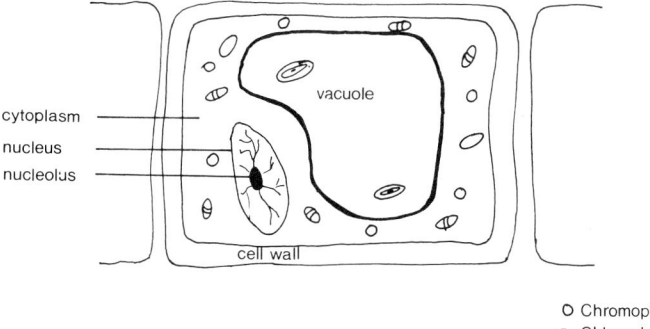

cytoplasm
nucleus
nucleolus
vacuole
cell wall

O Chromoplast
⊖ Chloroplast
⊕ Mitochrondria
⊛ Starch grain

Figure 4. Cell structure—diagrammatic only.

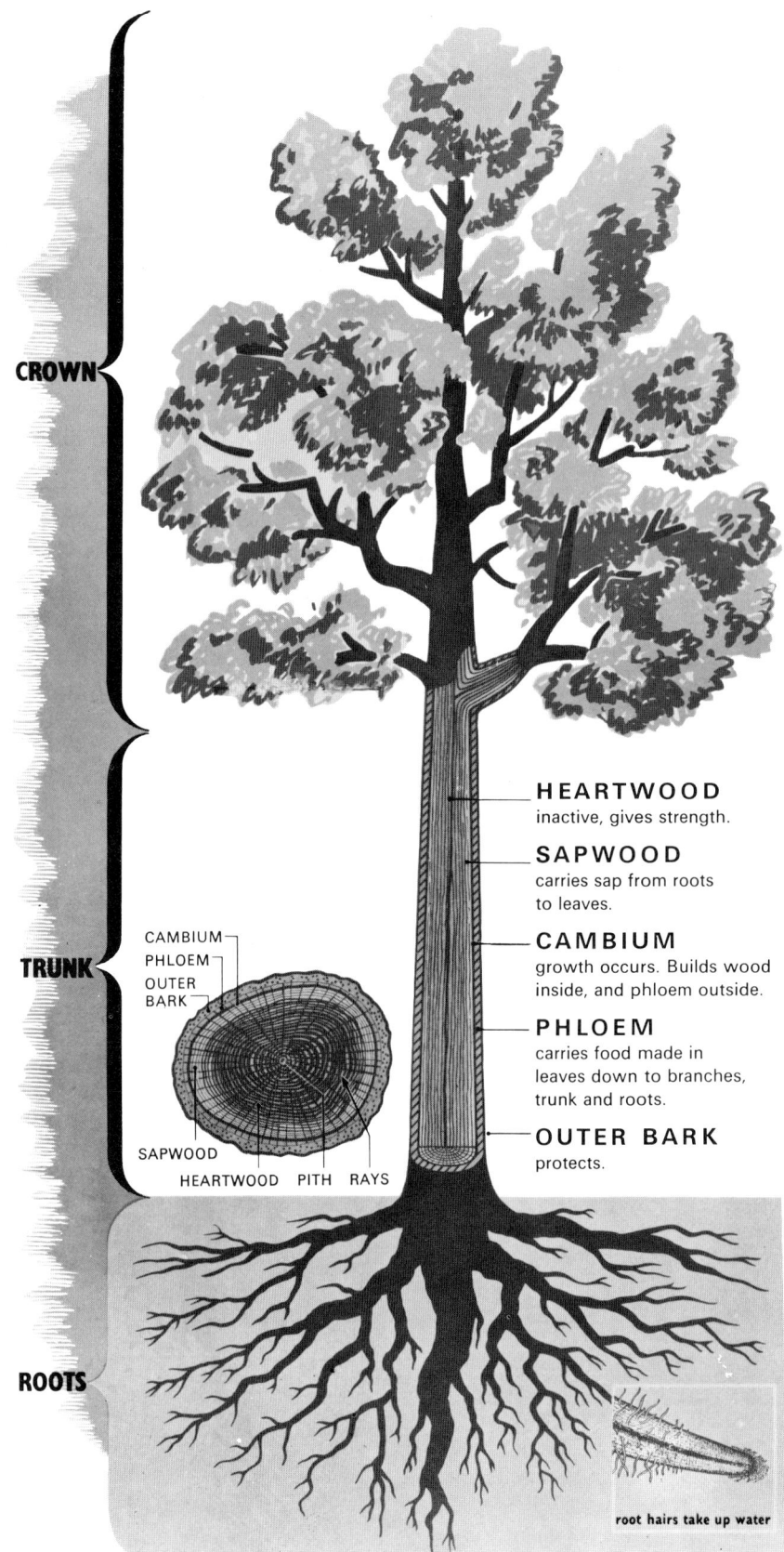

**CROWN**

**TRUNK**

**ROOTS**

**HEARTWOOD**
inactive, gives strength.

**SAPWOOD**
carries sap from roots
to leaves.

**CAMBIUM**
growth occurs. Builds wood
inside, and phloem outside.

**PHLOEM**
carries food made in
leaves down to branches,
trunk and roots.

**OUTER BARK**
protects.

CAMBIUM
PHLOEM
OUTER
BARK

SAPWOOD

HEARTWOOD    PITH    RAYS

root hairs take up water

Figure 5. Tree structure. By kind permission of the Timber Research and Development Association U.K.

(c) *Plants and Animals*

Before completing this section it may be of interest to point out the main differences between the cell structures of plants and of animals:

(i) In plants, the cell walls become thickened and rigid with maturity; in animals, they remain soft and pliable.

(ii) In plants, a vacuole develops, filled with sugary cell sap; in animal cells, the protoplasm fills the whole cell even at maturity. The protoplasm being primarily composed of protein, rather than sugars, explains the greater protein nutritional value of animal tissue.

(iii) The protoplasm of plants contains more operative structures since it needs to manufacture foods from the basic elements; animal cells receive nutrition from the breakdown of elaborated foods.

With an appreciation of the structure and growth of plant cells we can discuss the form of structure of a woody plant. For convenience, a woody tree or shrub can be considered to consist of a root, stem and crown. (Figure 5)

## ROOT

The function of the roots is to provide mechanical anchorage in the soil and to absorb water and nutrients for growth. The first seedling root normally grows downward as the primary root and in some species develops into a well-formed tap root. From the primary root, secondary and lateral roots arise by branching. Lateral roots are frequently horizontal and close to the surface and absorb both water and nutrients from the topsoil whereas the deeper roots primarily absorb moisture. Many drought-resistant plants have both well-developed surface and deep root systems, for example, sugar gum (*Eucalyptus cladocalyx*), the former system takes advantage of any light fall of rain and the latter taps resources of moisture deeper in the soil.

Root systems are much branched and permeate most of the soil near a tree. They extend much further than is commonly supposed and will grow far beyond the crown of the tree, even past adjoining trees. Under forest conditions individual roots may extend beyond several trees, so that the soil becomes permeated by a mass of roots. Intermingling root systems also reinforce the tree's anchorage in the soil.

All root systems comprise fibrous and woody roots. Some species, such as birches (*Betula* sp.), have a predominantly fibrous system while others, like eucalypts, are basically woody and relatively sparse. Those with fibrous systems are usually more easily transplanted since they have a much greater mass of fine absorbing roots to assist in re-establishment after the shock of transplanting.

Little work has been done on calculating the root mass of fully grown trees probably because excavating the root system in detail would be an almost impossible task. Any work that has been done has usually been on grasses and similar small plants which can be handled more easily. Although these results need to be applied cautiously to trees, they do serve to illustrate just how large root systems can be and how thoroughly they permeate the soil. A hectare of wheat may have 100 000 kilometres of roots and maize 80 000 kilometres. Fibrous root grasses may have a root system of one hundred or more kilometres in length and produce new roots at an average rate of 3 to 5 kilometres per day;

root hairs may be produced at the rate of several million per day while the total root surface area per plant can be up to about 0·1 hectare. These figures are large but serve to illustrate that a root system may grow vigorously and can rapidly saturate a given volume of soil.

Absorption of water and dissolved nutrients does not occur over the whole root system but only by the root hairs and soft root tips. Root hairs are small unicellular hair-like extensions arising from the epidermal (surface) cells just behind the growing root tip. They are usually less than 0·1 millimetre in diameter and 5 millimetres long and are generally too small to be seen by the naked eye. The roots that can be readily seen do not absorb much water as they have already become suberized and relatively impervious to moisture. They act mainly as an anchorage for the tree and as conducting channels to transport water and nutrients to the stem and leaves.

Root hairs survive in an active condition for a few days only but are produced rapidly and prolifically; one hundred per square centimetre is common. When an open-rooted plant is lifted for transplanting most of the root hairs are stripped off and the speed with which they are regenerated largely determines how readily a particular species will transplant; the faster they are produced the better the chances of survival. Mahogany gum (*Eucalyptus botryoides*) transplants fairly readily and will survive under adverse conditions, while messmate (*E. obliqua*) is notoriously difficult to establish under any but good conditions. Most of this difference is because the former has about eight times the root regenerative capacity of the latter. Further when root-rot fungi (such as *Phytophthora*) are present it is obvious that good root regeneration capacity will improve the chances of survival.

Unfortunately no practical means of stimulating root hair production has been discovered so far. The most practical means is to tube, pot, etc. so that on transplanting the root hairs remain in the tube or ball of soil and disturbance is minimal.

The internal structure of a root is complex and differs in many respects from the stem. The essential feature is the fibrovascular system which furnishes mechanical strength and serves as conducting tissues to the stem. It contains similar structural elements (xylem, phloem, cambium, etc.) to the stem but whereas in the latter they are arranged in concentric rings, in the root they are predominantly star-shaped in cross-section. Well defined rings, therefore, are not as apparent as they are in the stem.

Increase in diameter occurs by division of the cambium cells (see page 38). In the stem, viewed in cross-section, these are in a continuous circle or band, but in roots they occur only in the 'points' of the star-shaped fibrovascular strand. That is they occur as longitudinal strips or bands along the root rather than forming a sheath as they do in the stem.

## STEM

The stem or trunk is often the most conspicuous part of the tree. It serves to support the crown in a favourable position for photosynthesis, and as conducting tissue between the roots and crown. The stem begins at the point where all the roots are joined together. At this point the root tissue structure becomes the stem structure; here the cambium layer (see page 38) is continuous around the stem. The vascular strands or conducting tissue in the individual rootlets are collected together in the stem. In the crown they disperse again to the branches and finally become veins in the leaves.

The trunk or stem comprises pith in the centre, heartwood, sapwood, cambium and, on the outside, bark. The pith is of minor importance in the adult tree and need not be considered further. The heartwood is usually the greatest part of the trunk and provides the mechanical strength of the tree. It is inactive and takes no part in growth, which explains why a tree that is hollow may still be growing vigorously. The sapwood surrounds the heartwood and conveys the water and dissolved nutrients collected by the roots to the crown, consequently it contains more moisture than the heartwood. Heartwood and sapwood constitute the xylem tissue. In hardwoods, the sapwood is usually lighter in colour and less than 2 to 3 centimetres wide; in softwoods, there is often only a slight difference in colour, and the sapwood may constitute a large part of the trunk. The cambium is the actual growing tissue and will be discussed later in greater detail (see page 38). The bark comprises the inner bark and the outer bark. The inner bark or phloem tissue carries the food produced in the leaves down the tree to the cambium of the trunk and roots, and crosswise to the sapwood where the surplus is stored (in the form of starch) for use in adverse periods. The phloem tissue is very thin but remarkably efficient at transporting food from the crown to the rest of the tree. The outer bark is for protection from injury.

The movement of water up the sapwood and of food products down the inner bark is loosely called the sapstream. The rate of flow varies widely but is usually from 0·5 to 3 metres per hour. This movement is fastest in the growing season, the spring (hence the expression 'when the sap is up'), and slowest in the winter (or 'when the sap is down'). These are inaccurate expressions as the sap is always in the tree, if it were not the tree would die; even in a leafless deciduous tree in mid-winter there is still slight movement of the sapstream.

In the old practice of ringbarking, the flow of manufactured food from the crown to the roots is interrupted which reduces the growth of roots and new root hairs. As the root hairs have only a very short life, the time comes when insufficient new hairs are produced to sustain the tree and it eventually dies of starvation. Sapringing or cutting through the bark and sapwood acts in the same way, but here the flow of water up the tree is also interrupted and death is usually quicker than by ringbarking alone. Cincturing is a modified form of ringbarking and restricts the flow of sugars from the crown to the roots. A larger proportion of the sugars are therefore retained in the crown and aid top growth, flowering, or fruit production (as in vineyards).

(a) *Wood Structure*

The stem, like the roots, branches and leaves, consists of a mass of individual cells. In stems, the individual cells are aligned longitudinally except for the medullary rays which are horizontal and arranged radially. These extend through the wood from the pith to the inner bark and serve to store elaborated food substances or transport them from the inner bark to the storage tissues in the sapwood and heartwood. In some species, such as silky oak (*Grevillea robusta*), the rays are very broad, and in radial sections gives the timber its characteristic figure.

In temperate countries when growth is at a maximum during spring, the wood cells produced are large in diameter with comparatively thin walls; in late summer and autumn, when growth is less, they are smaller with the proportionately thicker walls so that the wood appears darker in colour and denser than the spring wood. This difference gives rise to the familiar rings, or annual rings, the lighter part representing spring growth and the darker part the summer and autumn growth. Since this pattern usually follows an annual cycle it can be used to estimate the age of a tree. It is only an estimation as occasionally false rings develop during an abnormal year, such as a dry spring followed by a mild wet summer, which gives two or more rings in one year. However, false rings usually do not form a complete ring around the trunk. Where growth is more or less continuous throughout the year, as in tropical forests, annual rings in the timber may not be so distinct or may even be indistinguishable to the naked eye.

(b) *Softwood or Hardwood?*

These terms originated in Europe where the softwoods (pines, firs, spruces, etc.) were actually much softer than the hardwoods (oak, beech, etc,). When the terms are used in other countries the clear distinction in Europe is not always as obvious. This leads to some peculiar anomalies, for example, some pines (softwoods) are much harder than many hardwoods, while balsa wood, the softest and lightest wood known, is a hardwood.

The essential difference between softwoods and hardwoods is one of wood anatomy. Softwoods consist mainly of large thin walled cells, called tracheids, which are usually about 2 to 5 millimetres (but can be up to 10 millimetres) long and 0·1 to 0·5 millimetres in diameter. They have the dual role of conducting water and dissolved nutrients to the leaves and of providing mechanical strength to the tree. Hardwoods, on the other hand, have 'division of labour' between two basic cell types: (i) fibres, which are small thick-walled cells usually not more than 0·1 millimetres in diameter; and (ii) vessels, which are large in diameter (up to 1 millimetre) and may be up to 50 to 100 centimetres long. Fibres constitute the bulk of the wood and give mechanical strength while vessels act as the conducting channels. The vessels, which are formed by the

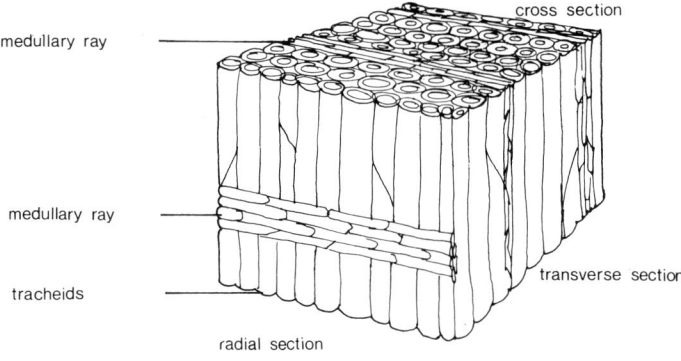

Figure 6. Softwood wood structure.

breaking down of the end walls of adjoining cells, are large in relation to the other cells and can usually be seen as fine hair-lines on the surface of the wood. Movement of the sap between adjoining tracheids is by diffusion through pits in the cell wall; a pit is a small, circular thinner area of the cell wall but with the primary wall still intact. In the fibres of hardwoods also, movement is via pits but the vessels have perforated ends, or sieve plates, to aid rapid flow.

Whether wood is classed as softwood or hardwood depends on the presence or absence of vessels; they are present in hardwoods, absent in softwoods. With a little practice the vessels (if present) can be seen on a planed or dressed surface of most timbers likely to be encountered in southern Australia. (Figure 6 and 7)

(c) *Composition of Wood*

Wood is a very complex substance principally consisting of cellulose, hemicelluloses, lignin and some extractives. All are compounds of high molecular weight. (Percentages given below are by weight.)

(i) Cellulose constitutes 40 to 50 per cent of wood and, as its long molecules are wound spirally, it is the main source of wood's tensile strength. As it is the main constituent of pulp and paper it is the component of most commercial interest. The composition of cellulose is 45 per cent carbon, 49 per cent oxygen and 6 per cent hydrogen. It is a very stable compound and is not attacked by chemicals, fungi or insects, except termites which still require the assistance of protozoa to achieve its digestion.

(ii) Hemicelluloses are a group of substances similar to cellulose, but are more gelatinous in nature. They are susceptible to chemical attack and can be dissolved in paper manufacture. Hemicelluloses are more common in hardwoods than in softwoods.

(iii) Lignin is the 'cement' which binds the cellulose together and provides the compressive resistance of timber. It is a highly complex substance constituting 23 to 33 per cent of softwoods, 15 to 20 per cent of hardwoods and comprises 65 per cent carbon, 29 per cent oxygen and 6 per cent hydrogen. Commercially it is a waste product in paper manufacture as it causes yellowing in paper, but chemically it is a raw material with the potential as a source of derivatives of coal and petroleum. It is very insoluble in solvents and is usually degraded in extraction.

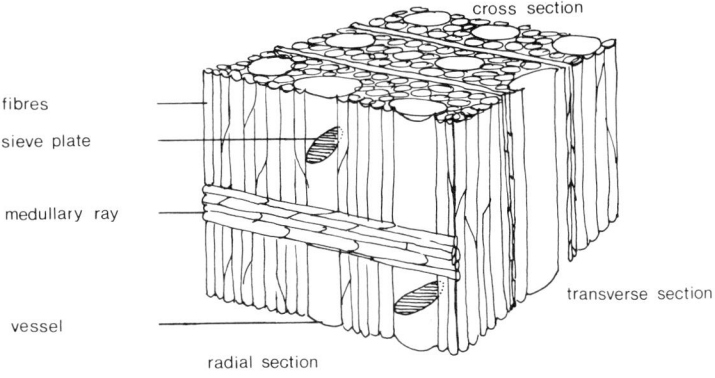

Figure 7. Hardwood wood structure.

(iv) Extractives represent all the rest of the various compounds found in wood in small amounts, such as, carbohydrates (up to 5 per cent), pectin, terpenes, phenols, tannins, proteins ($o_E2$ per cent), mineral ash (1 per cent) and silica (up to 1 per cent).

(d) *Reaction Wood*

Trees that lean are under an abnormal stress and to counteract this they develop reaction wood as a means of support. In softwoods it forms on the underside and is referred to as compression wood while in hardwoods it occurs on the upper side and is called tension wood. Compression wood is darker in colour, heavier, and contains a greater amount of lignin. Tension wood however is usually not very different in colour or weight to the surrounding wood but it does contain a greater percentage of cellulose.

Knots are a special type of reaction wood. As they are enclosed branches and, as branches have to withstand high wind-loading and flexing in relation to their relatively small diameter, wood structure is modified; softwoods having more thick walled cells and hardwoods more fibres. Consequently they are darker and harder than the adjoining wood.

## CROWN

The crown comprises branches and leaves. The branches are actually an extension of the trunk in that they have the same structure and sub-divide repeatedly to distribute the water and nutrients from the roots to the leaves. The trunk, and its extension the branch system, can be said to end at the leaf abscission layer. This is a layer of corky cells that develop at the base of the petiole (leaf-stalk) and is the point at which the leaf eventually falls off. The layer of corky cells produced seals the wound caused by leaf fall.

A leaf is a most remarkable structure. It is the region where the water and simple salts absorbed from the soil are combined with carbon dioxide by the action of sunlight to form more complex food materials. Leaves vary widely in size and shape from several metres long, as in some tropical plants, to the minute leaf teeth of the casuarinas about 1 millimetre in length. Shape varies from needle-like to flat and round, with flat, square or round cross-sections. Some are deeply lobed (oaks) or dissected (maples). Margins may be smooth (entire) or deeply and doubly serrated. Most leaves are simple but in others dissection has progressed to the stage where the leaf is divided into several leaflets, known as compound leaves (as in *Fraxinus* sp. and some *Acacia* sp.); the leaflets are joined to a common stem (*Fraxinus* sp. and *Acacia* sp.) or a common point (*Castanea* sp.). In most of the acacias true leaves are absent; the 'leaf' is a flattened and modified leaf-stalk (phyllode) which has taken over the functions of the true leaf. Leaf surfaces may be shining, smooth, hairy, rough, or coated in a glaucous bloom. Venation or pattern of the veins may be parallel or net-like.

The midrib and veins are actually the continuation of the vascular system of the trunk and repeatedly sub-divide until they are only one or two cells wide in the leaf. They are a double structure with xylem tissue in the centre surrounded by phloem tissue on the outside; both the xylem and phloem tissues are continuous with those of the stem and root. In a cross-section of a typical leaf the outer surfaces consist of a layer of rather thick-walled cells forming the epidermis. Over these there is usually a thin layer of a waxy substance, the cuticle. Within the leaf, and constituting the major part of it, are the mesophyll cells. On the upper surface they are tightly packed and arranged at right angles to the surface of the leaf to form the palisade tissue. These contain large numbers of the minute chloroplasts, up to several hundred thousand per cubic millimetre, in which chlorophyll, the activator of photosynthesis, occurs. In the lower part of the leaf the tissue is loose with large air spaces between the cells forming the spongy mesophyll. The surface exposure of these cells is five to twenty times that of the leaf surface and this is where the gas exchanges of the leaf occur. (Figure 8)

Gas exchange between the leaf and atmosphere—carbon dioxide into the leaf and water vapour and oxygen outwards—is via minute slit-like pores, the stomata. The stomata are very small and even though they occupy only about 2 to 3 per cent of the leaf surface they range in number, depending on the species, from a few hundred to several thousand per square centimetre. Each stomata has two guard cells which regulate the size of the opening to control gas exchange according to the plant's needs; when the guard cells are turgid the stomata are open, when limp (flaccid) they are closed cutting off any gas exchange.

In leaves that are not flat, such as the needle-like leaves of pines, some bottle-brushes, melaleucas, etc., the same basic structure and functions exist but in a modified form. Usually the veins of the vascular system run down the centre of the needle and are not visible externally, while the stomata generally occur on all surfaces either scattered or, as in pines, aligned in parallel rows.

Notable exceptions to the typical form are the eucalypts (*Eucalyptus* sp.) and some other Australian natives as most of them have stomata on both surfaces instead of the under side only. Since they frequently hang vertically there is really no 'under' surface but as there is often a colour difference between the two surfaces the darker is conventionally referred to as the upper. Structurally the eucalypt leaf is like a typical double leaf, i.e. epidermis and palisade tissue adjoining both surfaces with the spongy mesophyll in the centre.

## GROWTH

Growth is basically a process of cells dividing and then enlarging. There are three main forms in which it occurs although the division process is the same in each case.

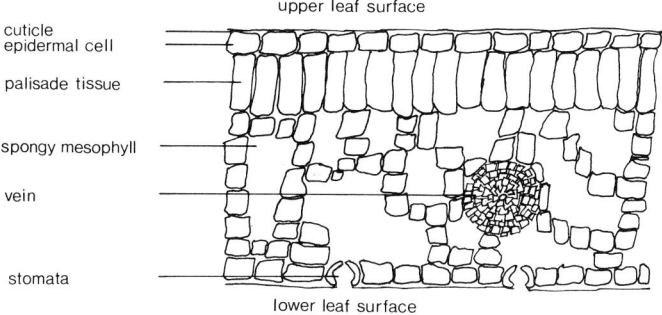

Figure 8. Leaf cross section.

*Determinate growth* is the type that takes place where the part has a definite limit to size, for example, leaves, flowers, fruit, etc. Each cell divides so that every part of the structure is expanding at the same time reaching a final size beyond which no further growth takes place.

*Terminal growth* is indeterminate in that it occurs at the tip of a root, stem, etc., and can continue almost indefinitely. Here only the cells at the tip divide, for example, with root or stem division it is restricted to the few layers of cells extending back for 2 to 4 millimetres from the tip. Immediately behind this is a zone of cell enlargement in which the cells produced by the tip enlarge to their mature size. Rate of shoot growth is related to the length of this zone of enlargement and not to the rate of production of cells, a process which in most species takes one and a half to two days to complete. In slow-growing trees the zone of cell enlargement may be only 1 centimetre long, but in fast-growing plants, such as bamboo, it may be more than 60 centimetres long. The cells further back are already mature and in their final form, hence the actual growing region moves 'outwards'. The pattern is similar for both roots and shoots; the former produces branch roots, and the latter produces branch shoots, leaves and often flowers.

*Cambial growth* increases diameter in roots, trunk, branches, etc. The actual growth occurs in the cambium layer which is between the sapwood and inner bark and extends over the whole tree (trunk, branches, roots, etc.). It completely encloses the woody aerial parts of the tree like a sheath while in the roots, although not continuous around the roots, it extends along them as longitudinal strips. It is only one cell in width (when not dividing) and does not increase in width during the life of the plant. Growth occurs by the cambium cell dividing into two smaller daughter cells which subsequently enlarge to normal size. It divides regularly and creates new cells to allow for expanding diameter, increasing height and lengthening roots. Being situated between the wood and bark, it gives rise to wood on the inside and bark on the outside. The youngest wood, therefore, is on the outside of the trunk and the youngest bark on the inside of the bark. Once the cambium has divided the newly-created cells will remain in the same position for the rest of the life of the tree, for example, those formed at 3 metres from the ground will remain at 3 metres, even if the tree grows to a 100 metres or more.

A specialized form of growth is shown by the cork cambium. It does not exist as a defined structure but can arise anywhere in the inner layers of the bark in response to a stimulus, such as injury to the bark, and produces a layer of corky cells which seals off the inner tissues from the atmosphere.

# PHYSIOLOGY

## ROOT ABSORPTION

Except for carbon dioxide a plant obtains all its nutrients from or by means of the soil. They cannot be absorbed in the solid state but must be dissolved in soil water. The root hairs penetrate the soil and come into intimate contact with the soil particles each of which has a capillary fringe of soil water containing dissolved nutrients adhering to it.

Water enters primarily by osmosis. The cell of the root hair contains a solution of sugars, etc., with a concentration that is normally higher than that of the soil solution. Water moves from the soil through the cytoplasmic membranes as it tends to equalize the concentrations, and continues to move from cell to cell in the root.

The entry of water is relatively simple but the mechanism of entry of solutes (soluble compounds mainly nutrients) is complex and not yet fully understood. Nutrients are first assembled and concentrated in the vacuole of the root hair cell. To reach the vacuole they must pass through the cytoplasm which is spread around the cell wall like a membrane, forming the cytoplasmic membrane. The cytoplasmic membrane of the cell exercises a controlling influence on entry and permeability; some solutes are admitted, others are rejected, some admitted at one time are rejected at another, smaller molecules generally find it easier to enter than larger ones, and so on. If a certain solute is being mobilized by the plant, its concentration in the cells decreases and more of it is preferentially absorbed by the root hairs. On the other hand, one element will 'antagonize' another preventing it from being absorbed in toxic quantities, for example, potassium prevents the uptake of calcium in toxic amounts.

When a solute in the soil is dissolved in soil water, it dissociates into two electrically-charged ions. The positively-charged (hydrogen and the metallic radicles) ions are known as cations and the negatively-charged ions (for example, sulphate, nitrate, etc.) are known as anions. These electrically-charged ions can be moved across membranes by an electrochemical potential. Electrical potential differences between the vacuole and the external solution of about 50 to 150 millivolts usually exist; the vacuole is generally negatively charged. A positively-charged cation, therefore, readily enters the cell but if the cell concentration of a particular cation is high the entry of more of the same cations is difficult and requires the expenditure of chemical energy. On the other hand, the entry of a negatively-charged anion (since like charges repel) also requires the expenditure of chemical energy.

Root hair cells have a life of only a few days. Once the cytoplasm is dead, there are no such restrictions on entry and any solute can pass freely. Since there is a time lapse between the death of the cytoplasm and the physical collapse of the root hair, this is a period of unrestricted entry for any solute.

The entry of nutrients, therefore, is not simple. There is a strong selective pattern with a marked differential absorption both of solute type and of occasion. Despite this, these variations act in a beneficial and regulatory manner so that the plant usually has a concentration and distribution of nutrients in the root hair cells which is fundamentally satisfactory for growth.

## MYCORRHIZA

While absorption by roots is regulated by many factors, superimposed on these is the effect of mycorrhizal relationships.

Mycorrhiza is the term given to a mutually beneficial relationship between the roots of a plant and a fungus. It is commonly seen in forest formations and is in fact necessary for the satisfactory nutrition and growth of many trees. Its importance to conifers has been known for about one hundred years but it is only in the last few decades that its occurrence in association with hardwoods has been recognized. Many Australian species require it and as more are investigated it is being found to be much more widespread than was previously realized.

Mycorrhiza in trees is usually seen externally on the roots unlike some species, such as orchids, where it occurs internally between the cells. On trees it can be seen either as a white, greyish, or light brown mass of fungal threads (hyphae) enveloping the roots, or as soft, light-coloured, thick, short branching structures which look like very short roots. It can be most easily seen on seedlings. The quantity produced by a mature stand is very large; *Pinus radiata* has about 1 tonne of mycorrhizal roots per hectare.

To those not familiar with mycorrhiza, the root system may look as though it is infected with a root disease, and it has often happened that plants with a good mycorrhizal relationship have been discarded as not suitable for planting. This is unfortunate as these are the plants which are the best and most vigorous.

Considerable research has been undertaken on the subject in the past fifty years. The relationship has many variations but, in essence, its function is that it gives the root a much larger absorbing area for water and nutrients while in exchange the fungus benefits by absorbing for its own nutrition some of the carbohydrates produced by the tree. This two-way transfer takes place via fine, short, hyphae of the fungus which penetrate the outer layers of the root cells.

Numerous fungi are involved. Species usually occur with only one or a few species of trees but there is not a strict inter-specific relation between them. Most are basidiomycetes (the toadstool and puff-ball type of fungi), for example with *Pinus radiata* two of the more important species are *Boletus granulatus* a large, brown toadstool, slimy on top and yellow underneath, and *Lactarius deliciosus* orange and white on top and often becoming slightly funnel shape when mature. Both are commonly seen under pine trees during autumn. It is sometimes necessary when establishing a nursery to introduce suitable fungi to ensure satisfactory mycorrhizal relationships. With native trees which usually are raised in containers of some kind, the soil generally has the fungus present and it is only on rare occasions that special inoculations are required. With introduced conifers, the soil is often satisfactory but where it is not inoculation is required. Laboratory produced cultures could be used but it is simpler and more positive either to grow plants from a site with mycorrhiza for a year or two, or to scatter over the soil partly decomposed litter taken from underneath older trees of the species concerned. If none of these methods are possible, such as with a completely new species, experience has shown that even in these cases a satisfactory relationship develops although it may take two or three years to do so. Apparently some species of fungus which are already present adapt to the new tree species or possibly some fungal spores may come in with the seed.

## TRANSPIRATION

### Ascent of Sap

Once soil moisture and the dissolved nutrients have entered the root hair cell they travel via the vascular system from the roots to the trunk, up to the crown and disperse again through the branches to the leaves. The ascent of this flow of sap is through the secondary xylem or wood and, since the heartwood consists of dead or mature tissues, for practical purposes the flow can be considered as occurring in the sapwood.

How sap ascends to the top of a tall tree has intrigued plant physiologists ever since they started studying trees. Many theories have been advanced but, while most of them may function in a contributory manner at some time or stage of growth, the Cohesion Theory is the most satisfactory and is now generally accepted. Some of the other theories and the objections to them are summarized below.

(i) Atmospheric pressure. The pressure of the atmosphere can only support a column of water about 11 metres high at sea level and much less at higher altitudes where many trees grow. Some eucalypts and redwoods can exceed 100 metres in height and, ignoring the frictional resistance of the cell to water movement, pressures ten times as great as atmospheric pressure would be required to reach the tallest known trees.

(ii) Osmosis. Osmosis can be a powerful force and could 'lift' sap to an almost indefinite height. But it is a slow process, a few centimetres per hour and is far too slow to explain the 2 to 3 metres per hour rate of the sapstream during spring.

(iii) Root pressure. The force which causes some cut hardwood stumps to 'bleed' for some time after cutting, was at one time widely believed to be responsible. Measurements with pressure gauges showed that maximum pressure occurred about sunrise but was much lower during the warmer part of the day when sap movement was at its greatest. Moreover, many conifers do not develop root pressure.

(iv) Biological pump. The existence of a pumping system has been postulated but this is not tenable as not only have no pumping cells been detected but sap movement can take place equally readily through dead cells.

(v) Capillarity. The rising of water in very narrow tubes is well known but in tubes of the same diameter as cells a rise of only about 1 metre is obtained.

The most widely accepted explanation of the ascent of sap is the Cohesion Theory, first suggested independently by Dixon and Renner as long ago as 1911.

It is based on the cohesiveness of water particles in a column and their resistance to being pulled apart; a concept analogous to tensile strength in solid materials.

Simply stated, as a seedling grows into a tree the 'threads' or 'columns' of water through the tissues of the stem are continually subjected to increasingly greater suction forces which tend to 'stretch' or 'pull up' the water column. The force required to pull this column up to the top of the tree comes from the evaporation (transpiration) from the leaves. Under favourable conditions water is evaporated via the stomata causing a 'pull' on the water in the tissues of the leaves and the trees. Since drying out, or the transfer of water vapour from any object permeable to water and moister than the surrounding atmosphere to the atmosphere, is one of the irresistable forces of nature, more than adequate suction force exists to cause sap to ascend.

Despite its liquidity and mobility, water has great cohesive strength, measured at well over 100 atmospheres, so that it is quite able to withstand the pull needed to draw it to the top of the tallest trees. For example, in a tree 100 metres tall, a force equal to about 10 atmospheres is required to raise water to an equivalent height. Add to this the frictional resistance of the tissues, which is a further 12 to 15 atmospheres, to give a total force of 25 to 30 atmospheres which is considerably less than the cohesive strength of water.

*Water Use*

The nutrients in the sapstream are utilized by the leaves in the manufacture of more elaborate compounds but less than 5 per cent of the water absorbed by the roots is used within the plant. Some heavy usage plants, such as maize and tobacco, retain less than 1 per cent. The remaining 95 per cent or more acts merely as a carrier of nutrients and, once it has delivered these to the leaves, it becomes surplus and is disposed of to the atmosphere via the leaf stomata. The 5 per cent or so of water remaining in the plant is used mainly (i) in the cell tissues which are 75 to 90 per cent water, (ii) as a carrier and distributor of elaborated foods from the leaves via the phloem to all parts of the plant, and (iii) a small part of it is broken down into hydrogen and oxygen during photosynthesis.

Evaporation from the crown is roughly proportional to the size of the crown. Wind is the most potent factor in that it removes the sheath of saturated air from around the leaf and increases the diffusion gradient from the leaf to the atmosphere. In high winds, the flexing of the leaf creates a 'pumping' action of the air in the tissues which increases water loss; if the weather should be cold, death of the leaf may follow as the transpiration stream may not be moving fast enough to replace the loss. Often death from this cause is blamed on frost.

During winter, transpiration is very small but in spring and early summer the quantities of water transpired by trees can be very large. It is influenced by many factors and varies widely from day to day; if the supply is good and conditions are favourable for transpiration, usage is high; if either the supply is limited or transpiration is restricted, usage is naturally much lower. On a sunny day in spring, individual trees may use 250 or more litres per day although this is much higher than the daily average. One hectare of forest in a medium rainfall area may use the equivalent of 1000 millimetres of rain; in high rainfall areas, such as some tropical forests, over 2500 millimetres per annum may be transpired and in low rainfall regions, less than 250 millimetres. (Five hundred millimetres of rain per hectare is about 5 million litres.) The amount used varies with the species; conifers and evergreens require less water than deciduous trees, even though in winter deciduous trees use only about 1 per cent of their annual consumption compared with 18 to 20 per cent for conifers and evergreens. Low-rainfall or arid species are much more economical in their use of water than those occuring in wetter climates but, as a general rule, it takes 200 to 300 litres to produce 1 kilogram of dry wood so the better the water supply, the better the growth. This may sound a large amount of water but it is not as large as the amount required by most agricultural crops; for example, maize requires 3500 to 4000 litres per kilogram of dry matter.

An actively growing productive forest needs to transpire water equivalent to perhaps 500 millimetres or more of rain a year, or probably more than the evaporation as measured by the usual meteorological methods. In the higher rainfall areas where some of the finest forests grow, usage may be higher but total annual evaporation is often less than 500 millimetres. This may seem to be a paradox or at least a finely balanced system but the answer lies in the small size of the leaf stomata. Evaporation through numerous stomata-sized small pores is about fifty times that from a single pore of the same total area (analogous to the open tank type evaporimeter) so the tree has more than adequate capacity to cope with its transpiration needs.

Figures like these are fairly general although there is a degree of consistency in them. Many species have evolved an anatomy and physiology which enables them to survive and grow in low rainfall areas, but the growth rate and ultimate total growth are usually proportionately less. Planting where a good supply of water is available is an advantage, as is shown by the improved growth made by many dry-climate plants when planted in higher rainfall zones, but even so there is a limit to the response which can be obtained.

## DROUGHT RESISTANCE

When the water supply is inadequate the first reaction (other than wilting) is for growth to slow down. If the shortage continues during the summer and the summer (or dry season) rainfall is below average, transpiration is reduced by progressive shedding of leaves, continuing if necessary until the tree is leafless. Usually the older leaves are lost first but if the leaf-shedding reaction is inadequate to maintain internal water relations, the tree may not survive, particularly in the medium to high rainfall zones. Once the crown is dry the tree is dead in most cases; there are only a few species which will re-shoot and then only if the main stem and dormant buds do not dry out. With species which coppice, felling them promptly before the stump and lower trunk are dry will give coppice growth in the autumn. Coppice growth, having an already established root system to draw on, will make quicker growth than a seedling, and from it stems can be selected to replace the lost tree.

In areas of low rainfall, the growth and metabolism of trees is the same as in better-watered regions but how the trees are able to survive is a problem which has absorbed physiologists for a long time. Succulents, which are more common in dry areas, are different from trees as they can store large volumes of water in their tissue and their stomata are only open at night. They have a modified form of metabolism, succulent metabolism, in which carbon dioxide is taken in at night and, with organic acids, is accumulated for conversion to carbohydrates by photosynthesis during daylight. As far as is known, no trees have these characteristics. (Figure 9)

To the question 'How do trees survive in low rainfall and arid zones?' there is no simple and direct explanation. Species native to dry areas have evolved a number of adaptations to cope with their situation. Many have low transpiration rates or can greatly reduce the rate under conditions of severe moisture stress. On the other hand, most may transpire just as freely as species from higher rainfall zones when the water supply is adequate. Anatomically, they usually have fewer stomata and these are often sunk into the surface of the leaf to reduce evaporation. Pubescene, or 'hairy' surface, of the leaves is common and reduces evaporation by restricting air movement close to the leaf surface. Leaves are usually thick, 'hard' and almost waterproof which reduces cuticular loss and insulates inner tissues against excessive heating and evaporation. The leaf shape may be modified to reduce the transpiring surface as in needle-like leaves, or reduced in area as in the minute leaf teeth of *Casuarina*. The actual number of leaves may be small and the crown thin and sparse; features which are common to many dry-climate species.

Physiologically, water loss can be controlled very significantly by the stomata. They may open only late at night or early in the morning when humidity is higher or when dew is likely to occur, and close by mid-morning to stop further transpiration loss. If the stress is prolonged, stomata may be closed for long periods and open only for short periods to discharge gaseous waste (mainly water vapour and oxygen) and take in carbon dioxide. Growth is then almost at a standstill as arid-zone trees still require about the same amount of water to produce the same amount of dry matter as trees in better watered areas. However, contrary to what is often believed, they do not have the capacity to recycle water within the tree. (Figure 9)

Many species tolerating dry conditions have both a surface and a deep root system (soil conditions permitting). The surface roots take advantage of light

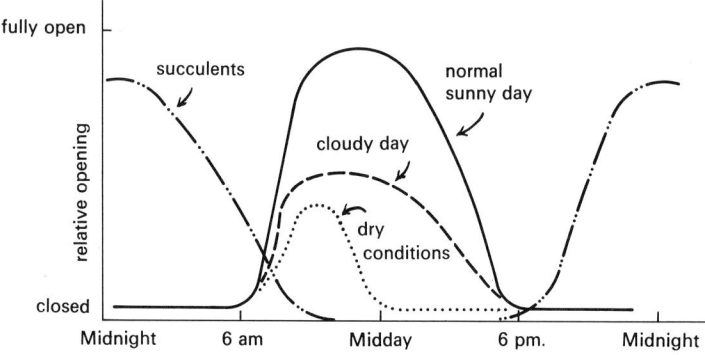

Figure 9. Typical pattern of stomatal opening.

falls of rain and the deeper roots tap water caught in the lower levels from previous heavy rainfall. Sugar gum (*Eucalyptus cladocalyx*) is a good example of this class. Others have a strong, deep root system which may reach the actual ground water and obtain an adequate supply regardless of the atmospheric conditions above ground, or the tap root extension downwards may be sufficiently rapid to keep ahead of the progressive drying out of the soil from the surface.

Sandy or light soils, common in many of the drier districts, do not have a large water-holding capacity (see Soils, page 64) but even a low annual rainfall will give a large subsoil reserve. Much of this gradually seeps through to the natural drainage lines which give trees growing along dry watercourses an adequate supply. Red gum (*Eucalyptus camaldulensis*), which, in south-eastern Australia normally grows along streams, is found in many parts of the inland on dry watercourses where it obtains water from subsoil reserves.

Species which have root systems enabling them to tap the water stored deeper in the soil frequently do not have the anatomical modifications referred to above. Except for their root system, they are usually similar to or may even be the same as species growing in higher rainfall areas.

## COLD RESISTANCE

Resistance to cold may seem to be a long way from drought resistance but physiologically they are both concerned with water shortage. It is only very rarely that the protoplasm of a cell freezes under natural conditions. In cold or frosty weather, ice crystals form in the spaces between the cells and water moves out of the cell to build up the ice crystals. This results in the cell contents being under the same kind of water stress as in drought conditions. While water expands on freezing the tissue as a whole contracts with cold.

On thawing, water moves back into the cell by osmosis. The degree of resistance to frost damage is related to the speed with which this takes place. If the flow back into the cell is fast, little damage occurs and the plant is regarded as frost-resistant, but if it is slow, metabolism cannot resume and damage follows so the plant is frost-sensitive. If chilling continues and water removal reduces the water content of the cell to less than about 75 to 80 per cent of its normal level, cells of most plants are killed. However, some alpine species can withstand a reduction to about 30 per cent of normal water content.

The damage by frost happens in the thawing and not in the freezing cycle. Rapid thawing of any plant will cause injury. The leaf surface temperature does not rise above freezing point (0°C) as long as there is frost on the leaf but, once this has been melted by the sun, a comparatively rapid rise in temperature follows. Washing off the frost by hosing or spray irrigation about sunrise enables leaf temperature to rise more slowly and reduces the risk of injury.

A certain degree of frost-hardiness can be induced in the nursery but there is not much that can be done to planted-out stock. Nitrogen encourages soft growth which is more liable to damage, so nitrogenous fertilizers should not be applied after mid-summer where there is a likelihood of frosts in autumn and winter.

Frost-sensitive plants are less likely to be damaged as they grow taller. This is not because they develop frost resistance (their resistance is the same throughout their life) but because with increasing height they grow above the frost level. In a frost, ground or grass temperature is usually 3 to 5°C lower than the temperature at 1 to 2 metres above ground and once plants are above this height the risk of frost damage is reduced considerably. (In the settled areas of southern Australia frosts of −5°C grass temperature are uncommon.)

Where practicable, covering plants gives good protection. The heat absorbed by the ground during the day is radiated back to the atmosphere at night and the covering functions as a blanket which reduces the rate of heat dissipation to the atmosphere. It may also prevent cold air settling around the plant particularly if the plant is completely covered. This method also traps the radiated heat from the soil which maintains the temperature inside the covering up to several degrees above the general air temperature.

In alpine regions, plants often survive below the snow but the part above it is killed. This is due to desiccation as the atmosphere, even though it is cold, frequently has a low relative humidity and is dry as the moisture has been 'frozen out' by the cold. As the state of metabolism in the plant is low during the winter, water movement is not rapid enough to replace that lost by evaporation and the part above the snow dies from lack of moisture, just the same as it would under arid conditions.

Frost-hardiness is a very variable factor. In the nursery, substantial hardiness can be induced in plants native to temperate regions but most tropical plants do not seem to respond. The survival of plants in colder areas depends greatly on the extent to which frost-hardiness develops naturally or has been induced, but some nurseries nowadays are promoting rapid growth without giving the plant an opportunity to harden-off. While this is acceptable for indoor plants and those planted in frost-free areas, heavy losses may be experienced in colder situations.

Frost-hardiness is not constant throughout the year. In mid-summer plants may be killed at 5°C (if these temperatures occur) yet may withstand cold to −30°C in winter. The annual killing temperature range is usually at least 10°C

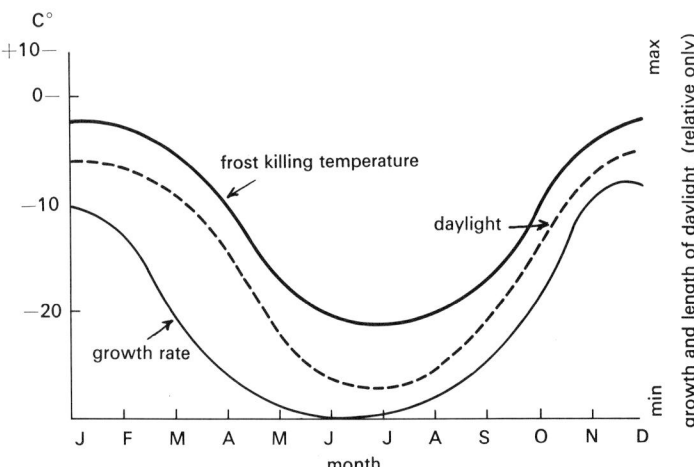

Figure 10. Relation of frost hardiness to growth rate and length of days.

but in many species, particularly deciduous trees and conifers, it may exceed 40°C.

The induction of frost-hardiness occurs gradually through the autumn and is influenced mainly by growth rate and length of daylight (Fgure 10). In actively-growing plants, hardiness is poor and only develops when growth slows down, as it normally does during autumn. Growth and hardiness are related inversely; fast growth gives poor hardiness, slow growth induces greater hardiness. Therefore, for hardening-off, growth should be checked. Good nursery practice, from late summer, is to cease any fertilizing program and withhold watering to just when incipient wilting appears. Obviously, then slow-release fertilizers which maintain growth through the autumn should not be used in raising plants intended for areas where there is any risk of frost or snow.

Checking growth rate on its own will not induce very much hardiness; it is also closely linked with the length of daylight. The shorter days of autumn promote it naturally and while dark-shading in the nursery would reduce the length of daylight, it adds little to the effect of the naturally shortening days. There is, therefore, little practical application in this direction. However, the opposite effect can occasionally be seen in city trees where there is an absence of branches around street lights. Even though the light is weak, it is a sufficient extension of daylight to prevent frost-hardiness developing to the same extent as in the rest of the crown; the foliage and twigs nearest the light are frost-killed.

Hardening-off cannot be commenced under warm conditions. There is a threshold temperature of about 5°C to 10°C above which it will not begin, irrespective of growth and light conditions. Continuous low temperature is also not required; the alternating cool and warm temperatures during frosty weather will induce hardiness as well as continuous cold. As the period of hardening continues, plants build up greater resistance but there is a limit. Even though cold treatment may be maintained during winter, when the warmer conditions of later winter to early spring arrive the plant loses its hardiness. Inducing frost-hardiness is therefore essentially an autumn activity.

## PHOTOSYNTHESIS

The primary function of leaves is photosynthesis or the manufacture of food from simple inorganic materials. For thousands of years Man has appreciated the value of animal fertilizers on crops but it has only been in the last 200 years that the role of the leaf has begun to be understood. Yet by far the greater mass of plant growth originates in the leaves; for every 100-kilogram increase in the weight of a plant the soil in which it is growing will decrease in weight by only a few grams (excluding water).

The first clue to photosynthesis came in 1732 when Priestly found that air in which animals could no longer live was 'restored' when green plants were grown in it. Further experiments showed that only the green parts of the plant were involved but it was not until about seventy to eighty years ago that the general outlines of photosynthesis were evolved. Even today it is far from being fully understood.

Chlorophyll, the green pigment of leaves, is the activator of the process and is itself synthesized in the leaves by a little known process. It is a complex protein of carbon, hydrogen, oxygen, nitrogen and magnesium with the formulae $C_{55}H_{72}O_5N_4Mg$ for chlorophyll a, and $C_{55}H_{70}O_6N_4Mg$ for chlorophyll b.

The general principle of photosynthesis is that the chlorophyll in the leaf uses the energy of sunlight (see also under Shade, page 105) to combine carbon dioxide from the atmosphere with water to form sugar and release oxygen. The simplified general equation is:

$$6CO_2 \;+\; 6H_2O \xrightarrow{\text{light energy}} C_6H_{12}O_2 + 6O_2$$

carbon dioxide     water                 glucose    oxygen

The energy for the reaction comes from sunlight. Less than 3 per cent of the light falling on a leaf is used. (Most plants could grow under shade with only 10 to 20 per cent full sunlight provided other factors such as root competition for

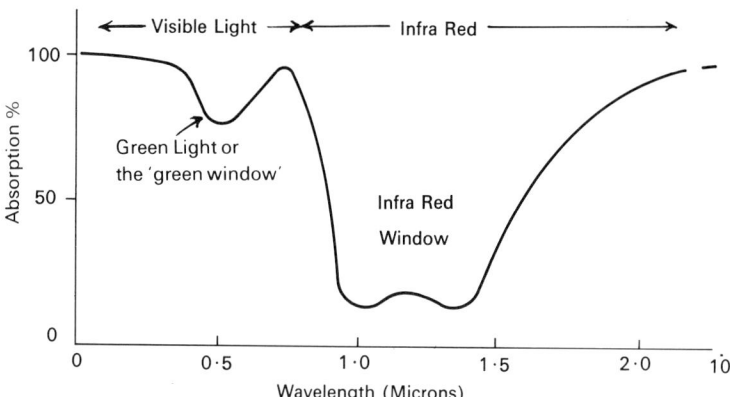

Figure 11. Absorption of light by leaves.

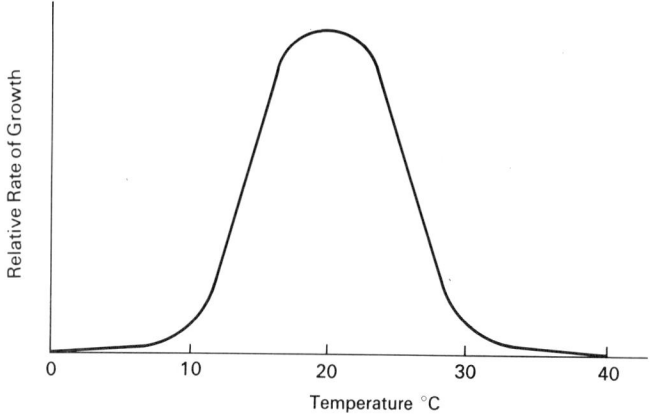

Figure 12. Growth and air temperature.

moisture and nutrients, soil temperature, etc. are the same as in the open, but such conditions can usually only be obtained in the laboratory.) Carbon dioxide, which is present in the atmosphere to about 0·03 per cent, enters the leaf via the stomata in the course of general air circulation within the leaf. Water from the soil is already in the leaf. Growth proceeds best at 18°C to 22°C for plants from temperate regions and at 24°C to 28°C for those of tropical origin. As temperature increases or decreases, growth declines and virtually ceases at about 32°C to 34°C or 5°C to 7°C. (Figure 11 and 12)

It is only since World War II, and with the availability of radioactive tracers, that some knowledge of the mechanism of photosynthesis has been obtained. The initial events happen rapidly and concurrently; at least twelve major steps take place in 6 to 7 seconds before glucose phosphate, the first stable compound, is formed. Chlorophyll, using the energy of light, splits water molecules into their constituent atoms of hydrogen and oxygen. Most of the oxygen is released to the atmosphere, but some of the split molecules re-unite, releasing energy which is used to manufacture adenosine triphosphate (ATP). ATP provides the energy for the subsequent reactions but is not itself consumed in them. Carbon dioxide taken in by the cell from the atmosphere unites with another compound stored in the plant, ribulose diphosphate (RDP). The hydrogen from the split water molecules now combines with RDP to form phosphoglyceric acid (PGA) molecules which then react to form sugar phosphates, the first stable products of photosynthesis.

The formation of sugar is continuous in all green leaves in light and sugar is present in the sap of almost all cells. The presence of a high level of sugar in a chlorophyll-containing cell would inhibit further manufacture. Some of it is used in other reactions but most is converted into more complex sugars and finally into related large starch molecules $(C_6H_{10}O_5)_n$ in which form it is stored for future use. Little sugar and starch are stored in the leaf and then only temporarily during the day, almost all is translocated to other parts of the plant where the concentration is lower or for storage in the cells of the sapwood region. Sugar translocation is via the phloem (inner bark) tissues which usually contain 5 to 10 per cent sugar. It is extremely fast but the mechanism by which it is achieved is still one of the unknowns of plant growth. When in storage, starches are very stable products, but when they are required as sources of energy for all the processes of growth and metabolism they are brought into a highly reactive state by ATP.

The rate of sugar and starch production by leaves in sunlight is very rapid, about 0·5 to 2·0 grams per square metre per hour. Calculations have been made by some authorities which show that plants produce 150 000 to 250 000 million tonnes of sugar a year, which is much more than the 300 million tonnes produced commercially.

Using carbohydrates (sugars and starches) as a basis, the other substances within plants are manufactured from them. Cellulose $(C_6H_{12}O_5)_n$, the main constituent of the cell wall, is in the same homologous series as starch. Lignin is a complex compound, consisting mainly of aromatic hydrocarbons, whose molecular structure still needs to be precisely determined. Fats are generally similar to carbohydrates except that they are the esters of the fatty acids and glycerine $(C_3H_5(OH)_3)$. Amino acids are modified from carbohydrates by the addition of nitrogen to form the basis for proteins. Proteins are a large group of

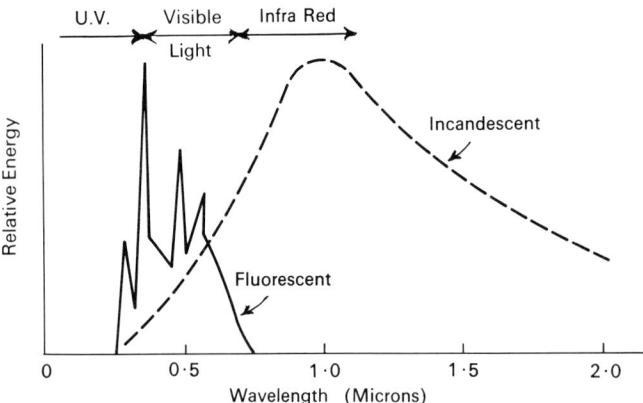

Figure 13. Emission spectra of electric lights.

very variable and often unstable compounds generally consisting of approximately 51 per cent carbon, 25 per cent oxygen, 7 per cent hydrogen and 16 per cent nitrogen, sometimes with 0.4 per cent sulphur or phosphorus.

Resin, waxes, aromatic oils, gums, kino, turpentine, terpenes, etc. are all derived, often by a long series of reactions, from sugar. Each stage of every reaction has a particular enzyme to trigger off the sequence of events, consequently hundreds of different enzymes are involved in the growth of one plant.

Different plants may respond differently to the colours of the sunlight spectrum. Artificial light is sometimes advocated as a means of stimultaing growth but its effect will depend on its colour components and whether they are the optimum for a particular plant. Attempts to use artificial light as a stimulus to growth on a commercial scale, as in nurseries, have met with mixed success. In a very general way a plant will respond to those colours which are complementary to its own colour as the colour of the plant as we see it is due to reflected light. The other colours are absorbed and provide the energy to activate the chlorophyll and promote photosynthesis. (Figure 13)

Stimulating growth by the technique of increasing the available raw material, carbon dioxide, has met with some success in recent years. It is becoming increasingly used in nursery practice where good increases in growth rate have been obtained by enriching the atmosphere in a restricted space with commercial carbon dioxide. Concentrations as high as 2 per cent, or nearly seventy times that occurring naturally in the atmosphere, are commonly used.

## RESPIRATION

Plants require a source of energy to carry out their normal function of growth. In photosynthesis, glucose (and other carbohydrates) are formed by the energy of sunlight acting on carbon dioxide and water. The energy thus contained in glucose is released in respiration by the reverse of the photosynthetic process:

$$C_6H_{12}O_6 + 6O_2 \rightarrow 6CO_2 + 6H_2O + \text{chemical energy}$$

glucose      oxygen      carbon dioxide      water

Sugars when combined with oxygen, and under the control of ATP, can release chemical energy (1 gram of glucose releases 3·74 kilogram calories of heat) and this is used by the plant in its other processes of metabolism. In this reaction oxygen is used up and carbon dioxide is released, just as in the respiration of animals. To keep things in perspective, the amounts of carbon dioxide and oxygen involved are very small in relation to those in photosynthesis. Moreover, most of the compounds concerned in respiration are either obtained from or used within the plant and have little effect on the atmosphere.

Both respiration and photosynthesis proceed at the same time although the latter, because it requires light, is limited to daylight. Respiration is also aided by a large number of enzymes, such as the zymases which convert sugars to carbon dioxide and alcohol, but each such group of enzymes is specific to one series of reactions.

## TEMPERATURE

Growth, photosynthesis and respiration are very largely dependent on temperature. For plants from temperate regions, growth commences at 5°C to 7°C, reaches a peak at 18°C to 22°C and declines at higher temperatures, finally ceasing at 32°C to 34°C. Within this range, optimum temperature for development (emergence, growth, flowering, seeding, etc.) varies over about 4 to 6°C; optimum temperature for flowering and seeding is about 5°C higher than that for vegetative growth. (Figure 14)

Growth varies from year to year: some years are good growing years while others are average or poor. Qualitative terms like this, as they are essentially subjective, are not very satisfactory in making comparisons from year to year. To overcome this problem the concept of total heat units for the season is frequently used, particularly in ecological work. For a system to be useful, readily available units of measurement are necessary and, in this case, the common meteorological ones of maximum and minimum temperature are used. Briefly, seasonal total heat units are the sum of the mean of maximum and minimum temperature for each day minus 5°C, as 5°C is about the temperature at which

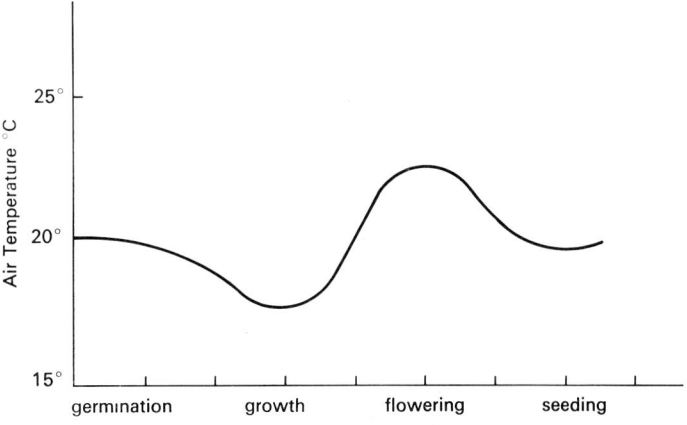

Figure 14. Relative optimum temperature at various stages of growth.

growth commences in south-eastern Australia. (In tropical regions, a correction of 7 to 8°C may be more appropriate.) Expressed as a formula:

$$\text{total heat units} = \left(\frac{T_M + T_m}{2} - 5\right)$$

where $T_M$ and $T_m$ are the maximum and minimum temperatures respectively. If the mean of $T_M$ and $T_m$ is less than 5°C the term inside the brackets is rated as zero.

Like all systems, this one is not without fault, for example, during summer heatwaves the mean temperature (less 5°) may be above the optimum for growth and affects the result disproportionately. As these days are only a few in the eight to nine months of the spring, summer and autumn growing season the distortion is not large. Even so the heat unit concept expressed as 'growing degree-days' is a vast improvement on the subjective method of 'a good growing season' or 'a poor growing season' and deserves much wider use.

## NUTRIENTS

The elements essential for growth are carbon, oxygen, hydrogen, nitrogen, potassium, calcium, magnesium, phosphorus, sulphur, iron, chlorine, manganese, zinc, boron, copper and molybdenum. The first two are obtained from the air as carbon dioxide, hydrogen is obtained from water absorbed by the plant, nitrogen indirectly from the air via or directly from the soil, and the remainder from the soil. The first nine are sometimes known as the major elements and the last seven (from iron) as the minor or trace elements.

Many other elements are sometimes found in plants, such as silica, aluminium, sodium and titanium but their exact roles, if any, are not yet clear. Silica is absorbed in large quantities by some hardwood trees, for example, a few Australian rain-forest trees have up to 1 per cent silica in their tissues. (Over 0·5 per cent silica quickly blunts any tools being used on the tree.)

Relative concentrations of nutrients in plants vary according to species but the following table will serve as a guide:

| Element major | ppm* | Element minor or trace | ppm* |
|---|---|---|---|
| Carbon | 450 000 | Iron | 100 |
| Oxygen | 450 000 | Chlorine | 100 |
| Hydrogen | 60 000 | Manganese | 50 |
| Nitrogen | 15 000 | Zinc | 20 |
| Potassium | 10 000 | Boron | 20 |
| Calcium | 5 000 | Copper | 6 |
| Magnesium | 2 000 | Molybdenum | 1 |
| Phosphorus | 2 000 | | |
| Sulphur | 1 000 | | |

\* ppm—parts per million
(Adapted from *Plant Physiology* by Salisbury and Ross.)

Carbon, oxygen and hydrogen are the basic materials for all tissues and constitute about 95 to 96 per cent of the plant.

*Nitrogen* is essential for chlorophyll production and for amino acids which are used in protein synthesis. Nitrogen deficiency is shown by a general yellowing of the foliage; the younger leaves remain green longer since nitrogen is remobilized from older tissues and transported to them. Plants with over-supply of nitrogen are dark green in colour and have dense foliage. Excess nitrogen delays flowering and seed-set.

*Phosphorus* is necessary as the activator of energy relations and conversion of sugars. It occurs in flowers and seeds and in small amounts in proteins. It is readily remobilized and transferred from older to younger tissues when in short supply. A strong interaction exists between nitrogen and phosphorus. Deficiency is shown by stunted growth and dark green foliage.

*Potassium* does not form a permanent part of the structure of the plant even though it is required in relatively large amounts. It acts mainly as an activator (or co-enzyme) of many enzymes which need to be 'triggered' before they can initiate their particular metabolic process. It is also required in large amounts for protein synthesis but does not become part of the protein structure. It is easily redistributed in the plant from older to younger tissues. Deficiency symptoms are a yellowing of the leaves, lesions on the leaf surface, and a drooping of the smaller and younger shoots.

*Calcium* is found in cell walls in association with the bonding substance pectin, in cell vacuoles as crystals of calcium oxalate, and as a part of a few enzymes. It is not very mobile in the plant. Recent work suggests that it may be used only in small amounts by plants, one of its main roles possibly being to reduce, by competition ('antagonism'), the entry of more toxic elements into the plant.

*Magnesium* is an essential part of chlorophyll and activates all of the enzymes (so far as is known) utilizing energy available from the ATP process. It is not very mobile in the plant. Deficiency symptoms are a yellowing of the foliage between the veins, occurring first on the older leaves.

*Sulphur* is a part of some proteins. It appears to be absorbed in greater quantities than necessary and incorporated into various compounds. It is not easily remobilized and so deficiencies usually show first in the younger tissues as a general yellowing. However, a shortage of sulphur is uncommon in southern Australia.

The minor or trace elements are involved mainly as activators of enzymes or as an essential part of some metabolic process. Only rarely are they incorporated into any stable part of the tissues.

*Iron* is necessary for chlorophyll production, although not incorporated in it, and occurs in some colouring pigments. It is very immobile and once lodged rarely moves. Deficiency is a yellowing between the veins as for magnesium, but in the younger leaves.

*Chlorine* has only been recognized as necessary for plant growth since about 1953. Its role appears to be as an activator for the splitting of water molecules in the early stages of photosynthesis. Deficiency symptoms are a wilting of growing tips followed by yellowing or, in the roots, a stunting and thickening of the tip.

*Manganese* is an enzyme activator for chlorophyll synthesis and for fatty and nucleic acids formation. Deficiencies are very uncommon in trees but fairly common in agricultural crops. Yellow spots and lesions are typical.

*Zinc* is an enzyme activator for cell metabolic processes and the formation of growth substances. It is fairly mobile in the plant. Symptoms are reduced leaf size and shoot growth leading to rosetting. It is not known to be connected with chlorophyll formation hence affected tissues remain green. It is a common deficiency of sandy soils in southern Australia but can be controlled easily and cheaply by spraying the foliage of trees with a 2·5 to 5·0 kilograms of zinc sulphate in water per hectare.

*Boron.* Little is known of its role but it appears to be connected with sugar transport. It is immobile in the plant. Deficiencies occur in trees but are not common. Symptoms are a reduced growth of the growing tips leading to a bushy habit. It can be controlled by spraying with borates.

*Copper.* Like boron, its role is not well understood other than that it is connected with a few enzymes and seems to be involved in the transport system. Deficiencies in tree growth are rare, it is doubtful if they have ever been positively identified in the field. However, nutrient solution cultures show it to be necessary and, if absent, the leaves turn dark green and twist and distort rather severely.

*Molybdenum* is necessary for the enzyme involved in the nitrate to ammonium conversion. Deficiencies are uncommon in trees but well known in agricultural crops. Symptoms are a yellowing of the older leaves, progressing to the younger.

## GROWTH REGULATORS

The processes of growth do not proceed haphazardly. Growth regulators exist in plants which determine that the growing tips grows upward, roots grow downwards, flowering occurs at a certain time, and so on. Some of these are conditioned by temperature, soil moisture, nutrient supply, photoperiodism, etc. but practically every facet of growth seems to be under some degree of control by growth regulators.

The presence of growth regulating substances has been known or suspected for about fifty years. However, it is only in the last thirty years or so that they have been identified, and the extent of their influence has begun to be understood. All are produced in extremely small quantities; amounts of one hundredth part per million being adequate to control growth. Several groups are recognized.

### (a) *Auxins*

These are produced in the growing shoot, a few millimetres from the tip. They stimulate cell growth and move downward through the stem, partly because they are polarized, i.e. move physiologically towards the roots, and partly because they appear to be under the influence of gravity. They exert an inhibiting influence on buds lower down the stem and restrict their development but if the uppermost growing tip is removed, i.e. the source of the auxin and its inhibiting effect is removed, the lower buds then develop at a faster rate. This is seen in practice when the leading or apical shoot of a tree is removed or damaged and the lateral buds can then develop and promote branch development. In a bent stem, auxins collect on the underside stimulating enlargement of the cells in this region so that the stem gradually straightens up to an upright position. Their effect on roots is the opposite to stems (possibly connected with the absence of light) as they restrict growth. Any accumulation on one side of a root retards growth on that side so the root bends downwards. The reasons for these different reactions between stems and roots are not yet fully known.

The first auxin identified was indole acetic acid (IAA) which was found to stimulate root initiation in cuttings. Since the extraction of natural auxins would be a formidable task because of the very small amounts produced in plants, synthetic products have been developed. The related compounds indole butyric acid (IBA) and naphthalene acetic acid (NPA) have been found to be very effective and are now widely used as powder and solution dips for cuttings. When mixed with lanoline and smeared on a cut stem, they promote callus formation.

The synthetic auxin-like phenoxyacetic acids, mainly 2,4-dichlorophen-oxyacetic acid (24D) and 2,4,5-trichlorophenoxyacetic acid (245T) are widely used in their amine and ester forms to control the growth of broad-leaf weeds, blackberries, silver wattle, etc.

## (b) *Gibberellins*

These were first discovered in fungi but at least twenty-four different ones have now been identified in the higher plants. They promote stem elongation; an excess amount applied to a plant will give elongated stems and a shortage results in dwarf plants. Not all dwarf plants are due to a shortage of gibberellins as many are of genetic origin. Their action is to increase the length of cells rather than the number. Gibberellins also interact with the effect of daylight in the initiation of flowering and with cold temperature in the breaking of dormancy in seeds.

## (c) *Photoperiodism and Phytochrome*

Most plants remain in the vegetative stage until they have been subjected to an alternation of light and dark periods of certain lengths. This characteristic is known as photoperiodism and initiates flowering. Plants requiring twelve to sixteen hours darkness in each twenty-four hours for flower initiation are short-day plants, those requiring only eight to twelve hours are long-day plants. Some, such as tomatoes, are not affected by the length of the dark period and are neutral-day plants. The relative length of day and night (light and dark) varies with the season and latitude and accordingly causes flowering to take place at certain times of the year irrespective of the size of the plant. But, if the dark period is interrupted by a flash of light for a brief period, the flowering cycle is upset and inhibited; light such as ordinary household lighting, or a minimum of one two-thousandth of full sunlight for one minute is sufficient to upset the cycle.

The substance involved in controlling flowering and fruit (seed) production is phytochrome, a blue pigment produced in minute quantities in the leaves. It occurs in two interconvertible forms, each of which can reverse the effect of the other. Each form is motivated by light in the red portion of the spectrum; red light of about 660 millimicrons wavelength stimulates flowering but far-red of about 730 millimicrons, which is near the limit of human vision at 760 millimicrons or almost into the infra-red band, inhibits flowering. If the two wavelengths of red light are repeated alternately, the last one determines whether flowering is stimulated or inhibited.

Phytochrome is believed to be in all plants, including most algae, although its presence in conifers is only suspected rather than proved. It initiates the formation of growth-control hormones and consequently has an effect on many physiological processes in addition to flowering and fruit production. Normal white light, such as sunlight and incandescent light, contains both red and far-red and, being somewhat richer in far-red, it inhibits flowering and promotes vegetative development. On the other hand fluorescent light, which is almost lacking in far-red, favours flower production. In commercial production of plants, as in glasshouses, the form of lighting appropriate to the objects of management needs to be used; if growth is sought use incandescent light, if flowers, use fluorescent light.

### (d) *Miscellaneous*

Cytokinins stimulate cell division and so delay the onset of senescence. Synthetic cytokinins are sometimes used to prolong the life of cut flowers, vegetables, etc. There are also numerous other compounds which control specific functions but as they have not yet been developed to the stage of being of use in practice, no further reference is made to them.

## COLOUR

Colour is one of the most distinctive and sought-after features in plants. In almost all cases it is due to compounds within the cell and only rarely to substances in the cell wall or leaf cuticle. (The bluish tone of many leaves, particularly eucalypts, is not an integral part of the leaf but is due to a waxy bloom on the leaf surface.) Many hundreds of pigments are found in plants but they may be grouped into three main classes (a) chlorophyll (b) carotenoids and (c) anthocyanins.

### (a) *Chlorophyll*

Green colour is due to chlorophyll which strongly absorbs the red and blue wavelengths for photosynthesis. Coloured foliage, such as golden, purple and red varieties, still contains chlorophyll but the green colour is masked by the more abundant compounds of other colours. The presence of these does not prevent the green chlorophyll performing its normal function of photosynthesis.

### (b) *Carotenoids*

Carotenoid pigments are contained in small bodies, chromoplasts, in the cell. In some species they exist in prolific numbers and impart their colour to the plant. Other than giving colour their exact function is not accurately known, but they are believed to be linked with chlorophyll in photosynthesis. They comprise a large number of pigments which fall into two groups: (i) carotenes with yellow tones and (ii) xanthophylls with red to red-brown colourings.

(i) Carotenes are yellow and their relative abundance affects the shade of leaf green, for example, light green to golden varieties of foliage. They are also responsible for the colour of yellow flowers. As they are chemically very stable, they remain in the leaves of deciduous trees after the chlorophyll has disappeared which results in the characteristic yellow autumn leaf. Their stability enables them to withstand digestive processes and they give eggs and butter their colour.

(ii) Xanthophylls are red to red-brown and cause the red and brown tonings of some variegated leaves. They occur more frequently in seaweeds to which they give the typical brown colour. However, the brown colour of autumn leaves is not due to xanthophylls but to an oxidation process following the death of the leaf.

(c) *Anthocyanins*

Unlike chlorophyll and carotenoids, these do not occur as particles in the cell but are dissolved in the cell sap. They give the red, blue and purple tones to flowers and leaves but otherwise their role in the plant is not known. Anthocyanins require high light intensity and high sugar concentrations for their formation. A few frosts in early autumn cause a drop in plant temperature and in the rate of metabolism which leads to more sugar being retained in the leaves. With a light intensity in early autumn which is higher than a month or two later, more anthocyanins are produced giving a much better display of autumn leaf colours. They are affected by acidity and turn red in an acid medium and blue as the medium becomes less acid. Hence flower colours may be determined by cell sap acidity which in turn may be modified by adjusting the soil acidity or alkalinity. However, cell sap acidity alone does not determine colour as colour may be influenced by the presence of other pigments (principally tannins and flavones) and by the state of anthocyanin in solution. If it is present in true solution, it is red; in colloidal dispersion, it gives blue colours. The action of 'bluing' mixtures is believed to be by precipitation of anthocyanins in colloidal form and not by reducing soil acidity. (Aluminium sulphate is commonly used in such mixtures; it increases soil acidity which favours the production of red rather than blue tones.)

## SEED

Seed is the living link between one generation and the next, and contains all the hereditary material to transmit the characteristics of the species. A seed is a developed embryo; it is a miniature plant comprising seed leaves (cotyledons), bud (plumule), stem (hypocotyl) and a rudimentary root (radicle). Monocotyledons such as palms, lilies and grasses have one cotyledon, dicotyledons or the flowering trees and shrubs have two, while conifers have from two to fifteen depending on the species. Cotyledons may manufacture food for the developing seedlings until true leaves are formed or may serve as storage organs to support the developing seedling.

On ripening, the simple sugars, amino acids and fatty acids of the seed are converted into more complex carbohydrates, fats, oils and proteins. These are for the initial nourishment of the future plant and are stored either in the cotyledons or endosperm. In plants where nutrients are stored in the cotyledons, these are large and fleshy and comprise most of the seed; when stored in separate tissue, the endosperm, the cotyledons and other seed parts are much smaller (often microscopic) in size.

The chief function of the seed coat (testa) is to protect the seed. It may be soft, hard, or occasionally leathery, and has a surface which may vary from smooth to rough. In a general way hard-coated seeds remain viable for much longer than 'soft' seed. Seed longevity may be only a day or two as in poplars, or up to about one hundred years as in *Acacia* (although at this age the percentage germination may be very low). Normally after three to five years the germination drops considerably even with species of noted longevity. Since a seed is an embryo plant some respiration is always taking place; reducing respiration by storing at just above freezing point will prolong seed life to at least twice that of seeds stored at ambient temperature. Hard-coated seeds often need treatment to ensure germination; acid treatment, immersion in boiling water, or scarifying are the most usual methods. Under natural conditions fire or abrasion are the main means of seed coat breakdown.

Inhibitors are present in the seed coats of many species. They comprise a wide range of compounds from common salt (sodium chloride), ammonia or cyanide releasing substances, to more complex compounds. While they prevent seed germinating in the fruit where moisture, temperature and other conditions are ideal their main interest is ecological. These substances must be leached from the seed coat by rain before germination can occur; if sufficient rain falls to remove them then there will be sufficient moisture in the soil to ensure the survival of the germinate. Several Australian species from the drier regions have common salt in the seed coat and good soaking rains are necessary to leach out most of it and wash it from the immediate surroundings of the seed before germination and growth can take place. Consequently some native seeds need to be soaked in several changes of water before sowing.

Dispersal of seeds may be by wind, water, birds or animals. Many seeds have wings (pines, maples) or hair-like structures (pappi) for wind dispersal. Water-dispersed seeds have a waterproof coat and sometimes an air sack to aid flotation. Birds and animals aid dispersion by eating the seed and passing it through the digestive system or the seed may be sticky and adhere to the feathers or fur. Hooks, spines and ejection mechanisms also occur in some species. On the whole, Australian native species are not well equipped for dispersal by wind; most are limited to free fall from the plant which restricts their spread to two to three times tree height. Long-range dispersal is mainly via birds and animals.

Large quantities of seeds are produced by most plants but only a small proportion actually germinate and produce new plants. Losses are high; birds and animals use them for food while with the smaller seeds, particularly natives, a large proportion is harvested by ants. Germination occurs once moisture is available, but contact with soil is necessary for growth. Litter on the ground prevents most seeds making contact and it is only those which actually fall on to favourable soil conditions that are able to grow into a new plant. Of the thousands or possibly millions of seeds produced by some trees only a few actually end up as a new plant. If a good crop of seedlings is desired, the ground litter needs to be reduced; in large areas a light fire with a flame height of about 0·3 to 0·5 metres is the most effective and least damaging means. Otherwise, cultivation is necessary.

Dormancy in seeds is common in many species, reducing or delaying germination. It can only be broken by a certain period of cold; in days gone by seeds were packed in layers or strata in moist soil, sawdust, etc., and left in the open for the winter. By spring, dormancy had been broken and good germination was obtained. The technique was known as stratification and is widely practised but nowadays the seed is usually soaked in water for one to two days then stored in a refrigerator at 2°C to 4°C for several weeks. The length of cold storage varies from three to fifteen weeks with different species. Stratification also reduces the germination and emergence time to about one-third or half the time for untreated seed.

Seeds must be dry for storage; drying in summer sun is adequate. They should be kept in air- and light-tight containers, preferably under refrigeration at 1°C to 2°C for maximum life. Nut-like seeds (acorns, walnuts, chestnuts, etc.) must be kept in moist sand, sawdust, peatmoss, etc. to prevent excessive drying out otherwise they will remain viable for only a few days after falling.

# SOIL PHYSICS

Soil provides trees with mechanical support and anchorage, acts as a reservoir of water, and provides some of the nutrients. Excluding water, it includes mainly mineral fractions of sand, silt, clay and re-formed materials such as buckshot gravel, limestone nodules, etc., together with the organic matter and its population of micro-organisms.

The physical condition of soil is of fundamental importance since this determines moisture relationships, drainage and aeration, and the extent to which the root system can penetrate. If the physical conditions are good, then solutions to many other growth problems usually follow without much trouble. All too frequently the chemical aspect is over-emphasized but, while the nutrient level is important, any deficiencies can usually be overcome simply by adding commercial fertilizers. The physical aspects, however, are not so easily modified but the effort and attention are worthwhile. If the physical environment is favourable, roots can occupy and utilize a greater volume of soil which enables them to adjust more readily to less favourable nutrient and moisture levels. Consequently, the physical conditions of soil are of such primary importance that they are dealt with in greater detail than is usual in most publications on plants.

## FORMATION

All soils have their genesis in the weathering of rocks, but the effect of climate, re-deposition, etc. can lead to the development of different soils from the same basic materials. For example, the same bedrock may give in a warm 600 to 700 millimetre rainfall area a yellow-grey soil with a sharp boundary between the top soil and underlying clay, and in a cooler over 1000 millimetre rainfall area a red soil with a gradual increase in clay with depth, and no sharp boundaries in the profile.

Alluvial soils are a special case and were obviously not developed from the underlying bedrock, although they may have been deposited by wash from adjoining hillsides. Their texture can vary from fine to coarse depending on the velocity of the originating load-bearing stream.

With soils developed in situ, climate, particularly rainfall, is the major formative factor. Rain-water soaking through the soil carries the finer clay particles down in suspension and always dissolves something from the soil. In wetter climates, the clay particles are gradually filtered out further down in the soil so the clay content of the surface decreases and that of the lower layers increases, while the soluble materials are eventually lost in drainage and seepage from the soil. Compounds like calcium sulphate, sodium chloride (common salt), and calcium, magnesium, and potassium bicarbonates are very mobile, while iron and aluminium are very immobile. Iron and aluminium, therefore, eventually become the main elements as the others are removed by leaching. Where leaching has continued for some time and wet conditions have evolved following the formation of a relatively impervious clay layer, buckshot gravel may form above the clay. Buckshot gravel comprises 10 to 20 per cent (up to 35 per cent) iron cementing together grains of sand, quartz, etc. Its presence indicates that poor vertical drainage has evolved and most of the nutrients have been lost by leaching, i.e. the soil is old and degraded.

In drier climates the position is different. Soluble compounds and clay particles are only carried down as far as the winter rain penetrates so clay layers or buried clay pans are formed. The soluble components, like salt, are also deposited while calcium carbonate precipitates out as limestone nodules or massive limestone. In general, limestone is likely to occur in the soil where annual rainfall is less than about 500 millimetres.

Not all water movement and leaching are vertically downwards. In hilly country lateral movement down the slope is of some consequence and can lead to enrichment of the soils further down the slope; or, where these have poor drainage characteristics, to water-logging or salting.

Laterite soils, common in the drier inland areas, are a special case. Usually they comprise an ironstone capping over soft clay and are believed to have been formed in the Tertiary period under the influence of past wet climates. However, the term laterite is often loosely used to describe any soil with ironstone present as nodules.

Crab-holey or 'gilgai' country is common in many inland areas. Crab holes only form on heavy clays which expand considerably on wetting and shrink on drying. When dry, large cracks develop and crumbs of soil fall down them. These crumbs swell on wetting causing a pressure which can only be released upwards which gives rise to the typical puff.

## CLASSIFICATION

The soil type is the basic unit, analogous to the plant name. A vertical section through the soil is called a profile and each layer of soil, that is where the soil appearance changes, is called a horizon. Each horizon in a profile is briefly described, for example, 0 to 15 centimetres grey clay loam, 15 to 40 centimetres light grey silty loam, 40 to 70 centimetres yellow-grey clay, over 70 centimetres yellow clay with mottling. The soil type is named by the texture class of the surface layer and the locality where it is found, as in the Gardiner clay loam above. (The practice of referring to the layers above the clay as the A horizon, the clay layer as the B horizon and the bedrock as the C horizon, is tending to be discontinued.)

Soil types are loosely grouped into classes such as podzol, chernozem, krasnozem, solenetz, etc. The use of such terms is not strictly accurate as many, being of Russian origin, refer to the soils formed under a winter-snow, continental type of climate which does not occur in Australia. Their use can be justified only on the grounds of convenience but not of accuracy.

An internationally acceptable system of soil classification has still to be evolved. While a particular soil type can be described, it refers just to the soil in that locality and means little to anyone not familiar with the location. This is a severe limitation on the use of the information elsewhere in Australia or overseas. Of the many systems proposed, the one being evolved in Australia (mainly by Northcott and his associates) is finding good acceptance and is worthy of a brief mention. The primary classification is on the texture pattern, a feature which is easily identified in the field. Three main classes are recognized: (a) uniform, i.e. uniform texture through the profile, (b) gradational or a gradual increase in clayiness with depth, and (c) duplex or where a sharp change to clay over about a range of 5 centimetres occurs. Supplementary classifications to cover unusual soils, such as peaty and alluvial soils, are included. The major divisions are progressively sub-divided by visual features such as calcareous and non-calcareous, presence or absence of ironstone, etc. as appropriate. The ultimate description is for the soil type as referred to above.

## MINERAL CONSTITUENTS

In the soils of southern Australia organic matter is usually less then 5 per cent (commonly less than 1 per cent) the other 95 per cent or more comprising the mineral fraction. All the mineral constituents came originally from igneous rocks, even though in the course of geological history they may have been eroded, deposited as sedimentary strata, uplifted, and weathered again to form soil. Movement by wind or water may transport soil far from the location of its first appearance. Once in situ, the type of formation developing is controlled mainly by the climate but there are many consequential reactions which can further modify its characteristics.

Sand, silt and clay are the main parts of soil. Sand and silt are particles derived from the parent rock but clay does not occur in unweathered rock; it is re-formed predominantly from minerals which are the products of weathering. It is not surprising, therefore, that there are numerous forms of clay and that the study of clay mineralogy is a science of its own.

In Australia, the mineral components, or the mechanical composition of the soil is graded according to the International System. This system recognizes five fractions, each being one-tenth of the next larger, viz:

| | |
|---|---|
| gravel | over 2 millimetres diameter |
| coarse sand | 2 to 0·2 millimetres diameter |
| fine sand | 0·2 to 0·02 millimetres diameter |
| silt | 0·02 to 0·002 millimetres diameter |
| clay | less than 0·002 millimetres diameter |

Gravel is not found in every soil but some of the last four types usually occur in most soils. The relative proportions of sand, silt and clay give a soil its characteristic feel or texture and most soils can be described in these terms, or combinations of them. Some of the commoner texture classes are:

(a)  Sand—the individual sand grains can be seen and felt.
(b)  Sandy loam—like a loam, but sand grains can be felt.
(c)  Silty loam—loam with a smooth or 'silky' feel.

(d) Loam—holds together when moist, but sand cannot be felt, and it will not 'ribbon'.

(e) Sandy clay loam—like a clay loam but sand grains can be felt.

(f) Clay loam—friable yet plastic, will 'ribbon', but tends to break.

(g) Sandy clay—clay in which sand grains can be felt.

(h) Clay—plastic, works out to a long ribbon.

Other combinations, especially those involving silt, are also used but as silt is not so commonly the dominant fraction, they are less likely to be met in practice. A more precise break-up is obtained by mechanical analysis of a sample; representative percentages of a range of texture classes are:

| Texture class | coarse sand | fine sand | silt | clay |
|---|---|---|---|---|
| sand | 40 | 50 | 6 | 4 |
| sandy loam | 35 | 40 | 15 | 10 |
| loam | 15 | 45 | 20 | 20 |
| clay loam | 5 | 35 | 30 | 30 |
| clay (heavy) | 2 | 8 | 25 | 65 |

With each texture class there are also differences between the surface levels and lower levels. In general, the latter become 'heavier', i.e. more clay is found; increases of from 20 per cent clay in the surface soil to 70 to 80 per cent at 40 to 60 centimetres depth are common.

The specific gravity of the mineral constituents is 2·5 to 2·6 but the bulk density, or the density as it exists in the field, is much less. An easily worked soil is about 0·9 to 1·2, and a difficult soil 1·3 to 1·6. The remaining space consists of pores of varying sizes, from large to very small, and is occupied by soil air, water, and roots. Roots do not enter the solid particles but ramify through the pores and make only surface contact with the particles. As soil is compacted, pore space decreases, bulk density increases, and root penetration is more difficult. Once compaction reaches a bulk density of 1·9 to 2·0 then roots can no longer penetrate.

## SOIL WATER

Water is essential for the growth of plants even though more than 95 per cent of that taken up is eventually transpired by the leaves. It acts mainly as the carrier of nutrients from the soil to the leaves, although some is required for growth and moisture balance within the plant.

Soil acts as a reservoir of water on which the plant can draw, and the importance of a moist soil is appreciated by everyone who has ever grown anything. When water is added to soil (by rainfall or irrigation), it saturates the surface layers before moving to lower layers. A well defined, though often irregular, wetting front is present behind which the soil is saturated but in front of which there is no change in moisture content, i.e. water saturates the soil through which it passes before moving further on. Eventually the soil becomes fully saturated and surplus water either drains away as seepage to the natural drainage system, or replenishes the ground water. Above the ground water a capillary fringe of moist soil exists, which, if within the root zone, can be used by trees.

Many species native to arid areas have strong root systems which can penetrate to depths of 8 to 10 metres to reach water in the capillary fringe.

Soil particles attract and hold water but after a day or so water in the larger pores will have drained away under the influence of gravity, i.e. gravitational water. The soil is then at field capacity which represents the maximum amount of water which the soil can hold without restricting plant growth through excessive wetness. It may appear a vague measure but is actually a fairly precise one for each particular soil. As the water is used, the soil becomes progressively drier until wilting point is reached, that is the soil moisture content at which plants will not recover from wilting. The soil would then be rather dry but the remaining water is so tightly held by the soil particles that the roots cannot extract it; it is known as hygroscopic water. Drying will still continue until the soil is air-dry, that is in equilibrium with the atmosphere.

Thus the water available for plant growth is that between field capacity and wilting point. Soils vary in the amount of available water that they can hold and also in the rates at which they can absorb it. Typical relative values are:

|  | Available water (millimetres) | Rate of infiltration per hour (millimetres) |
|---|---|---|
| sand | 18 | 15–18 |
| sandy loam | 40 | 13 |
| loam | 48 | 10 |
| clay loam | 54 | 7 |
| clay | 62 | 5 |

The lighter soils, therefore, cannot store as much usable water as the heavier ones. On the other hand they can absorb it faster which is an advantage in storms. Lighter falls of rain will penetrate further into sand than into clays; a light shower may barely wet the surface layer of clay but in sand may penetrate several centimetres into the surface root zone.

As the soil dries out it becomes increasingly difficult for the plant to extract water since it needs to apply a greater 'suction' to break the attraction between water and soil particles. Measuring this force provides a system of measuring the moisture available for growth which is expressed in terms of the pF scale. The pF scale represents the logarithm of the height of a column of water (in centimetres) needed to exert a pressure equivalent to the suction required to extract water from the soil. Completely saturated soil requires no suction and so has a pF value of 0. At field capacity (i.e. after the soil has drained for a day or so), a suction of about one-third of an atmosphere (300 to 400 centimetres) is required or pF 2·5 to 2·6. A suction equivalent to 1000 centimetres or about 1 atmosphere is pF 3. At wilting point about 15 atmospheres or the force of a column of water 15 000 centimetres high is required, giving a value of pF 4·2. Quite strong 'suctions' are therefore exercised by a plant in obtaining water. Air-dry soil has a pF value of 6·0.

Water in the range of about pF 2·6 to pF 4·2 is available water, or water that is available to the plant. Best growth can be expected in moist soils or at about pF 3. Soil with lower values is too saturated (aeration is inadequate) for growth and with higher values the water is more tightly held by the soil. Both field capacity and wilting point vary with soils and the species of plant, so need to be regarded as representing a narrow range rather than a precise value.

The value of the pF scale is that it is applicable to all soils and is independent of the actual amount of water in the soil; at wilting point a sand may have only 2 to 3 per cent water and a heavy clay 20 to 25 per cent but in each case the water is equally tightly held and requires the same 'suction' force by the plant to absorb it or, in terms of pF, a value for both soils of pF 4·2. The pF scale therefore forms a convenient means of reference.

Unfortunately pF values are not easily measured in the field and the method is not suitable for everyday use. Some moisture measuring devices based on the electrical conductivity of a water film are available but they indicate dry to wet soil conditions rather than available moisture. Gas pressure instruments which relate the pressure of acetylene gas, produced by a mixture of soil and calcium carbide, to the water content of the soil are rapid and reliable, once they have been calibrated for a particular soil type. The resistance to an electric current between two electrodes embedded in gypsum or nylon/fibreglass blocks, which are buried in the soil, is a widely used method for repeated measurements at one site over a period. More sophisticated methods which involve neutron-scattering devices are costly in equipment although rapid and comparatively convenient in use.

Once soil has drained for a day or two, the movement of water is very slow. It moves as vapour from warmer to cooler layers. During a warm day when the surface is warmer, water vapour in the soil moves downwards to the cooler layers; at night or in the early morning when the surface is cooler, it moves upwards, that is, the dew rises. Moisture is evaporated from the surface but drying out, even in summer, does not extend to more than a few centimetres depth (except in cracks) provided vegetation is not present to reduce it by transpiration. Hence, the value of fallow in conserving moisture is by eliminating weeds, not by breaking up the capillary system as was once thought.

In the liquid phase, again after free drainage, movement is so slow that for practical purposes it can be ignored. In other words, water does not move towards the roots or redistribute itself to give a uniform moisture level; there is usually drier soil around a root tip with moister soil a fraction of a centimetre away. What actually happens is that the plant taps the water by extending its root system towards it, not by the water moving to the root. By continual extension of its roots a plant will eventually absorb all the available water, unless the supply is replenished by rain or irrigation.

## STRUCTURE

Structure refers to the aggregation or otherwise of soil particles into crumb-like pieces. It is the main factor determining infiltration and drainage characteristics. Soil colloids, particularly clay and organic matter compounds, are the prime agents of cohesion and bind the particles together into stable units.

Sands, which have little clay and organic matter, consist of single grains without any cohesion between them. This single-grain structure is very stable under all conditions and as each particle is an individual unit, this class of structure is different from that in other soils.

Most other soils contain a fairly large amount of colloids and are usually capable of forming some kind of aggregates. Well-structured soils have particles which have coalesced together into crumbs up to about 2 millimetres in diameter. Therefore, they have good drainage characteristics and allow cultivation under a wide range of moisture conditions. Poorly-structured soils do not have a stable crumb-like organization and when wet, disperse into a fine muddy suspension. Soil pores are sealed as the suspension moves downwards and, on drying, sets to a massive state; breaking up such a soil leaves a cloddy and lumpy mass. Between the limits of good and poor structure there is the range of gradation from one to the other.

Structure can be assisted by treating with lime or gypsum, the calcium content of which flocculates the clay particles. Flocculation in itself will not give good structure but is an essential pre-requisite. A large content of organic matter is also vital and the addition of green crops to soil is a well-known and proven method of aiding its development. Soil conditioners are available and do an effective job but the high cost precludes their general use; green-cropping or adding other organic matter as mulch, etc., is still the best and most economical way. There appears to be an irreversible colloid produced from organic matter by microbic action which 'cements' the particles into crumbs. The finer rootlets also exert a mechanical influence by binding particles together.

Well-structured soils maintain their stability under cultivation but any form of cultivation does have some destructive action. The more traditional implements generally have the least effect but some modern powered types, such as rotary hoes and similar 'beating' types of implements, are not so kind to the maintenance of good structure, and their repeated use should be avoided as much as possible.

## AERATION AND DRAINAGE

The importance of good drainage cannot be over-emphasized. Structure, aeration and drainage are interrelated and complementary; poor structure leads to poor aeration which implies poor drainage.

Respiration by plant roots and micro-organisms releases carbon dioxide in the soil. Concentrations of it are much higher than in the atmosphere (0·03 per cent) commonly reaching 1 per cent, or up to 5 per cent where drainage is poor. Plants differ widely in their sensitivity to soil carbon dioxide; those that can tolerate larger amounts grow more readily on poorly aerated soils.

In most soils of southern Australia, the pore volume or 'air space' is usually 35 to 45 per cent, but the trend towards increasing use of heavier machinery is leading to more compaction and reduction of the pore space. Breaking up a compacted soil has little effect on aeration but it does make penetration by rainfall and roots easier, and reduces the risk of surface run-off and washing. On the other hand, a poorly-structured soil disperses under the influence of rain or irrigation leading to a sealing of the surface which is equivalent to compaction. Although the surface seal may only be a few millimetres thick it impedes the exchange of carbon dioxide and oxygen between the soil and atmosphere and needs to be broken by cultivation.

The development of a tree's root system occupies pore space and increases compaction by the increasing diameter growth of the larger roots. But the process of growth and decay of the finer rootlets and root hairs, together with the burrowings of the associated soil life which develops under a tree crop and the addition of organic matter from leaf fall, counterbalances this so that over a period of time the total pore space actually increases.

Poor drainage can be improved by mechanical means, particularly on duplex soils, i.e. those with a well-defined heavy subsoil horizon overlain by a lighter surface soil. Ripping to about 30 to 50 centimetres depth with heavy machinery is widely used, although the risk of initiating tunnel erosion needs to be watched. With a heavy horizon close to the surface mole drainage is effective but it is not as widely used as it could be. Artificial drainage by agricultural pipes (clay or plastic) is very effective but costly, and is normally only used in high-value areas such as orchards. On wet sites, the planting of trees which can tolerate wet conditions is effective as trees transpire large amounts of water and will gradually remove the excess water.

There are large areas in the medium to higher rainfall areas where tree growth on duplex soils could be much improved by more attention to drainage. By late winter and early spring the soil above the clay is saturated by winter rainfall and is drained only slowly by downward or lateral seepage. Therefore, a seasonal perched water-table occurs each year presenting a wet and cold environment to the roots in spring just when maximum growth is, or should be, occurring.

## TEMPERATURE

Radiation from the Sun warms the soil but has little direct effect on air which cannot absorb the Sun's relatively short waves (less than 0·0015 millimetre). (See also the section on Shade, page 104.)

Consequently it is not unusual for surface soil temperatures in early afternoon during summer to reach 60°C to 70°C, or well above air temperature. In fact, during the day in the warmer seasons, the surface soil is usually warmer than the air a metre or so above it, particularly while the soil is dry. During the night, heat continues to be radiated and while cloud may have some blanketing effect, soil surface temperatures usually fall below that of the air. The effect of the daily fluctuations is transmitted downwards through the soil but, with a lag in the downward movement of heat, maximum temperature at 15 centimetres is not reached until about 6.30 p.m. (daylight saving time). Lower levels are affected correspondingly later and to a lesser extent; the effect of daily fluctuations disappears at about 40 to 50 centimetres depth.

As well as daily fluctuations, temperature varies with the season. Under a tree canopy variation is least at the lower levels, the 45-centimetre and 75-centimetre temperatures being cooler in summer and warmer in winter, for example:

|        | *Temperature at Depth (centimetres)* | | |
|--------|------|------|------|
|        | 15   | 45   | 75   |
| Summer | 19°C | 18°C | 17°C |
| Autumn | 12°C | 14°C | 15°C |
| Winter | 4°C  | 7°C  | 9°C  |
| Spring | 12°C | 12°C | 12°C |

The variation is greatest at 15 centimetres and decreases with depth, there being less variation in the 30 centimetres between 45 and 75 centimetres than between 15 and 45 centimetres. In winter, the lower levels are warmer; in spring, they are about the same; in summer, the surface is warmer; but in autumn, the surface cools quicker with the lower depths retaining more of their heat and remaining warmer. The figures given are very general as variations between localities, particularly due to the blanketing effect of the density of tree cover, can be very large.

Annual average temperatures under trees do not vary much; at 15-, 45- and 75-centimetre depths the values are about 12°C, 13°C and 14°C respectively.

Evaporation cools the soil but the greatest effect on temperature and growth is caused by rain, particularly during the active growing period of spring. Spring rains can be cold or warm, a difference of 8 to 10°C is not uncommon. Cold rain may cool the soil several degrees to its depth of penetration, similarly warm rain may warm it. This effect of rain on growth can be very marked where most of the trees' roots are still only a few centimetres from the soil surface, such as with young trees or in nurseries.

Frost can cool the surface very rapidly but its overall effect is not great. Except at the higher elevations, soil is rarely frozen to more than 2 to 3 centimetres depth or for more than a few hours, which is too short for much heat transfer to lower levels to take place. A succession of severe frosts can cause frost heave but it only affects young plants whose roots are still in the top few centimetres of soil. In nurseries several severe frosts will check growth since the night temperatures of the upper few centimetres may fall to about 5°C or slightly less, that is the temperature at which root growth practically ceases.

## ORGANIC MATTER

Organic matter, sometimes loosely called humus, is found in all soils. It is normally dark brown to black and occurs mainly in the top few centimetres. Under trees three stages are usually identifiable; the uppermost layer which consists of fallen leaves, twigs, etc., is underlain by a partly decomposed mass of the surface layer permeated by fungal hyphae, and the broken-down products which are no longer recognizable in their original form incorporated in the top layers of the soil. Under natural conditions the cycle of life and decay is continuous, the soil supports the growth of trees which on their death are broken down again and reused by the new crop.

The agents of decay are the micro-organisms of the soil, chiefly bacteria and fungi, although insects play their part in assisting to break down some of the larger components. Soil is not solely an inert mass of mineral particles but contains literally thousands of bacteria of many species per gram. In the upper 10 to 20 centimetres of a good, fertile soil they amount to about 0·1 per cent by volume but are less in poorer soil and in the subsoil.

Reference has already been made to fungi and their role in decomposition. As they do not contain chlorophyll they obtain their food from other vegetation. Bacteria, on the other hand, comprise two main groups (a) autotrophic, which like plants can build up complex molecules from carbon dioxide using sunlight as a source of energy, and (b) heterotrophic which live on plant remains.

From the viewpoint of tree growth, two groups of autotrophic bacteria are important. Firstly, the nitrifiers which convert ammonia to nitrous acid and then nitric acid and secondly, the sulphofiers which convert sulphur and its compounds to sulphuric acid. Both acids react with other compounds in the soil to give nitrates and sulphates respectively. Each sequence comprises numerous stages and there are bacteria specific to each stage.

Heterotrophic bacteria break down organic residues. The number of steps and reactions is large, with bacteria specific to each stage. Not only do they release nutrients which are of value to the tree but they also detoxify toxic compounds produced as by-products (such as phenols) or may produce substances which restrict the growth of other organisms, e.g. penicillin. Some are also anaerobic, i.e. can live in the absence of air and are frequently found in wet conditions but most are aerobic or require oxygen for their survival.

Bacteria are most active between 20°C and 30°C and virtually cease functioning below 10°C. They also require moisture so they are usually most active in spring and autumn. Following a cold winter and heavy rains on well-drained soil, slight deficiency symptoms may appear in trees since microbial activity will have been very low and the supply of nutrients released from organic matter will have been largely leached from the soil. This position, however, normally corrects itself rapidly in the spring as ground temperatures rise.

In addition to being a source of nutrients, organic matter has two important physical effects. Its colloidal nature, in the broken down form, increases the water-holding capacity of soil since in this form it has two to three times the water-holding capacity of clay. Equally important is its effect on soil structure and its ability to bind together soil particles to give the desirable crumb structure. Peaty soils may have 50 per cent or more organic matter while 10 to 20 per cent has a marked effect on structure. Many soils in southern Australia are well below these figures (less than 5 per cent is common) but even so with a good supply of calcium the structure can still be good.

The value of organic matter has been appreciated for a long time. Virtually any decomposable vegetable matter is effective but in some cases, such as large quantities of sawdust mixed with the soil, the addition of a nitrogenous fertilizer is necessary to counteract induced nitrogen deficiency (see Carbon/Nitrogen Ratio later, page 79). Animal manure has been used for centuries while green cropping with legumes (field peas, dun peas, vetches, lupins, etc.), rye and sudax grass, oats, and pasture grasses is well established in practice. Volunteer weeds, grasses, etc. are also useful and can be ploughed in.

Green crops, whether sown or natural, should be ploughed in at flowering as (a) most of the growth has taken place and the crop hardens off and gradually loses its succulence, (b) some nutrients within the plant are remobilized and used in seed production, and (c) seed should not be allowed to form, otherwise re-growth in the soil can be a problem. The success depends on the volume of material added; several tonnes per hectare is preferable. Legumes and sudax grass are usually more satisfactory as the other grasses and oats, because of their relatively large cell size, do not produce so much solid matter; in practice they maintain the level rather than increase it.

Krasnozem

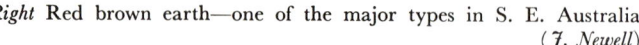

*Right* Red brown earth—one of the major types in S. E. Australia
(*J. Newell*)

Solodic soil

*Below* Heavy black clay on basalt—typical of the Western District of
Victoria

Severely leached sandy soil

Montane soil, typical of the
Highlands

*Above right* Terra rossa on limestone

*Left* Typical duplex soil—note sharp change to red clay
(*J. Newell*)

*Left* Typical alluvium
(*J. Newell*)

Sand on calcareous clay

*Below right* Red sand of reasonable quality for tree growth

*Below* Gilgai country

In medium to high rainfall zones sowing a green crop the year before planting is an advantage in aiding establishment, although not often done in practice. Thorough ploughing-in before planting is necessary to avoid grass and weeds competing with the seedlings. Broadcasting 400 to 600 kilograms of sulphate of ammonia per hectare at ploughing-in will hasten decomposition. Perennial lupins have been found suitable on sandy soils as they provide a source of nitrogen for tree growth and are eventually suppressed by the tree crop.

In lower rainfall areas, green crops will still produce the results as in more favourable localities but a conflict of interests develops. During its growth the green crop transpires moisture to the atmosphere and, as moisture is more likely to be limiting to tree growth than the disadvantage of not green cropping, fallowing for a year before planting is preferable.


## MULCHING

Allied to organic matter is the practice of mulching. Although vegetable materials used for mulching are gradually incorporated in the soil the objective is not primarily to add organic matter but to reduce weed competition and conserve moisture.

Moisture is conserved in two ways. Firstly, by shading the soil surface and reducing the daily fluctuation in surface temperature, the movement of water vapour from below the surface is reduced or almost eliminated. At night, dew is more likely to condense on the mulch and, while this may not add materially to the soil moisture, it has to be evaporated before any significant heat transfer to lower levels takes place. Secondly, the mulch prevents the growth of weeds which would themselves transpire moisture. This is by far the greater effect since the roots withdraw moisture which would otherwise be available to the plants. Weeds are aggressive plants and many have fairly strong root systems which extend further and more quickly than the favoured plant, thus occupying the soil in advance of it. Eliminating weeds reduces the competition for both moisture and nutrients.

A wide variety of materials can be used. Any cut vegetable material such as grass, lawn clippings, straw, old hay, leaves, crushed bark, sawdust, seaweed, etc. can be used. If available, wood chips, waste pulp, spent grains or grain husks are also satisfactory. A depth of 8 to 10 centimetres for about 25 to 30 centimetres around the stem is about the minimum necessary to be effective as mulch. More can be added if available. The mulch need not be dry but green material should not be placed against the stem as the heat generated in the first day or two may cause injury. With mulch materials like sawdust, it is not necessary to add a nitrogenous fertilizer (see Carbon/Nitrogen Ratio, page 79) as the material is only slowly incorporated into the soil and the tree's roots will not enter it to any extent. Old sawdust is preferable as extractives leached from green sawdust may retard growth. If seaweed is used, it should be stockpiled until 50 to 100 millimetres of rain have fallen on it and washed off most of the salt. Small quantities can be washed sufficiently free of salt by hosing thoroughly and allowing to drain.

Inert materials such as subsoil sand, washed or river sand, fine gravel, etc. are also satisfactory. Surface sand or soil should not be used as it usually contains weed seeds which will grow and defeat the aim of mulching. If only a few trees are involved, opaque polyethylene film, cardboard or several layers (fifteen to twenty) of newspaper placed for about 25 to 30 centimetres around the plant and weighted with soil are satisfactory.

## COLLOIDS

Colloids are particles of matter small enough to remain in suspension indefinitely. In colloid chemistry, they are defined as particles ranging from 2 to 200 millimicrons in diameter. They range from easily deflocculated to very stable in water, that is, hydrophilic such as starch to hydrophobic such as clay.

Clay is the most important colloid in the soil although organic colloids from organic matter are also important. The significance of colloids lies in the fact that, while very small, they present a large total surface area, much larger than if in a single mass. For example, a cube of clay one centimetre in each dimension has a surface area of 6 square centimetres, if divided into half along each of the three dimensions, its surface area is 12 square centimetres. By the time subdivision has reached colloidal size, the surface area is many thousands of times greater than the area of the original cube.

The value of clay lies in its role of holding nutrients against loss through leaching. When a nutrient, say potassium sulphate, is dissolved in soil water, it dissociates into positively-charged cations ($K^+$) and negatively-charged anions ($SO_4^{--}$). The cations attach themselves to the negatively-charged clay and are held (adsorbed) against leaching and loss from the soil. Consequently in soils low in clay, such as sands, fertilizers should be applied in light but frequent dressings compared with a heavier soil where one large application may suffice.

Although clay is classed as a hydrophobic colloid, there are many types ranging from those which are easily dispersed to very stable aggregates. One which is easily dispersed washes down into the soil, is filtered out, and eventually fills up pore space and impedes infiltration and vertical drainage. Stable clays, on the other hand, retain their form when wet and allow good infiltration and drainage.

The degree of natural stability of a clay can also be affected by sodium and calcium salts. Sodium ions, usually from common salt (sodium chloride) in cyclic salt in the rainfall, cause dispersion so that the clay particles are filtered out blocking the soil pores. Calcium ions have the opposite effect and flocculate clay, that is, they cause the individual particles to coalesce; a feature which is widely used in the liming of heavy soils. Lime or gypsum can be used, the latter having the advantage that it does not affect soil acidity or alkalinity to any degree. Gypsum is sometimes added in irrigation water.

## ACIDITY AND ALKALINITY

Acidity or alkalinity of the soil affects trees to varying degrees. For some it is very critical, for others of little consequence. Acidity is usually measured in terms of the pH scale, a negative logarithmic scale expressing the concentration of the acidic hydrogen ion. It ranges from pH 0 to pH 14; a value of pH 0 is strongly acidic, pH 7 is neutral, and pH 14 is strongly alkaline.

Most soils are in the range of pH 5 to pH 7·5, although pH values of less than 4 are known for some strongly acidic soils, and values up to pH 9·5 are known for

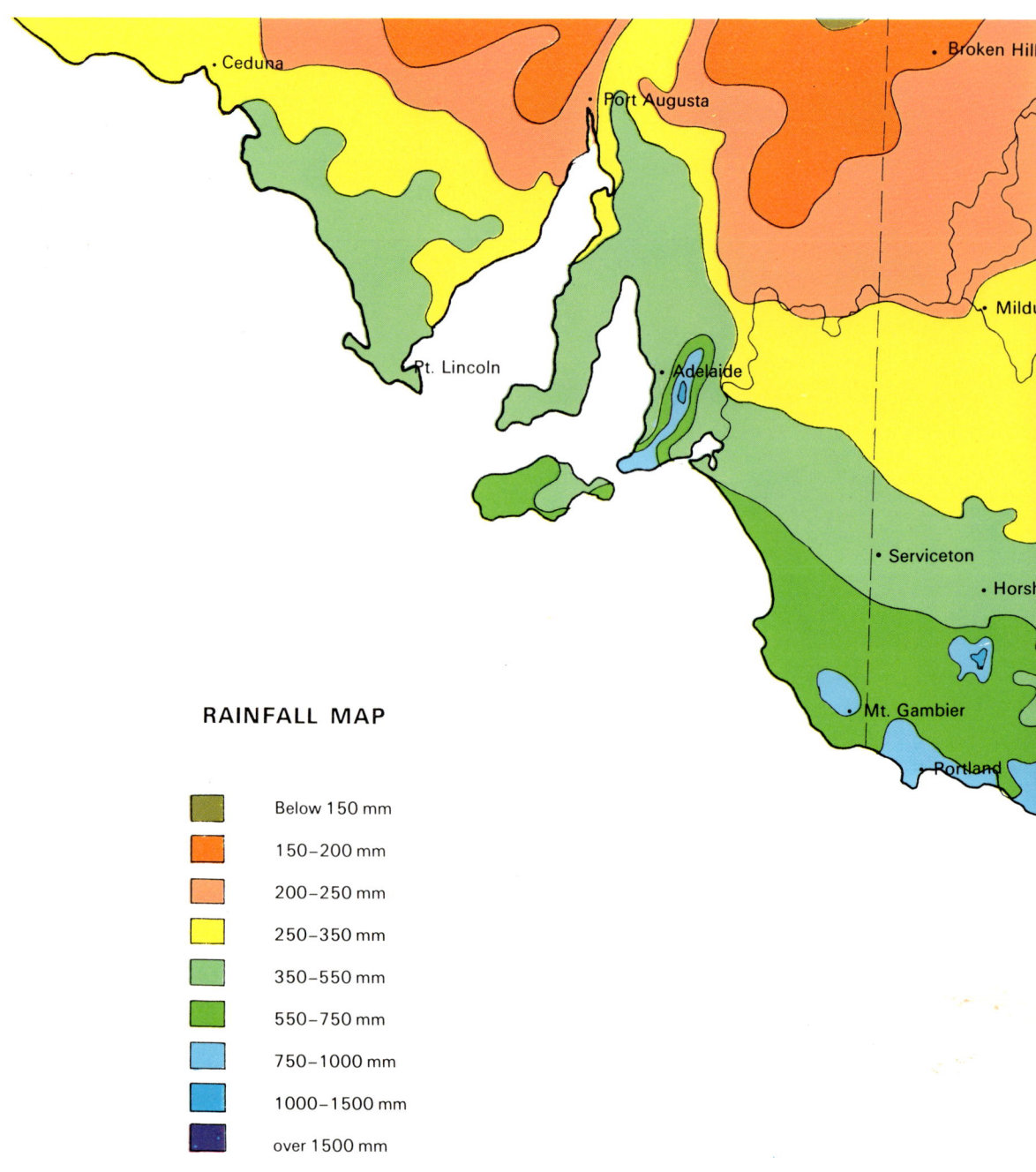

## RAINFALL MAP

| | |
|---|---|
| | Below 150 mm |
| | 150–200 mm |
| | 200–250 mm |
| | 250–350 mm |
| | 350–550 mm |
| | 550–750 mm |
| | 750–1000 mm |
| | 1000–1500 mm |
| | over 1500 mm |

Scale of kilometres

100    50    0    100    200    300

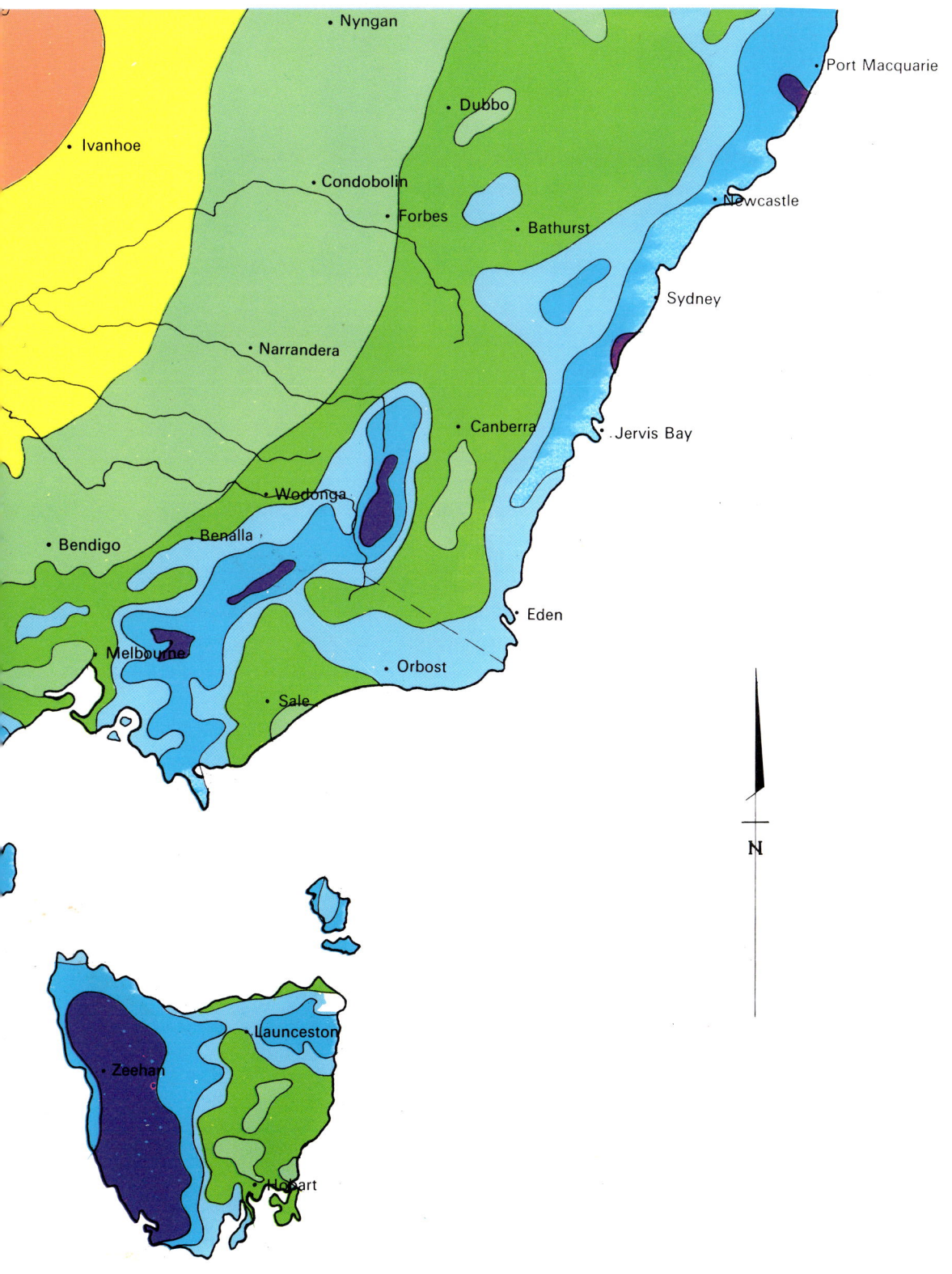

calcareous soils (as in some of the Mallee soils of southern Australia). A range of pH 5·5 to pH 6·5 is probably the optimum. Most trees grow satisfactorily at these levels of acidity.

Acidity in itself has little effect. Trials in water cultures indicate that most plants will grow well from pH 3·5 to pH 8·5, that is over the range usually found in practice. The difference between water cultures and natural soils lies in the differing solubilities of some nutrients or toxic compounds with varying pH values of the soil.

## (a) *Acidity*

Strongly acidic soils are heavily leached soils so they are naturally low in some essential elements, particularly phosphorus, calcium and magnesium. Aluminium and manganese are usually relatively abundant and are toxic in excessive amounts. Most plants can tolerate only a few parts per million (ppm) of aluminium but a well known exception is the hydrangea which can have as much as 1 per cent aluminium (dry weight) in its tissues, hence aluminium compounds (usually aluminium sulphate) can be used in 'blueing' mixtures. Aluminium and manganese are least soluble in neutral soils (less than 1 ppm) but increase in solubility as acidity increases. Excess aluminium also reduces phosphate (phosphorus) uptake, producing a deficiency which cannot be corrected by adding additional phosphate such as superphosphate. Aluminium and manganese troubles can be overcome by adding calcium because of the 'antagonistic' effect in root absorption between calcium and these two elements. Molybdenum, and possibly potassium, supply may also be improved by the addition of lime.

Bringing a very acidic soil nearer to pH 5·5 to pH 6·0 by liming also improves micro-organism activity on organic matter, leading to a quicker production of nitrates. Lime (as calcium carbonate), calcium oxide (which is converted to calcium carbonate in the soil), or dolomite may be used to reduce acidity. Heavy dressings should not be applied to young trees on sandy soils as the more open texture of sands may lead, by leaching, to the formation of pockets of high alkalinity. Gypsum (calcium sulphate) does not affect the pH level but adds calcium and improves the soil structure more effectively than lime.

## (b) *Alkalinity*

Soils with pH values up to about 8 do not usually suffer from alkalinity problems. Calcareous soils rarely exceed pH 8·5 unless sodium salts are abundant when pH values up to 9·5 can occur. These, however, are uncommon.

High alkalinity causes shortages of phosphorus, iron, boron and manganese as they are also comparatively insoluble at high pH, so are unavailable to the plant. Lowering the pH is the only solution and can be achieved by adding sulphur or aluminium sulphate. Sulphur is converted to sulphuric acid in the soil which reacts to give sulphates. It is safer and more effective than aluminium sulphate which may also cause trouble by causing excess aluminium to come into solution. The safest way is to grow and plough in green crops. Their breakdown by micro-organisms increases the carbon dioxide content of the soil and the initially high pH is lowered by the increasing carbon dioxide content of the soil. A carbon dioxide concentration of 1 per cent around the roots can reduce pH 8·5 to about pH 7·0. However, it should be noted that this reaction only occurs at high pH; in neutral or acidic soils the carbon dioxide concentration has less effect on pH.

(c) *Measurement of pH*

Measuring soil pH is difficult. The usual techniques of making a soil solution and determining the pH with a glass electrode or by colour solution kits do not necessarily give the pH in the critical zone around the roots. Differences of pH 1·0 are common. Anyway there is usually little point in determining pH to an accuracy greater than 0·5. For most soils in southern Australia it is only at the extremes of the range—and such soils are not common—that pH measurement is likely to be of practical value in tree growth.

# SOIL CHEMISTRY

## NITROGEN

Of the nutrients obtained from the soil, all except nitrogen initially came from igneous rocks. Nitrogen is in a class of its own in that it is obtained from the atmosphere via the soil.

It is a necessary component of protein and protoplasm so it is required by all living organisms. Plants require it in relatively large quantities but can only absorb it as ammonium and nitrate; all other forms of nitrogen compounds must be converted to these to become available to the plant. Nitrogen is very mobile in the soil. It rarely remains for more than a few months as most forms of it found in the soil are very soluble, and easily lost by leaching. Some is also lost to the atmosphere as ammonia during the ammonification of plant residues.

A good replacement system is therefore necessary and supplies are obtained from several sources. A small quantity, probably only 4 to 5 kilograms per hectare per annum, is added by rainfall, particularly by electrical storms which synthesize nitric acid from atmospheric nitrogen (air contains 78·5 per cent nitrogen), but the remainder is derived from bacterial action.

(a) Organic matter, through its breakdown which releases nitrogen from the protein and protoplasm, is the main source. Adding organic matter as dead material or as a green crop will improve the supply. Breakdown is rapid and, provided a small amount of nitrogen (perhaps only 10 to 20 parts per million in the surface few centimetres) is initially available, the rate of decay is usually sufficient to maintain growth. Since bacteria are inactive below 10°C and most active between 20°C and 30°C, the supply is low in winter and early spring but increases to a peak during summer before decreasing again in autumn as consumption gradually overtakes production. In drier climates, nitrogen accumulates over the summer but in the wetter areas it may be leached out of the soil, depending on how wet the summer is. The nitrifying bacteria involved are *Nitrosomonas* which convert ammonia to nitrites and *Nitrobacter* which convert nitrites to nitrates.

(b) Fixation of gaseous nitrogen in the soil is accomplished by *Azotobacter* species (aerobic bacteria), *Closteridium* species (anaerobic bacteria which function only as long as adjoining bacteria remove any free oxygen formed) and blue-green algae of the *Nostococcaceae* family. Some authorities believe that nitrogen is excreted by bacteria to the soil as complex protein but when they die their remains contribute to the organic matter and are utilized as in (a) above.

(c) Symbiotic fixation (or symbiosis), in association with plants, is a major source. Many plants, mainly the family *Leguminoseae* or legumes, carry root nodules containing bacteria of *Rhizobium* species which live in harmony (or symbiotically) with the plant. They fix free nitrogen from the soil atmosphere and on the death of the root this is released for the use of other plants via the organic matter cycle as in (a) above. Representative leguminous species amongst trees are *Acacia*, *Cassia*, *Albizia*, *Robinia*, *Laburnum*, and *Gleditsia*; amongst shrub genera the leguminous species are the native *Pultenaea*, *Daviesia*, *Dillwynia*, and *Goodia*, and the introduced *Cytisus*, *Genista*, *Ulex*, and *Lespedeza*; and amongst ground plants *Platylobium*, *Bossiaea*, *Hardenbergia*, and *Kennedia*, and pasture plants such as clovers, medicks and lupins. Some non-leguminous species particularly *Casuarina* and *Alnus* are also able to fix free soil nitrogen.

Numerous forms of nitrogenous fertilizers are available. The main ones are sulphate of ammonia, nitrate of soda, ammonium nitrate, urea, dried blood, calcium ammonium nitrate, calcium cyanamide, blood and bone, and anhydrous ammonia. All are very soluble and easily lost by leaching, except anhydrous ammonia which volatilizes readily and can be lost rapidly to the atmosphere. Unlike most other fertilizers, nitrate nitrogen (and to a lesser extent urea and zinc) can be absorbed via the foliage as well as by the roots, consequently it is one of the main constituents of foliar fertilizers. Some forms, particularly sulphate of ammonia, will acidify the soil if applied in quantity but about 2·5 tonnes per hectare per annum for a few years is required to reduce the pH value by 1·0. Levels of this order are not required for trees.

Adding nitrogenous fertilizers to the soil tends to suppress the growth of nitrogen-fixing plants since the extra nitrogen then available encourages the growth of other plants. This principle is applied when sulphate of ammonia is used as a control measure to get rid of clovers (nitrogen-fixing plants) in lawns.

## CARBON/NITROGEN RATIO

The ratio of carbon to nitrogen in most soils is about 10 to 12 : 1, although in peaty soils it may be as high as 50 : 1. The bacterial population normally rapidly adjusts any variation to about this ratio so that average applications of organic-matter, fertilizers, etc. do not cause much change.

However, if a large volume of carbon is suddenly added it may take months for equilibrium to be reached. For example, if straw, sawdust, etc. (which contain a large amount of carbon in the cellulose and lignin) are mixed with the topsoil to lighten it, an increase in the bacteria which attack cellulose is stimulated. As these require nitrogen for their growth, nitrogen is removed from the soil and locked up in the bacterial cells. It is then not available to plants and nitrogen deficiency symptoms (yellowing) may develop. The nitrogen locked up does not become available until the cellulose has been oxidized and the bacterial population decreases and decomposes. Consequently when mixing straw, saw-dust, etc. with soil a nitrogenous fertilizer should also be added. If sawdust, etc. is added as a mulch the problem is not so likely to occur as only the lower layer is in contact with the soil and while all of the mulch is gradually incorporated there is more time for the carbon/nitrogen ratio to adjust itself.

When fresh material such as a green crop is added, the deficiency is unlikely to occur as the green crop, while representing a substantial increase in carbon compounds, still has nitrogen available in the protein and protoplasm of its tissues. Following this principle young crops, because of their more succulent nature and less carbon, are better than mature or hardier ones, particularly with cereals such as oats, rye-corn, and grasses such as rye, sudax, etc.

A parallel bacterial reaction can also occur if too much nitrogen is added to the soil but as the amount added is usually small in proportion, it is unlikely that the carbon/nitrogen ratio will be substantially changed this way. It is rarely a serious problem and for practical purposes can be ignored.

## PHOSPHORUS

Southern Australian soils are usually deficient in phosphorus, mainly because they are generally old and the natural phosphorus has been lost by leaching. The only soils with adequate natural supplies are a few volcanic soils of very recent origin and some of the better alluvial flats. Native Australian trees have adjusted to these low phosphorus levels and can recycle phosphorus within the plant, from old leaves to younger tissues, more readily than most introduced species. However, contrary to a still fairly common belief, native trees will respond to fertilizers, particularly phosphorus and nitrogen.

Phosphorus is applied as superphosphate which is rock phosphate treated with sulphuric acid to make it more water soluble. (It is actually a mixture of calcium phosphate and gypsum.) When it is applied, most of it reverts in the soil to give a less soluble form which helps to retain it in the surface 3 to 5 centimetres of soil. Most of it is insoluble in water but a gradual breakdown by the soil solution slowly makes some of it available. If it were not for reversion most of the usable superphosphate would be lost in a few weeks by leaching.

A more serious problem is fixation in which nearly all of the superphosphate is precipitated in an insoluble form. Fixation happens most frequently on soils which are high in aluminium or iron. Excess aluminium is usually not visible but excess iron gives red soils ('krasnozems') their characteristic colour. (Iron can also be high in yellow soils.) Red soils are commonly regarded as good soils but this is due more to their good physical characteristics, especially structure and drainage, than to their fertility. They are usually very deficient in available phosphorus and, because of fixation, require fairly large amounts to be added before a response is obtained. In other acid or neutral soils, calcium fixation may occur in which superphosphate reacts with the calcium to give a less soluble form which, over a few weeks, gradually becomes more and more insoluble. In an alkaline soil, it can become insoluble within a few days; such soils frequently exhibit strong phosphorus deficiency.

Despite a widespread belief superphosphate does not acidify the soil; its effect, if any, is more likely to be towards alkalinity. In any case, the quantity required for trees is too small to have any effect.

Nitrogen and phosphorus when present in adequate quantities, as well as making their individual contribution, interact to produce a bonus of growth. The amount is variable but is commonly 20 to 25 per cent extra growth.

## POTASSIUM

Plants require potassium in fairly large quantities. Reserves in the soil are adequate in the inland lower rainfall areas but deficiencies commonly occur in the wetter coastal regions, especially on the red volcanic soils.

When applied to the soil, very little of the potassium remains in solution to be lost by leaching as most of it is adsorbed by the colloids and so remains in a readily exchangeable form. In soils with illite clay (the common form in southern Australia) some may be incorporated in the clay lattice and plants are able to use this source by a mechanism not yet fully understood. No technique of chemical analysis has yet been devised to determine these forms; analysis often gives mis-leading results as it may indicate a severe shortage while plants show no signs of deficiency or response to potassium fertilizers. On the other hand, soils with kaolinite clay (such as the red volcanic types) are usually low in potassium since there is no place for it in the clay lattice. Consequently good responses to potassium fertilizers are commonly obtained on these soils.

## OTHER NUTRIENTS

The other (or metallic) nutrients were also initially derived from mineral particles of the original igneous rocks, even though they may have been re-sorted as sedimentary rocks or recombined in other forms, such as clay minerals. Most of these particles or primary minerals occur in the fine sand fraction, since smaller particles are rapidly decomposed because of their large surface to volume ratio.

Primary minerals are usually less than 10 per cent of the fine sand, but can be as high as 30 per cent in recent volcanic soils. Felspars provide potassium and calcium; olivene provides magnesium and iron; micas provide potassium, magnesium, and iron; apatite provides calcium and phosphorus (also fluorine); tourmaline provides boron; and the calcium magnesium silicates provide calcium and magnesium. 'Trace' elements or the minor nutrients generally occur as impurities in some of these minerals or as separate minor components in the original rocks. A good supply of primary minerals virtually guarantees an adequate supply of these nutrients.

## CYCLIC SALT

As well as the natural degree of stability of a clay, the stability can be affected by the addition of sodium or calcium salts. Sodium disperses clay so that it washes down more readily and fills the soil pores, a process known as solonizing. The most common source of sodium is common salt (sodium chloride) brought in by rainfall as cyclic salt. Cyclic salt originates as ocean spray mixed with the atmosphere so near the coast the concentration is naturally much higher than it is further inland. Where rainfall is high, most of the salt is washed through the soil without any serious effect on the clay; in drier areas, which in southern Australia are generally further inland, there is insufficient cyclic salt in the rainfall to cause much trouble. In between these two zones there is a fairly wide belt where the supply is sufficient to cause clay dispersal, which has led to the formation of soils in which the upper few centimetres or so of the clay subsoil is so finely dis-persed that it is relatively impermeable to vertical drainage. Temporary perched water-tables are common in such soils at the end of winter.

The problem of 'salting' which is common in many areas is not caused directly by cyclic salt but by a gradual accumulation of salt from drainage (or irrigation) water. Salt patches on lower slopes and flats arise where ground seepage from higher slopes comes to the surface. The salt, in this case, usually originates from cyclic salt and is concentrated by increased seepage following the clearing of the original tree cover. Vegetation on the higher slopes is not sufficiently dense to use the water available and the surplus percolates to lower levels washing out the salt at the same time. Eventually the salt-enriched water reaches streams by natural drainage.

## CHEMICAL ANALYSES

Chemical analyses of soil are useful in tree nutrition but their value is all too often overrated. Enough has been said to indicate that the immediate surroundings of the root and root hairs (or 'rhizosphere'), and the manner in which nutrient elements may be held in the clay structure, make duplication or reproduction of these conditions difficult in a laboratory analysis.

Analyses of the foliage are more useful but even these are subject to many reservations. Precise sampling is important as nutrient concentrations vary with season, time of day, age of the foliage and position on the crown; samples must be collected under identical conditions to give information of meaningful value. They show only the amount absorbed by the tree and may indicate a marked deficiency but, if supplies of one element are in abundant excess, uptake or 'luxury consumption' may confuse the picture. However, these analyses do show the quantity removed by the tree and thus the minimum amount that needs to be provided, either naturally or by adding fertilizer, to prevent depletion of the soil reserve. On the other hand, they give no indication of how much has been lost through leaching or consumed by the soil population of insects, bacteria, fungi, etc.

If results are interpreted with understanding, soil and foliar analyses do supply useful information and guidelines in tree nutrition, but they are not the infallable answer that many people who are unacquainted with the difficulties seem to believe. Despite the considerable amount of research that has been devoted to this field, the results have not yet lived up to the high expectations of the original analysts. Results must still be applied with the discretion and understanding derived from experience.

## FERTILIZING

Under southern Australian conditions, fertilizing improves the initial growth of trees. It is not vital but it is an advantage. In regions of high rainfall, nitrogenous fertilizers give a good response; in poorly-drained or wet sites, superphosphate is very effective. Both these soil types have an abundant supply of moisture and, as indicated earlier, uptake of fertilizer elements depends largely on the available soil moisture; the nearer it is towards field capacity the better the uptake and response. Conversely on dry soils, poor or nil results can be expected. In fact on naturally dry soils, fertilizing is of little benefit in an average year; it is only when a wet spring and summer occur that any response may result.

The essential nutrients can be supplied by ordinary commercial fertilizers or organic matter. Organic matter has the added advantage that it improves the physical condition and water-holding capacity of the soil, but there is no sound evidence to show that the nutrients (elements) in it are any more beneficial than those obtained from artificial fertilizers. On the other hand, it frequently does not contain a sufficient supply of all the elements which are desirable for maximum growth, and supplementary fertilizing is often an advantage. Organic matter, because its decomposition takes time, gives a slower release of nutrients than artificial fertilizers which, being immediately available, are more likely to be leached out and lost much sooner.

Of the nutrients not previously referred to, sulphur is rarely required by itself, possibly because it is present in quantity in superphosphate and the normal levels of application of superphosphate are more than adequate to supply sulphur requirements. Magnesium fertilizers are usually soluble and are eventually lost by leaching. Iron is difficult as it oxidizes rapidly to the stable and insoluble red ferric oxide. (In wet or waterlogged soils it may be reduced to ferrous oxide giving these soils their characteristic grey colour.)

Trace elements are generally present as impurities but, where deficiencies are known, they may be mixed with superphosphate to the required level. Of the trace elements, zinc can become unavailable in the soil and is usually applied to trees as a foliar spray. Addition of copper, molybdenum, and manganese for tree growth has so far not been found necessary. Boron deficiency has been detected but the natural decay of grass after each summer seems to release a sufficient quantity.

Farm tree plantings will usually benefit from application of fertilizers to adjoining crops or pastures. Consequently there is no need to take special precautions to keep fertilizer away from them.

The percentage of major elements in some of the more common fertilizers is approximately as follows:

*Nitrogen* Dried blood 12%, hoof and horn 12%, blood and bone 5–7%, nitrate of ammonia 34%, nitrate of soda 16%, sulphate of ammonia 21%, urea 46%, calcium ammonium nitrate 23–26%, anhydrous ammonia 82%.

*Phosphorus* Ground phosphate 16%, superphosphate 8–10% (18–22% $P_2O_5$), blood and bone 5–8%, bone meal 12%, 'double' superphosphate 18–20%.

*Potassium* Muriate of potash 48–50%, sulphate of potash 39–42%.

*Sulphur* Superphosphate 10–12%.

*Calcium* Agricultural lime 40%, ground limestone 30–35%, dolomite 16–18%. (Amount of calcium can vary widely between different trade brands depending on the source of the limestone rock.)

*Magnesium* Dolomite 11%, Epsom salts 20%.

Blood and bone has 5 to 7 per cent nitrogen, 4 to 8 per cent phosphorus and usually most of the trace elements. The lower concentrations of elements makes it a safe fertilizer as the quantities applied need not be critical. 'Complete' fertilizers containing N,P,K (nitrogen, phosphorus and potassium) in varying proportions and often with trace elements are available for specific purposes. Ammoniated superphosphate, calcium ammonium nitrate, magnesium ammonium phosphate and 'slow release' fertilizers are good but expensive.

Amounts of nitrogen, phosphorus and potassium in animal manure are much lower; it is only because it is applied in such large quantities that a satisfactory supply is available. Horse, cow and sheep manure contain nitrogen about 0·4 to 0·6 per cent, phosphorus 0·3 per cent, and potassium 0·3 to 0·8 per cent.

Fowl manure is richer containing respectively 3·5 per cent, 2·5 per cent and 1·5 per cent.

Most of the fertilizer mixtures available commercially are prepared for 'a specific crop or set of conditions and, while good for the intended conditions, they may be little better than the general types when used for other crops or conditions. As far as trees are concerned, superphosphate, sulphate of ammonia or urea, and blood and bone would be adequate in most cases, although there is no reason why pre-mixed complete fertilizers or the more expensive ammonium based and 'slow release' fertilizers should not be used. A mixture of five parts superphosphate, two parts sulphate of ammonia (or one of urea), and one part each of sulphate of potash and blood and bone is a good all-round mix. Alternatively, a good response may be obtained from 50 to 100 grams of super-phosphate or blood and bone, or 30 to 50 grams of sulphate of ammonia per plant at planting. Nitrogenous types can be applied to leguminous species even though these can fix atmospheric nitrogen, as some additional source will accelerate growth. If blood and bone is used in quantity, dogs should be kept off the area for a few days as the smell suggests an ancient bone to dogs, and the vigorous scratching which follows may uproot the tree. Lime should not be mixed with nitrogenous fertilizers nor applied several weeks before or after since it reacts with them releasing most of the nitrogen, often as ammonia, which is then lost to the atmosphere.

The quantity applied depends on the fertilizer and size of the plant, but roughly 2 grams per centimetre of height of the plant is safe for seedlings. It should be placed in a heap on the low side rather than spread around the plant as being in a heap reduces the rate at which the soluble constituents are leached out and possibly lost before the plant can use them. After several years broadcast applications can be given at the rate of up to 5 kilograms per hectare for super-phosphate and 1 to 2 kilograms of sulphate of ammonia.

While fertilizing will stimulate the growth of the plant, it will also stimulate the growth of competing weeds and these will need to be removed to reduce competition. (See next section.)

# PLANTING

Various aspects of planting have already been referred to and will not be repeated here. Reference should be made particularly to the sections Soil Moisture (page 63), Soil Structure (page 65), and Fertilizers (page 82). It is assumed that most people will obtain plants from a nursery rather than grow their own as techniques for raising vary with species. However, there is no reason why the interested person should not propagate his own plants even though, in most cases, it will be more convenient to purchase them.

In the medium to high rainfall areas, sowing a green crop in the late summer before planting is an advantage. In dry areas, fallowing over the previous summer and autumn is very desirable, if not essential. Thorough cultivation a few weeks before planting should be undertaken in all areas. Ripping to 30 to 40 centimetres depth along planting lines with a rabbit ripper will loosen the subsoil and aid root and moisture penetration. Planting can be done any time after the first good rains in April or May and can normally continue to August. Later planting is possible with container-grown stock (tubes, pots, tins, etc.) particularly if it is practical to water them once or twice should the summer be dry. Except for plants in thin paper tubes which can be planted as they are, all plants need to be removed from their containers. Wooden veneer tubes can be unrolled, plastic tubes and bags should be cut vertically with a sharp knife, tins cut away with tin-snips, hessian or other wrapping material removed, and knock plants out of earthenware pots. If the plant is 'pot-bound', that is the roots are coiling around the inside of the container, cut the coiling roots in two or three places and ease them out. Loosen slightly (to about a quarter of the distance towards the centre) the outer soil, otherwise keep soil disturbance to a minimum. Protect the roots from drying out at all times. Be generous with the size of the planting hole (at least two or three times the 'root volume'), plant to the soil collar on the stem, firm well, and, if practical, water thoroughly to ensure good root contact with the soil. Leave a slight depression on the surface to collect rain water.

Failures need to be replaced the year after planting. Keep weeds under control at all times. Cultivation around the plant is best; where mechanical cultivation is used between the trees some hand work will still be necessary immediately around the plant. A contact weedicide can be used on weeds but take care that none of the spray contacts the plant. Several good weedicides are available but a strong solution (equivalent to 100 grams or more per square metre) of sulphate of ammonia will kill any foliage it contacts as well as acting as a fertilizer when washed into the soil. If weedicides are preferred, a strong mixture of the ester form of 24D or 245T is satisfactory. These are not very

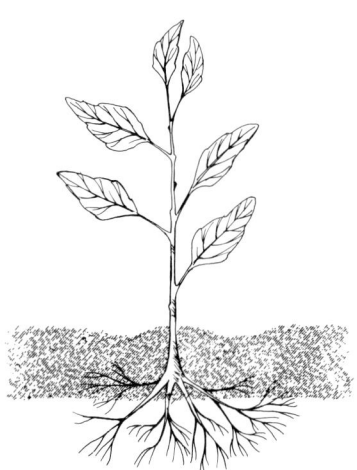

Do not dig into clay—it only forms a water sump

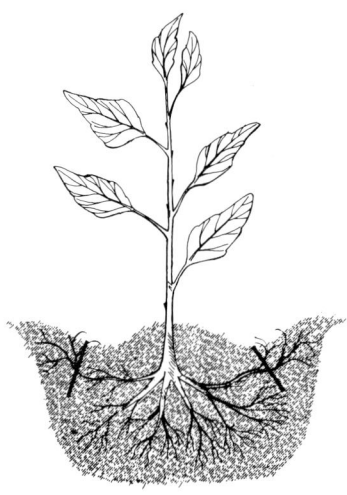

Cut back abnormally long or damaged roots

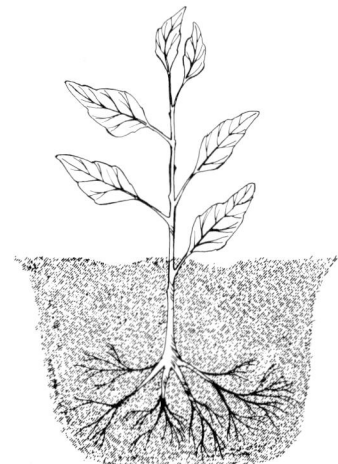

Planting too deeply

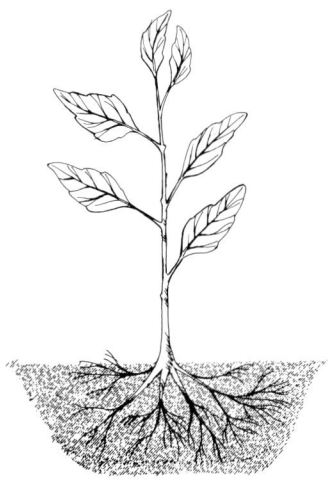

Planting too shallow

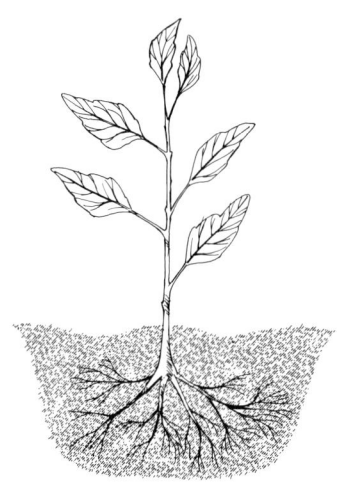

Spread the roots, and be generous with the size of planting hole

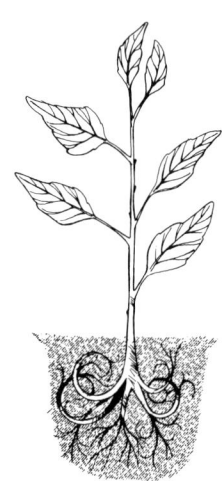

Planting hole too small.
It should be at least twice the loose root area

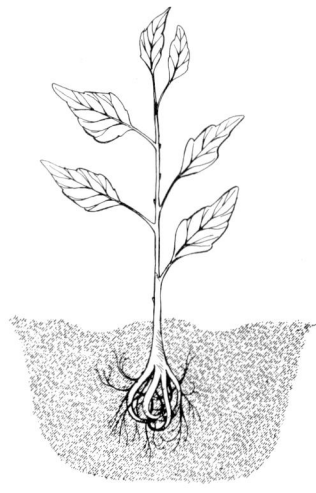

Separate tangled and upturned roots

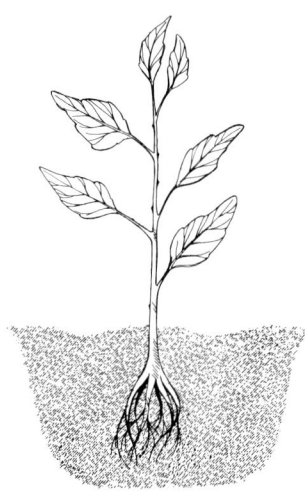

Bunched roots—spread them before covering with soil

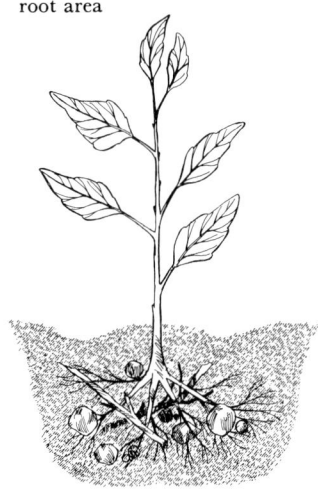

Do not leave stones and sticks etc. in the back fill soil

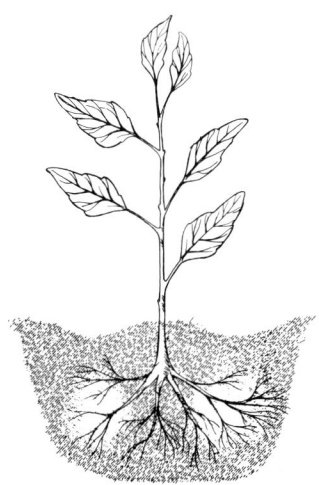
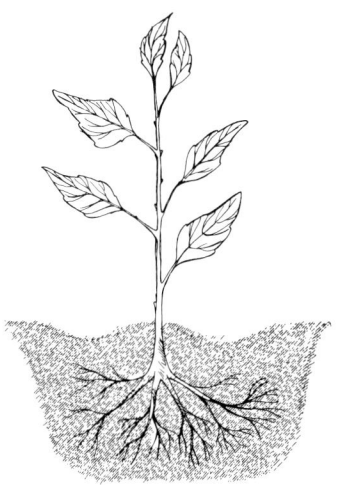
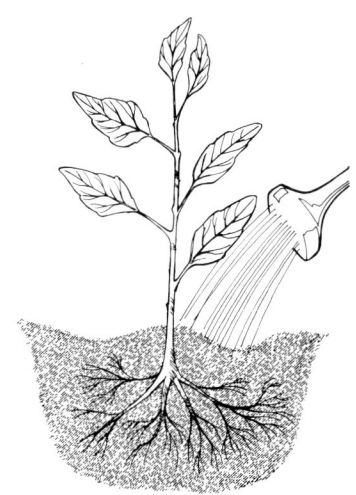

| | | |
|---|---|---|
| Eliminate large air pockets by firming well | Leave a shallow depression around the plant to collect the rain water | Water well immediately after planting, to give good root contact with the soil |

persistent in the soil and break down in a few weeks. With larger woody species, spraying the butt to run-off point with 245T gives a slow but effective kill. 'Tordon' can also be used but it is more hazardous. In vine and tobacco growing areas, the use of 24D or 245T may be prohibited at certain times so enquiries should be made beforehand on any local restrictions. Persistent type weedicides can be used on surfaces free of weeds but care is required in their selection and application to prevent damage to the plant. The use of weedicides is not encouraged as their effect on soil micro-organisms and as a chemical additive to the soil cannot be controlled. This also applies to the contact types since the dead plant tissue is eventually going to be added to the soil or the weedicide itself may be leached out and washed into the soil. Weedicides should only be used as a supplementary measure when the other techniques fail or are impracticable.

Regular watering is usually not necessary but, if the planting project is not too large, it is certainly an advantage. It should not be commenced too early in the season as, even though the surface may be dry, some moisture will be available at lower levels. The aim should be to allow the plant to grow for some time as, particularly if the site has been ripped to 30 to 40 centimetres or deeper, the roots will rapidly extend downwards in search of moisture. If the soil is loose and does not present a barrier to the roots, most seedlings about 20 to 50 centimetres tall can extend their roots to a greater depth before the first summer. The significance of good and deep soil preparation before planting is therefore clear.

# PROTECTION

## INSECTS

A large number of insects can attack trees but because means of natural control exist it is rare that serious damage is caused and even rarer for numbers of trees to be killed. The chances of suffering severe damage from insect attack are not great but it is just as well to be aware that the possibility does exist. Insects which are most likely to be troublesome to plants are briefly described below.

### (a) *All Trees*

Wingless grasshopper (*Phaulacridium vittatum*). This is one of the common grasshoppers occurring throughout southern Australia where the rainfall exceeds about 500 millimetres per year. Most have non-functional atrophied wings, hence the ability to disperse is limited and swarms are usually local. Winged forms do occur but rarely in sufficient numbers to form flying swarms. It does not appear to have any particular preference for trees but eats whatever is in its path.

### (b) *Conifers*

(i) Sirex wasp (*Sirex noctilio*). This wasp attacks most pines but principally *Pinus radiata*. Eggs are laid in the living wood beneath the bark and after hatching the larvae tunnel through the wood. Associated with the larvae is a fungus (*Amylosterium areolatum*) which predigests the wood for the larvae. It was first located in 1961 and is confined at present to central and southern Victoria but, as it seems to be established, in time it can be expected to spread to other areas. The adult is a steel blue wasp, but without the typical narrow wasp waist. Its length is 1·5 to 5·0 centimetres with a distinct horny tip at the end of the abdomen. Males have a broad tan band across the abdomen and females are uniform in colour. Larvae are cream legless grubs with a brown horny spine at the rear.

Vigorous control measures aimed at eliminating it are in hand. Experience in New Zealand, backed by experimental evidence in Australia, suggests that only the weaker trees, that is the ones that should be removed in thinning, are damaged or killed. However, fairly large trees may succumb if weakened by a poor growing season such as drought. Good management, involving the early removal of rapidly dying trees and using or burning them during winter, appears to be the most promising method of keeping it in check.

Introduced insect parasites are proving effective. Good control in large-scale field trials has been obtained with the wasp *Megarhyssa nortonii* from

North America (which is parasitic on the *Sirex* larvae), and the nematode *Deladenus siricidicola* from Britain (which enters the testes and eggs so that sterile eggs are laid). This nematode has a two pronged attack as the young, on emerging from the egg, also feed on the associated *Sirex* fungus.

(ii) Woolly pine aphid (*Chermes* (Pineus) *boerneri*). A minute insect which is difficult to see but is conspicuous because colonies form a white exudation (like cotton wool) around the needles. The colonies are found more often on young shoots of *Pinus radiata* and occasionally on older specimens of *Pseudotsuga menziesii*. It is not a serious pest and control measures are usually not required.

(iii) Moths. Various moth larvae such as painted acacia moth (*Orgyia anartoides*), light brown apple moth (*Epiphyas postvittana*), looper caterpillars (*Chlenias* sp.), case moth (*Hyalarcta* sp.), and tube moths (*Lichenaula* and *Clania* sp.) may attack the needles of conifers. Only rarely will attacks be severe enough to warrant control measures.

(iv) Bark beetle (*Hylastes ater*). Small, black beetles, 3·5 to 5·5 millimetres long whose colourless larvae feed on the inner, living bark of seedlings and saplings of pine, principally *Pinus radiata*. Attacked young trees may suffer die-back, loss of vigour or mortality. It does not generally cause serious damage.

(c) *Hardwoods*

These have many natural insect enemies but only a few are likely to cause trouble in southern Australia. They comprise four main groups—defoliators or leaf eaters, suctorial or sap-ingesting, leaf scale, and wood borers.

*Defoliators*

(i) Fire blight (*Paropsis* sp.). Small, brown or black, oval-shaped beetles, about 4 to 5 millimetres long. Both the larvae and adult eat the foliage; they drop off the foliage when it is touched so can be missed on a casual inspection.

(ii) Gum leaf skeletonizer (*Uraba lugens*). The larvae are hairy, cream underneath, and pink with yellow and dark brown markings on top. A fully-grown larva may reach 2·5 centimetres in length. Head capsules from previous moults are characteristically carried on top of the head. Eggs resemble small knobs and are laid on leaves in rows or clusters. They are commonly found in the Murray Valley, particularly on red gum (*Eucalyptus camaldulensis*).

(iii) Cup moth (*Doratifera vulnerans*). The caterpillar is brightly coloured but mainly light blue-green, about 2 centimetres long with four rosettes of reddish stinging hairs at each end. Occurence is mainly in the medium rainfall areas.

(iv) Sawfly (*Perga dorsalis*). Shiny black, slender caterpillar which is up to 5 centimetres long. They feed singly at night but during the day the larvae form a group like a black ball, on the smaller stems. They may seriously defoliate young trees.

(v) Christmas beetle (*Anoplognathus chloropyrus*). Both larvae and adult insects attack trees, the larvae attack the roots and the beetle the leaves. The beetles are an irridescent coppery colour, about 2·5 to 3·0 centimetres long, 2 centimetres broad, of 'robust' build, and occur mainly on eucalypts. Primitive, but effective control can be achieved by shaking small trees by hand, or larger trees with a tractor, and squashing the beetles on the ground. Control by sprays can be difficult and is often not very effective.

(vi) Emperor gum moth (*Antherea eucalyptii*). Large blue-green caterpillar (6 to 10 centimetres long, about 1 centimetre diameter) found mainly on eucalypts but it will attack other trees as well. Due to the spread of a virus, the caterpillars of this moth are no longer of much economic importance.

(vii) Pear and cherry slug (*Caliroa limaeina*). Black or dark green, wedge-shaped, slimy caterpillar up to 1 centimetre long which attacks a wide variety of trees, both native and introduced. It is seen most commonly in late winter to early spring.

(viii) Stick insect (*Didymuria violescens*). Large, grey-brown insect up to 20 centimetres long which attacks eucalypts, especially those in the high rainfall zones. It has a two-year life cycle and has caused severe defoliation and mortality of mountain ash (*Eucalyptus regnans*).

(ix) Leaf blister sawfly (*Phylacteophaga eucalyptii*). The larvae feed on tissues beneath the leaf cuticle and the foliage becomes blotched and blistered. Damage is rarely serious. A wide range of eucalypts and related species can be attacked. Control is by systemic insecticides.

### Sap-ingesting (Suctorial)

They are generally not very serious pests of trees. Two groups, other than the scale insects, occasionally attack older trees.

(i) Crusader bugs (*Cereidae* in general but particularly *Mictis profana*). Large bugs which are usually brownish, marked with a large and prominent St Andrew's cross on their backs, and about 8 to 15 millimetres long. Sometimes they attack the seedlings of *Eucalyptus regnans* and *E. delegatensis*.

(ii) Leaf hoppers (*Eurymellidae*, often called jassids). Small, wedge-shaped bugs which are generally less than 1 centimetre long. They are occasionally seen on eucalypts but may also lay their eggs in the soft stem tissues of seedlings causing small scars (egg scars), but these are rarely sufficiently extensive to cause serious damage.

### Leaf Scale

(i) Lerp. Greyish-white scale-like covering, common on eucalypts particularly under dry conditions. 'Scale' may be up to 1 centimetre in diameter, often of intricate design. They are produced by and cover a minute insect (family *Psyllidae*). Control is by systemic insecticides.

### Wood Borers

(i) Ambrosia beetles (*Austroplatypus incompertus* and *Platypus subgranosus*). Larvae and beetles tunnel in the tree leaving galleries 1 to 2 millimetres wide which are stained blue-black by the associated fungus. They are not easily detectable in the standing tree; infestation usually only comes to notice when the tree is felled and used, otherwise it may survive for 20 to 30 years. Neither death nor defoliation are caused.

(ii) Longicorn beetles (*Phoracantha* sp. and *Uracantha* sp.). Large beetles which are 2 to 5 centimetres long, with antennae longer than the insect. Larvae are large and vigorous, leaving tunnels up to nearly 1 centimetre in diameter and attacking most native woody species. Sugar gum (*Eucalyptus cladocalyx*) and blue gum (*E. globulus* and *E. st johnii*) in particular can be severely attacked when growing on unfavourable sites.

FUNGI

Some thousands of species of fungi are known to occur in association with trees. They are a part of the cycle of growth and decay and appear in the latter part of the tree's life. Healthy trees under good conditions are not usually affected, attacks being confined to moribund or dead trees.

(i) *Phytophthora cinnamomii*. A root-rot fungus which causes the death of trees from single specimens to stands several acres in extent. It requires warm, moist conditions and causes most trouble where soil drainage is unsatisfactory. Conifers and hardwoods are affected, principally Douglas fir (*Pseudotsuga menziesii*), *Eucalyptus astringens*, *E. baxteri*, *E. delegatensis*, *E. macrorhyncha*, *E. obliqua*, *E. radiata*, *E. regnans* and *E. sieberi* whereas *E. camaldulensis*, *E. cladocalyx*, *E. maculata*, *E. st johnii* and *Pinus radiata* are resistant.

The Family *Proteaceae* (*Banksia*, *Grevillea*, *Hakea*, etc. sp.) are also susceptible. Control is difficult and no practical means, other than improving drainage, has been found so far for field use. In nurseries, the fungicides 'Benlate' and 'Dexon' give satisfactory control.

Natural or introduced antibiosis is showing some promise as a method of control. It appears that some of the common basidiomycete fungi (toadstools, etc.) produce antibiotics which discourage the growth of pathogenic fungi; under laboratory conditions control of *Phytophthora* has been obtained but the results still (1973) have to be tried under field conditions.

(ii) Needle cast fungi. Several species are known to affect pines usually causing yellowing and partial defoliation during the warmer months. They rarely cause the death of the tree. Main species are *Naemacyclus niveus*, *Lophodermium pinastri* and *Diplodia pinea*. (*D. pinea* is sometimes found in moribund or dead shoots but its presence is secondary and is not the primary cause of death.)

(iii) Honey fungus (*Armillariella mellea*). This is a root fungus, the first evidence of attack being thinning of the crown. The inner bark at the base of the tree is usually heavily impregnated with white fungal 'threads' or hyphae; in severe cases the tree is girdled and death follows. In late autumn the fruiting bodies may be conspicuous, occuring at the base of the tree in clumps of twenty to fifty, with a brown-yellow cap, each one 4 to 10 centimetres across. Control is difficult; few effective measures can be taken but larger infected trees should be ringbarked and left standing for a year or two if replanting is proposed.

(iv) Cypress canker (*Monochaetia unicornis*). It is confined to cypresses, particularly *Cupressus macrocarpa* and its varieties. It is fairly widespread and is characterized by a swelling on the branches with cracking and splitting of the bark, and copious resin exudations. The distal portion of the branch usually dies. Little can be done to control it other than removing affected limbs.

ANIMALS

Young trees require protection from all animals—native and introduced. Damage may be by grazing or, with larger animals, by trampling. Grazing by larger animals, including kangaroos, wallabies, cattle, sheep, etc., is usually incidental rather than deliberate although when the surroundings are dry, stock will browse on young green trees. Rabbits may cause damage after planting as they are attracted by the smell of freshly disturbed soil. They nip the top of the tree, apparently more from curiosity than the need for food. In the wetter areas, native bush rats may gnaw the bark at the base of the tree until it becomes hardened while possums will climb a tree of any age and strip off the younger bark towards the top. Deer are likely to be troublesome until the trees are too large for browsing.

## BIRDS

Cockatoos, galahs, magpies and birds of a similar size may damage young trees by nipping off the top. Attacks usually occur when surroundings are dry. Scare tactics are the best method of control, although as strong solutions of fungicides such as thiram, captan, etc., are distasteful to most birds spraying with these gives some control.

## PLANTS

Mistletoe and dodder are two parasites which may be found on plants. Both, but particularly mistletoe, usually only develop on plants lacking in vigour either because they are (a) old and senescent, (b) on unfavourable sites, or (c) are otherwise diseased.

(i) Mistletoe usually *Amyema* or *Muellerina* sp. (commonly *M. eucalyptoides* and *A. pendulum*) is a pendulous leafy growth found on branches anywhere on the crown. The sticky seed is distributed by birds and deposited on the bark of the tree. On germinating, it attaches itself to the host plant and 'taps' the sapstream descending from the leaves with elaborated foods. Mistletoe has no root system as such, but it has a plate-like growth or haustoria from which 'hyphae' penetrate the host tissue and gradually divert the food supply from the leaves. A heavy infestation will kill a tree but one or two will only affect the branches to which they are attached. There is no effective cure other than lopping off the individual mistletoes or infected branches. Trunk injections of 24D and 245T may be used but the results can be variable—ranging from success to death of the host plant.

(ii) Dodder (*Cuscuta* sp., commonly *C. epithymum*) is a climbing plant with long thin wire-like stems and no apparent leaves. The seed germinates on the ground and the emerging shoot climbs the nearest upright growth. It forms a similar, but not so obvious, relationship with the plant as mistletoe, eventually severing all connection with the ground and living on the host. The growth forms a tangled mass of wiry stems which may completely cover and 'smother' the host plant. Control is difficult; pulling off the tangled mass of stems or cutting off badly infected parts of the host and burning them is about the only method. Spraying with 24D or 245T may be effective but sometimes the host plant suffers more than the dodder; generally these sprays are less effective on dodder than on mistletoe.

## MISCELLANEOUS

On young plants, especially for a few years after establishment, a variety of insect and fungal pests can be found. Mostly they are the common agricultural or garden pests of a locality and do not usually cause any serious trouble. They can be controlled by using the normal spraying treatment.

## CONTROL

In the last twenty years, there has been a tremendous increase in the number of pesticides available. Some are broad spectrum formulations effective over a wide range, others are specific to a few species or to special circumstances. In large-scale operations, it is more economical and efficient to use the most appropriate type but for the average planting, a fairly limited range will suffice. Use should always be in accordance with the directions and precautions on the label of the container.

(a) *Insects*

Insects are usually either defoliators, in which case the leaves are eaten, or the sucking type where the insect sucks out the cell sap. The former require a stomach poison for control, the latter a contact or systemic spray.

Arsenate of lead is a good all-round poison for defoliators. It had been used for decades without any marked side-effect but fell into disuse with the introduction of new compounds since World War II. Many of the latter have now been found to have undesirable side-effects on wildlife, and, as a result, arsenate of lead is tending to come back into use as a general-purpose spray. Arsenic is very persistent in soil and with repeated heavy spraying over a long period of time lead and arsenic concentrations can be built up in the soil, but investigations have so far not shown concentrations greater than those which occur naturally in many soils. Theoretically birds may be affected by eating insects killed with arsenate of lead although most birds prefer live insects so the risk of adverse effects is reduced. However, arsenate of lead is hazardous and care in its use is required.

For sucking-type insects white oil is still preferable because many of the more recent compounds are too persistent or have undesirable side-effects. Small insects, such as thrips, mites, etc., are to a large extent immune to the traditional nicotine sulphate but can be controlled by spraying with a liquid, biodegradable detergent of the type commonly used for household purposes.

Borers are difficult if not impracticable to control. If only a few trees of special value or interest are involved, injecting carbon bisulphide, formalin, chloroform, or methylated spirits into the tunnel and sealing the exit will give good control. Normally the tunnel exits are visible or can be located by the frass on the outside bark.

Amongst the modern insecticides maldison is one of the better general-purpose types especially for defoliators. It is of limited persistence and controls a wide range of insects. Being particularly lethal to bees it should not be used during flowering (pollinating) periods. Dimethoate and carbaryl are also satisfactory.

For sucking-type insects systemic insecticides containing demeton-S-methyl are effective but slightly more hazardous. Systemic insecticides enter the sap-stream and affect any insect sucking the sap. They remain effective for three to four weeks before being detoxified by the plant. However, as none of these is selective, there is the unfortunate risk of killing beneficial insects as well.

(b) *Fungi*

(i) Foliage sprays. Bordeaux mixture and lime sulphur are two old-established

## Control of Pests and Diseases

| APPEARANCE | SPECIES COMMONLY ATTACKED | PROBABLE CAUSE | CONTROL AND COMMENTS |
|---|---|---|---|
| **CONIFERS** | | | |
| Needles nipped off or chewed, caterpillars present | Mainly pines | Case moths | Spray with 0·1% maldison or arsenate of lead. |
| Foliage dirty in appearance, partly covered with a blackish mould | Spruces, firs, cedars | Aphides, followed by the development of a sooty mould fungus | Spray with 0·1% maldison, nicotine sulphate, or a good bio-degradable detergent. |
| Woolly, white, 'frothy' material amongst the branches | Conifers, mainly pines | *Adeleges* sp. (Chermes) | Spray with 0·1% maldison or 0·2% dimethoate. Rarely causes any serious damage–permanent control difficult to obtain as the susceptibility of individual trees to attack varies. |
| Foliage brown and dead in patches | Cypress | Probably thrips or aphides but may be root problems due to drought or excessive moisture | Spray with an all-purpose spray, 0·1% maldison, or a bio-degradable detergent. |
| Branches or branchlets swollen with resin exudations and bark splitting, outer part of the branch dead | Monterey and Lambert cypress and varieties | Cypress canker | Difficult to control–cutting off the infected portion and burning is about all that can be done. Spraying is usually not practicable, but copper-based sprays (bordeaux mixture, copper oxychloride) are effective if the affected part is accessible to the spray. |
| Branches or whole trees die, circular holes up to 5 mm diam. in the stem | *P. radiata*, rarely other pines | Sirex wood wasp | Fell infected trees and burn. (Wood can be used if certain quarantine conditions are fulfilled–contact State or Commonwealth quarantine authority.) |
| Needles pale green to yellow but otherwise not damaged, partial defoliation may occur | Pines | Needle cast fungi | Control not practicable, but attack rarely causes serious trouble. Older needles only usually affected, more common in wet years. Trees generally recover. |
| **BROADLEAF (HARDWOODS)** | | | |
| Leaves eaten completely or eaten from the margins leaving a ragged edge | Any species | Case moths, cup moths, emperor gum moths, sawflies, Christmas beetles, paropsis, phasmatids | Spray with 0·1% maldison, carbaryl, or arsenate of lead. |
| Leaf surface eaten, leaving the veins as a skeleton network | Any species | On eucalypts, usually gum leaf skeletonisers; on other species, pear slug | Spray with 0·1% maldison, carbaryl, or arsenate of lead. |
| Buds eaten, leaves rolled with most rolls containing a small green caterpillar, numerous spidery threads | Any species | Light brown apple moth | Spray with 0·1% maldison or arsenate of lead, if attack just commencing. If leaves are already rolled, use 0·2% dimethoate to reach the caterpillars in the rolls. |
| Interior of the leaf eaten, leaving the outside intact; may be transparent | Any species | Leaf blister sawfly, leaf miner or tunneller | Spray with 0·2% dimethoate or other systemic spray. |
| Small lumps, 'knobs' of various sizes and shapes on the leaf surface | Mainly eucalypts | Coccids, psyllids, or egg cases of various species | Difficult to control, but usually do not affect the tree or spread to any extent. Spray with white oil if control is essential. |

## Control of Pests and Diseases (*Cont'd.*)

| APPEARANCE | SPECIES COMMONLY ATTACKED | PROBABLE CAUSE | CONTROL AND COMMENTS |
|---|---|---|---|
| Brownish, irregular woody lumps or knobs on the branches and twigs | Mainly acacias | Various gall insects | Difficult to control—spray with $0.1\%$ maldison or white oil. Indicates that the tree is lacking in vigour and should be replaced. |
| Leaves pale green, unhealthy in appearance, but otherwise undamaged | Any species | Leaf hoppers, crusader bugs, (and other sap ingesting species) | Spray with $0.2\%$ dimethoate or other systemic spray. |
| White or light-coloured flaky scale up to 1 cm diam. on the leaf surface, leaf pale green, with pale yellow patches where the scale has fallen off. Aphids may be present causing a blackish sooty mould to develop | Mainly eucalypts | Scale insects (living under the scale) | Spray with $0.2\%$ maldison, white oil, or a good bio-degradable detergent. |
| Foliage being eaten, tree sometimes defoliated | Wattles with pinnate foliage, sugar gum | *Paropsis* sp. (fire blight) | Spray with $0.1\%$ maldison, carbaryl, or arsenate of lead. |
| Branches or large parts of the crown wilting | Mainly deciduous trees | Probably *Verticillium wilt* (a soil fungus infecting the roots) | Difficult to control. Heavy fertilization with nitrogen fertilizer may help, but removal and replanting with a resistant species is usually necessary. |

GENERAL

| | | | |
|---|---|---|---|
| Foliage eaten, generally irregularly, attacks erratic | Any species | Grasshoppers, usually the wingless grasshopper | Spraying with dieldrin is the most effective but often environmentally unacceptable. Using $0.2\%$ maldison will give a knock-down effect. Attacks are only likely to be serious when grasshoppers are present in plague proportions. |
| Branches or large parts of the crown appear unhealthy or dead | Any species | Probably wood borers | No effective control possible on a large scale. Fell and replant. |
| Holes on the outside of the bark, lumps of sawdust-like material (frass) may be seen on the outside bark | Any species but mainly hardwoods, particularly native species | Wood borers | No effective control possible on a large scale. Fell and replant. See also notes on Protection (page 93). |
| Whole tree dying all parts more or less at the same time. (Generally indicates root troubles.) | (a) Pines, eucalypts | Honey fungus (*Armillariella mellea*) | Control difficult. |
| | (b) Conifers (other than pines), some native species and broadleaf trees, but few species of deciduous trees | *Phytophthora cinnamomii* root rot. | Control difficult, improving drainage before replanting is desirable. |
| | (c) Any tree | Cause physical e.g. drainage unsatisfactory, root system disturbed, crown sprayed with pesticide, etc. | Improve drainage, avoid disturbance of root systems, take care when using pesticides, etc. |

(The assistance of Mr F. Neumann and Dr G. Marks *Forest Commission, Victoria* in preparing this Table is acknowledged.)

sprays which are still very effective against a wide range of diseases. Bordeaux mixture is a bit troublesome to use so a copper spray, such as copper oxychloride, is commonly used instead. Modern organic sprays (thiram, ziram, zineb, captan, etc.) can also be used but are more specific to particular diseases rather than suitable for general control.

(ii) Root rots. Control of root rots is difficult and usually uneconomic on a large scale. Improvement of drainage is the best approach. For small areas or single plants sterilizing the soil with formalin before planting is effective, or with 'Benlate' or 'Dexon' after planting. However, treatment with the latter two compounds can be costly on large areas. The use of soil fumigants, such as methyl bromide, on soil in situ is usually difficult and often not practicable.

### (c) *Animals*

Rabbits are normally the main problem. Fencing will restrict movement but effective control is most conveniently achieved with 1080 and carrots or prepared baits. Two poison-free feeds at about four-day intervals followed by a poisoned feed is usual. During the poison-free feeds, observe whether all baits are taken; if not, lessen the intensity of the poisoned feed, otherwise the poisoned baits may be taken by other animals or birds.

1080 is a tasteless, odourless and cumulative poison. It is dangerous to humans so is subject to various State laws; you should consult with your local vermin destruction agency before using it.

## PESTICIDES

A summary of a few modern pesticides is given below for general information. Withholding time is the minimum legal time (for agricultural crops in Victoria 1973) between the last spraying and sale, when used at the recommended strengths. It is a rough indication of the hazardousness to people of the substances.

| Group | Common Name | | Withholding Period (days) |
|---|---|---|---|
| *Insecticides* | | | |
| Chlorinated | DDT | | 30 |
| hydrocarbons | dieldrin | | 90 |
| | endrin | | 90 |
| Organic phosphates | demeton-S-methyl | ('Metasystox') | 21 |
| | dimethoate | ('Rogor') | 7 |
| | maldison | ('Malathion') | 7 |
| | parathion | | 21 |
| Carbamates | carbaryl | | 3 |
| | arsenate of lead | | 60 |
| *Fungicides* | benamyl | ('Benlate') | 1 |
| | copper oxychloride | | 1 |
| | dexon | | – |
| | lime sulphur | | 1 |
| | thiram | | 7 |
| | zineb | | 7 |
| | ziram | | 7 |
| | captan | | 7 |

Pesticides should only be used as a supplementary control measure since their effect on the chemical status of the soil and on soil micro-organisms is not well understood. Except for when the occasional epidemic occurs, most troubles can be avoided by good cultural practices as insects and fungi normally only attack trees lacking in vigour. However, when pesticides are necessary, they need not be used excessively, as happens all too frequently; using the recommended dosage is adequate, putting in a bit extra for good measure does not achieve anything except making it more hazardous to the user and wildlife.

There is usually a choice of pesticides available so that the use of the more undesirable ones can be avoided. The use of persistent compounds, like DDT and its derivatives, should be minimized as their breakdown by digestive processes is slight which leads to large biomagnification in the food chain. Similarly, the use of other chlorinated hydrocarbons (such as aldrin, dieldrin, and endrin) and mercury compounds should be avoided if possible. On the other hand, there are some which are reasonably safe on present evidence provided they are used wisely and cautiously, for example, maldison, dimethoate, demeton-S-methyl, carbaryl, 24D, 245T, and most of the fungicides are in this group. With the 'growth regulators' 24D and 245T, spray drift should be controlled to prevent possible damage to nearby susceptible plants.

During the past few years, 24D and 245T have been suspected of being a risk to human beings. This has not been satisfactorily proved but the suspicion seems to have arisen mainly from the effects of large-scale, aerial defoliation under-taken during some phases of the Vietnam War, even though these compounds had been in widespread use for the previous twenty years. It seems that output was increased (presumably because of the rapid and great increase in quantities required for the war defoliation program) by running the manufacturing process at a much higher temperature which led to the formation of the compound dioxin as an impurity of up to 2 per cent. It was the high dioxin impurity, not the compounds 24D and 245T, which was suspect. Since manufacturers have modified their techniques in the last few years to give a product with less than 0·5 parts per million of dioxin, the possible threat has been removed. Some countries, such as Canada, which previously had severe restrictions on the use of 24D and 245T, have substantially modified them to the extent that the restraints on their use are now of minor practical significance.

Contrary to the views held by many people, most pesticides are fairly rapidly degraded. Light, particularly the ultra-violet end of the spectrum, is a potent factor in pesticide degradation in air and water; in soil, the ordinary chemical processes of oxidation, reduction, and hydrolysis, both by the microflora and by normal chemical reactions, are very effective and rapid. Significant amounts are also adsorbed on soil particles and held in a non-reactive state. The environment has a large capacity to cope with pesticides and other introduced substances but it can be overwhelmed by careless or excessive use of them; a rational approach does not cause any lasting deleterious effect and, with reasonable rather than excessive application, there should be no serious cause for concern.

Biological control methods do not produce the undesirable side-effects of chemical methods which have been the cause of much criticism in recent years. But biological techniques are not without risks; there is always the possibility that an introduced predator will adapt itself to some local fauna or flora or, if a native species is involved, a marked increase in population increases the chances of a more virulent and adaptable strain evolving. Careful screening trials will reduce the risk to a very low level but mutations are an unpredictable and inherent risk in biological control. Each case must be thoroughly and tediously examined before general release can be accepted. This is a slow and time-consuming process, nevertheless the approach is sound and, with careful investigation, biological methods will be of increasing value in the future.

In recent years there has been an upsurge of interest in these methods. There have been many highly successful techniques for dealing with specific problems, such as prickly pear and St Johns wort, but more attention is now being given to broader-spectrum methods. One of the more promising ones is the use of bacterium *Bacillus thuringiensis* in controlling *Lepidoptera* larvae (moth and butterfly caterpillars). *B. thuringiensis* occurs naturally in the soil in most countries, including Australia, and is a natural parasite of the larvae. It is marketed locally as a wettable powder and is applied as an ordinary spray. On ingestion, it produces a crystaline toxin which is soluble in the gut of larvae that are highly alkaline (pH 9·0 or more), such as the *Lepidoptera*. The toxin disrupts the walls of the midgut leading to infection by the bacterial spores which causes septicemia and death. The larvae cease feeding within a few hours of ingestion and die within one to four days. Most of the experience (to date) has been on agricultural crops but there is no reason for believing that it would not be equally effective on lepidopterous tree pests if they feed on the surface and can be infected. In Canada, it has been very effective on the spruce bud-worm, one of their major forest pests. On the other hand, larvae which feed in concealed situations are unlikely to ingest the bacteria on the surface so are largely immune.

To conclude, it needs to be appreciated that no pesticide is perfectly safe in all respects—if it were it wouldn't be effective. When control is necessary, it is a matter of compromise between the threat and the risk and this is a matter on which the individual must make up his own mind.

# ENVIRONMENT

## GENERAL

Trees are a part of the environment. Their influence outside the forest is similar in many ways to a forest stand but differs in that it is local rather than regional. As they are local, they can be used to affect conditions on a single property, be it a suburban block or a large grazing property, and can be adapted to suit the ideals of the owner.

### (a) *Reduction of Wind*

The reduction of wind speed by trees is well known but the physical aspects are complex. As well as dissipating some of the wind energy, trees deflect wind around and over themselves creating their own 'field' of wind currents. Wind velocity is reduced and direction is changed; this aspect is dealt with in greater detail under Windbreaks (page 101).

### (b) *Rainfall*

Whether trees increase rainfall is still not clear. It is doubtful if even forested areas covering thousands of square kilometres have any effect on the cyclonic pattern of world-wide air flows. However, they do affect orographical (local) rainfall to the extent that the extra height of the trees and the friction of the wind reducing the ground speed does increase the likelihood of precipitation. In clearings, such as between windbreaks, increases of up to 10 per cent in the amount of rain reaching the ground may occur. The general rainfall is not increased but there is much evidence to show that more does actually reach the ground, particularly in windy weather. Moreover, dewfall can be increased in both quantity and frequency of dews, while fog drip interception is frequently twice as great as in the open. In alpine areas, snow may accumulate on the lee side to a depth of a metre more than on the windward side.

### (c) *Evaporation*

While trees may not increase the actual rainfall appreciably, they do make it more effective by reducing evaporation. Relative humidity varies during the day but the average is higher close to trees and decreases with distance to that in the open. Reduced wind velocity and high humidity mean that evaporation is less— figures indicate that evaporation close to trees is about 40 per cent of that in the open, gradually increasing to that in the open at a distance equal to twenty to thirty times the height of the trees. The net effect can be equivalent to about 100 millimetres of additional rain per year.

### (d) *Temperature*

While it is usually cooler underneath or in the shade of trees, average air temperature between or behind windbreaks is usually little different, perhaps up to $\mp 1°C$ from that in the open. This may seem contrary to most people's experience in that they feel 'warmer behind the trees' but the explanation is that the wind dissipates body heat and the loss of heat produces a feeling of cold; behind the windbreak, wind and loss of body heat are less hence the feeling of comparative warmth. The range of temperature, however, is decreased as trees, especially as windbreaks, reduce the maximum and increase the minimum by up to a total of 4 to 5°C.

### (e) Soil Moisture

The cumulative effect of windbreaks on rainfall, relative humidity and evaporation is an increase in soil moisture. Investigations on a wide range of crops show soil moisture to be about 20 per cent higher near the windbreak decreasing to that in the open at a distance equal to about twenty to twenty-five times the height of the windbreak. In a dry area of about 250 millimetres annual rainfall, conservation of soil moisture is equivalent to about another 50 millimetres of rainfall. In wetter areas where rainfall normally exceeds evaporation, the difference is unlikely to be so significant.

## WINDBREAKS

Although windbreaks are a common feature of country areas, the underlying principles of their operation are not widely known. For a country, such as Australia, with extensive open plains, surprisingly little work has been done on the subject and it is necessary to draw on overseas investigations. Most of the investigation has been undertaken in notoriously windswept regions, such as the Great Plains of America, the Steppes of Russia, the plains of north Germany and Denmark, and the Highlands of Scotland, but the general agreement in the results suggests they are also applicable in Australia.

### (a) Design

A windbreak functions by deflecting most of the air stream (wind) up and over the trees. The area protected depends on the airflow over the windbreak; a sharp upwards deflection is followed by a quick descent on the lee side while a gentle upwards movement on the windward side gives a more gentle descent on the lee. The area protected in the latter case is therefore greater than that with sharp deflections.

The reduction in wind velocity may extend to about five to ten times the height of the windbreak (abbreviated to 5–10H) in front and fifteen to thirty times its height (15–30H) to the rear, depending on windbreak density or permeability. A dense or relatively impermeable windbreak, such as a good cypress hedge, has a greater effect close in and reduces velocity to about a quarter, but the total area protected extends back for only about 10–15H at which point wind velocity is back to that in the open. With a more permeable type, such as a row of eucalypts with an understorey of smaller trees or shrubs, wind reduction is less (only about 50 per cent) but the area protected extends back to 25–30H. With a dense one, the zone of maximum protection is from the tree edge extending back to about 2H; with a permeable type it occurs slightly further behind, at about 3–5H. Turbulence and eddying behind a dense one is also

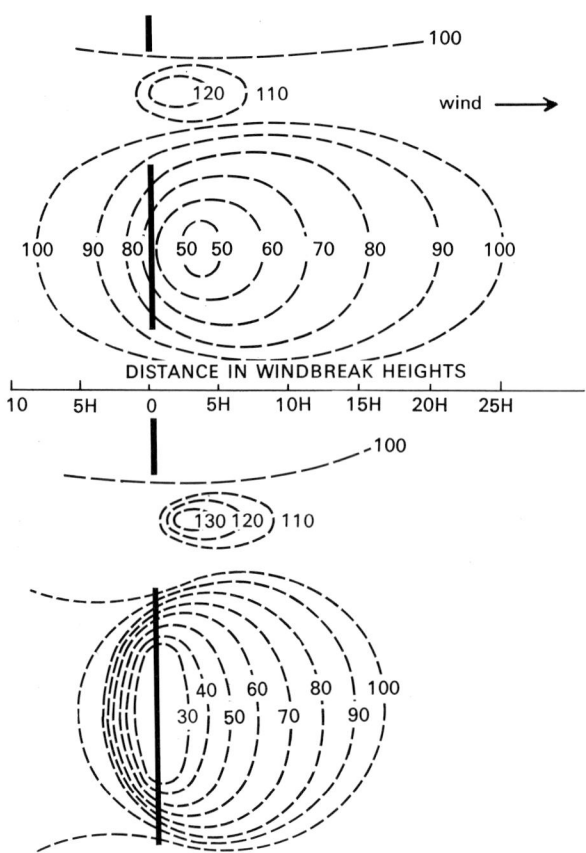

Figure 15. Simplified diagram of reduction in wind velocity by a permeable (above) and non-permeable (below) windbreak expressed as a percentage of open velocity.          Forest Commission Victoria

greater, wind currents on striking the break rise more sharply and consequently descend more rapidly on the lee side, whereas with a more permeable type some of the currents pass through and on the lee side 'cushion' the downward movement of the current which passed over the top of the windbreak. (Figure 15)

Which type is planted depends on the intention. A dense type is normally preferred around houses and buildings, or where intensive animal protection is desired while the permeable is more suitable for protection around paddocks for crops and open grazing.

The greater the height of the windbreak the greater the area protected. Tall-growing trees are obviously desirable. A single-row windbreak is very common and effective, but a three to five row design with the tallest trees in the centre and smaller trees in the outside rows presents an inclined slope to the wind on both sides giving a smoother airflow and less turbulence. With this design both the reduction in velocity and the area protected is greater than with a single row.

For general use, a three to five row design is the best. Wider breaks of seven to ten rows, or more if space permits, are desirable for noise reduction (see page

110). If the width becomes too great, say thirty to fifty rows as when a combined farm forest and windbreak are sought, the area protected is reduced as the wind currents travel along the top of the trees and at the lee edge descend fairly quickly to ground level.

The length of a windbreak needs to be about ten to twelve times the height if protection from one direction only is required, or about twenty to twenty-five times the height for all-round protection. Shorter lengths lose much of their effectiveness because of eddying around the ends. Gaps will cause increased wind velocity (up to 30 per cent more than that in the open) and they should be covered with a short windbreak a few metres away or, in multi-row plantings, the gaps should be set at an angle across the lines of trees. Gaps under trees will cause increased wind velocity by funnelling; in light soils, accelerated soil blow could occur. Under-planting with shrubs or bushy trees along the edges of the break will prevent it and should always be undertaken when using tall species which shed their lower branches, for example, eucalypts. For best protection, dense windbreaks should be about 10–12H apart and permeable windbreaks about 18–20H apart. On large properties, they should be planted at right angles to each other on a grid pattern. Under these conditions the protected area is increased because of the 'lifting' of the main air stream above the general ground level. In hilly areas, planting them along the ridge line will give the most 'lift' to the air currents and a larger protected area. Where planted along contours, gaps may need to be left in gullies, depressions, and other low points to allow cold-air drainage and prevent the creation of frost pockets.

(b) *Production*

Trees planted for shelter will affect adjoining crops but while the effect is plainly seen, the benefits are not so obvious. Adjoining each windbreak or tree is a strip about $\frac{1}{2}$–1H in which root competition reduces the yield appreciably. Not all the decrease is due to the windbreak since, in its absence, the open-field edge effect would also have depressed the yield. Beyond this yields will increase, more or less over the whole protected area but reaching a peak at about 5–10H on the lee of the windbreak. (Figure 16)

The greatest increases are obtained in the more adverse sites or years. When conditions are favourable, increases may be only slight but where conditions are extreme, or in a drought year, the increases can be substantial. Pasture grasses seem to respond more than cereals and these respond more than vegetables. It is obvious that increases can vary widely from year to year and site to site but local observation suggests that on the average an increase of 10 to 25 per cent can be expected. Much higher yields have been recorded, 30 to 50 percent is common, but during drought years on the plains in north Germany, Poland and Russia increases of over 300 per cent in winter wheat and lucerne have been recorded on several occasions over the last fifty years. It is not suggested that this would necessarily occur locally but it indicates that windbreaks can have a significant effect on yields, particularly in adverse years.

The effect on stock is indirect as a sheltered animal eats less than an exposed one, and less of its body weight is converted to heat energy for warmth. Under extreme conditions an exposed animal will loose several pounds more body weight over a few days than a protected one—another example of a cold cow is a hungry cow.

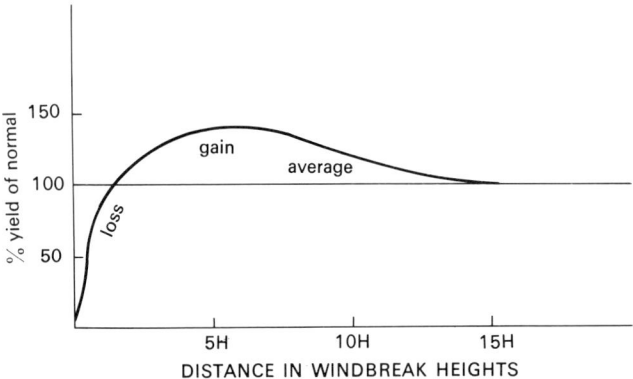

Fig. 16. Graph showing the effect of shelter on crop yield. Competition close to the plantation causes a substantial reduction in a limited area which is more than offset by the increase over the large sheltered area.
Forest Commission Victoria

The proportion of a farm which should be under trees is a difficult problem to resolve. Using the values already given it can be calculated that for most needs it should be a minimum of 3 per cent in average localities, and about 5 per cent in the more exposed or unfavourable sites. Even a single-row windbreak 20 to 25 metres high will give useful protection for about a half a kilometre.

### (c) *Heating*

Air is one of the poorer conductors of heat and, as such, is a good insulating material. Most household and industrial insulating materials function on the principle of a layer of air trapped within them preventing heat loss by air movement conducting heat away from the source.

A windbreak of trees functions in the same way; viz. by restricting air movement so that less heat is dissipated and lost. Although movement is not reduced to zero any reduction in movement will reduce heat loss. A saving in heat loss is reflected by savings in heating costs and, in the cooler parts of southern Australia, a saving of 20 to 40 per cent can be expected with good protection around a house. Protection on the cold and windy side only will also give a saving but not so much as all-round protection.

In the commercial field, heating glasshouses can be costly. An increase in wind from the south of 0 to 30 kilometres per hour commonly doubles the fuel or power consumption, indicating the quantity of heat dissipated by air movement. A windbreak on the cold side (usually south or south west) will restrict air movement and may reduce annual heating costs by 15 to 20 per cent.

## SHADE

The shading effect of trees is widely appreciated both for comfort in summer and its effect on the growth of plants. Temperatures under trees can be 10 to 15°C less than in the open, although mixing of air by the wind may reduce this.

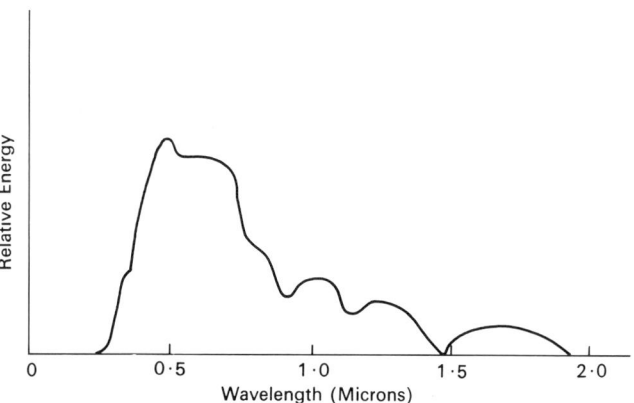

Figure 17. Emission spectrum of sunlight.

The temperature is cooler in shade because air is heated by heat reflected from the ground and not directly by sunlight. Solar radiation reaches the outer atmosphere at the rate of approximately 2 calories per square centimetre per minute. About one-third is reflected back into outer space and a similar amount is refracted through the atmosphere and absorbed by the ozone, dust, carbon dioxide and water vapour in the air, so that only about 0·8 calories actually reach the ground at sea level in southern Australia. Most of the absorption takes place in the atmosphere at lower levels; the radiation received increases rapidly with altitude, at 1500 metres it is 1.4 or nearly the same as in the outer atmosphere. The additional heat at higher altitudes is one reason why plants grow more rapidly in the short growing season of alpine areas; as well as why people get sunburnt more easily.

Radiation from the Sun consists of visible light, a small amount of ultra-violet, and the shorter infra-red rays. Wavelength is less than about 1.5 microns (0·0015 millimetres). Since air cannot absorb these relatively short wavelengths it is not heated by the sunlight passing through. They are absorbed by the ground (the surface often being heated to 65°C to 70°C) which re-radiates longer waves of 5 to 20 microns (0.005 to 0.020 millimetres) back to the atmosphere. It is these reflected waves that do the actual heating. (Figure 17)

Since the air is not heated by direct sunlight but by ground reflection, intercepting the sunlight with, for example, trees or screens as in nurseries, reduces ground heating and re-radiation. Consequently there is a lower temperature in the shade or underneath screens. If the atmosphere were heated directly by the Sun, wind movement would keep sunshine and shade temperatures about the same.

In built-up areas the same principle operates. Most external building materials, with the exception of timber, readily absorb the solar heat and re-radiate it to heat the surrounding atmosphere. 'Reflection' or re-radiation can be very great and is the main reason why towns and cities are often 5 to 10°C warmer than the neighbouring parkland or surrounding countryside. Whether the external wall is light-coloured (i.e. heat reflective) or dark (i.e. heat

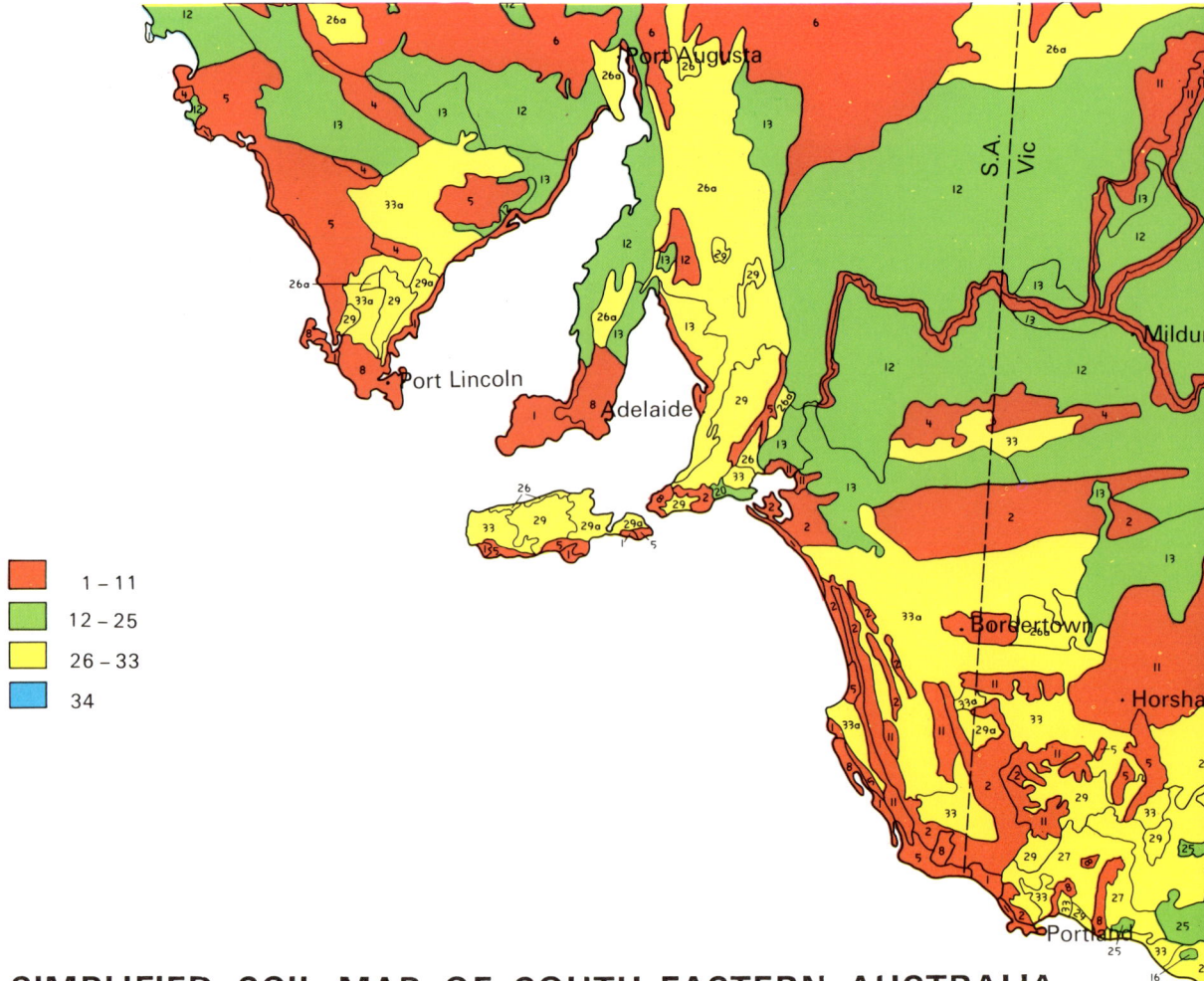

# SIMPLIFIED SOIL MAP OF SOUTH-EASTERN AUSTRALIA

(Adapted from the C.S.I.R.O. Division of Soils publication—K. H. Northcote and others—'Atlas of Australian Soils' —Sheets 1, 2 and 3, and the Bulletin accompanying each sheet.)

*Legend*

Use of the suffix 'a' e.g. 26a indicates that the soil is alkaline through the profile.

**A.  SOILS WITH UNIFORM TEXTURE PROFILES**

1   Sand soils with an unbleached $A_2$ horizon
2   Leached sand soils
3   Sand soils with an unbleached $A_2$ horizon
4   Sand soils with weak horizon formation
5   Coherent sandy soils
6   Loamy soils of minimum or weak horizon development.
7   Loamy soils with an $A_2$ horizon
8   Friable loamy soils
9   Organic loamy soils
10  Plastic clay soils
11  Cracking clays

**B.  SOILS WITH GRADATIONAL TEXTURE PROFILES**

12  Brown calcareous soils
13  Highly calcareous loamy earths
14  Red earths
15  Yellow earths
16  Yellow leached earths

17  Grey earths
18  Red friable earths
19  Brown friable earths
20  Dark friable earths
21  Leached friable earths
22  Red friable porous earths
23  Brown friable porous earths
24  Dark friable porous earths
25  Friable earths with a variety of loamy soils

**C.  SOILS WITH CONTRASTING (DUPLEX) PROFILES**

26  Hard setting loamy soils with red clay subsoils
27  Hard setting loamy soils with brown or mottled brown or dark clay subsoils.
28  Hard setting loamy soils with yellow clay subsoils.
29  Hard setting loamy soils with mottled yellow clay subsoils.
30  Friable loamy soils with red clay subsoils
31  Friable loamy soils with dark, brown, or mottled brown clay subsoils.
32  Friable loamy soils with gley clay subsoils
33  Sandy soils with yellow or mottled yellow clay subsoil.
34  Organic soils.

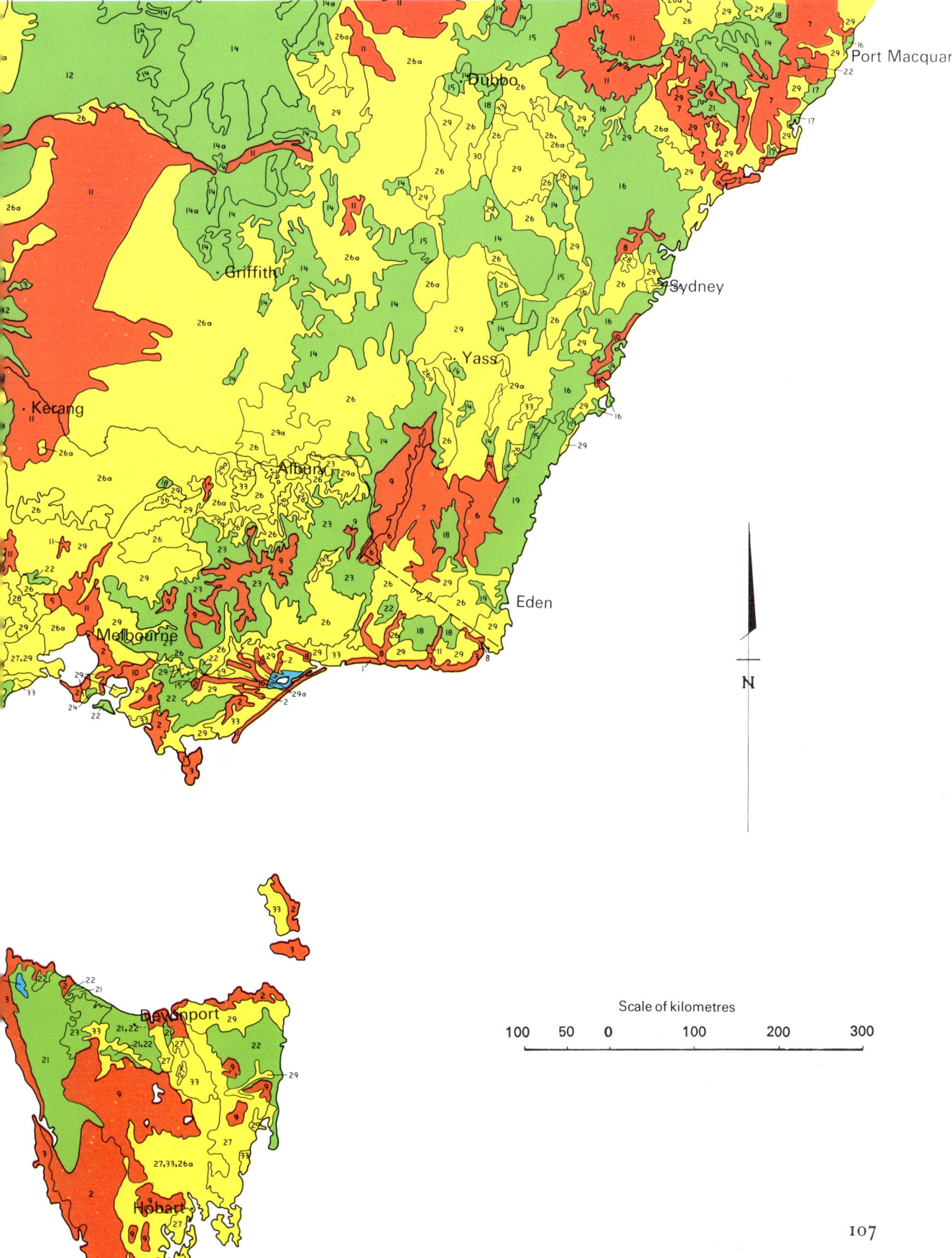

Scale of kilometres

100    50    0         100        200        300

absorptive) has an effect on internal temperature. It has less effect on the external temperature except that with a dark colour, because of greater absorption, heat is re-radiated to the air for a longer period. Bitumen pavements are a good example of this. The answer to this problem of heating in built-up areas is more trees and parkland in towns and cities.

## CARBON DIOXIDE

Much has been said about the possible effect of carbon dioxide ($CO_2$) on the Earth's climate (the 'glasshouse' effect) since it was first proposed more than seventy years ago. The use of the term 'glasshouse' alludes to the transmissivity of glass which only transmits the shorter wavelengths (up to 4·5 microns) but not the longer re-radiated wavelengths (5 to 20 microns) which are 'retained' in the glasshouse and create the heating effect.

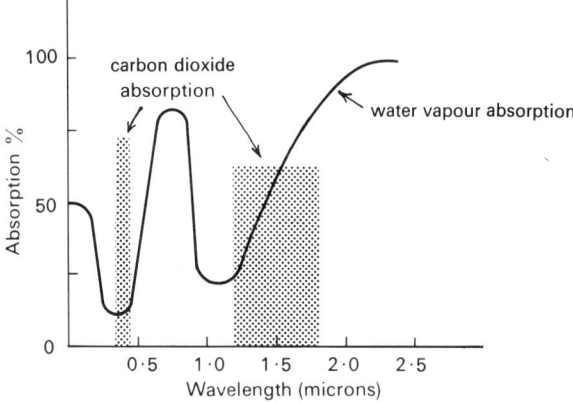

Figure 18. Absorption of solar radiation by water vapour and carbon dioxide in the atmosphere.

While carbon dioxide in the atmosphere absorbs some sunlight, it is a much stronger absorber of the longer wavelengths re-radiated from the soil surface, particularly in the infra-red section of 12 to 18 microns (0.012 to 0·018 milli-metres). When the soil-air difference is favourable the carbon dioxide acts as a blanket by back-radiating the heat to the lower air and surface, while some is lost to outer space. Consequently, if the carbon dioxide content of the air is increased, a slight warming of the Earth's atmosphere may be expected. On the other hand, if the temperature rises, some of the extra heat would be used in evaporating more moisture to bring the humidity into equilibrium How much the increase would be is complex to estimate; a recent estimate suggests a 25 per cent increase in the carbon dioxide content of the atmosphere would raise the mean temperature by a minimum of 0·6°C or a possible maximum of 4°C. (Figure 18)

The most significant effects of increased world-wide temperature would be:

(a) Potentially increased photosynthesis and production by plants, although other factors such as water and nutrients may be more limiting.

(b) A rise in the level of sea level by melting of the polar ice caps. Melting of all the ice in Antarctica would raise sea level by 30 to 40 metres which would have a catastrophic effect on most coastal cities. Sea level has risen 10 centimetres in the last one hundred years but a 25 per cent increase in atmospheric carbon dioxide would probably increase this rate to the order of a metre or so per century.

(c) Warming of sea water, especially the surface layers. A rise of 1 to 2°C during 1880–1940 in the northern Atlantic was sufficient to cause a redistribution of cod fisheries towards Greenland, and a retreat of the sea ice around the Arctic Ocean.

Although in the period 1880–1940 world mean temperature increased by 0.5°C, since 1940 a slight cooling trend (about 0·1°C) has become apparent in spite of the great increase in the last thirty years in the consumption of fossil fuels and in the carbon dioxide output. Possibly this reduction is due to the greater concentration of particulate matter in the air which increases reflection by the atmosphere thereby reducing the amount of sunlight and heat reaching the Earth's surface. Much has been said about pollution of the Earth's atmosphere but it is worth noting that Man's efforts, unfortunate though they may be, are small compared with Nature; the eruptions of Mt Krakatoa in Java in 1883, Mt Katmai in Alaska in 1912, and Mt Hekla in Iceland in 1947 contributed more particulate matter, gases, etc., to the atmosphere than the total contribution by Man throughout recorded history.

The position is not yet clear enough for any firm predictions to be made. It is only in the last fifteen years that some accurate measurements of the carbon dioxide problem have become available but on the whole they support the theory that the carbon dioxide content of the atmosphere is increasing and a rise of mean temperature is possible. Measurements suggest an annual increase of 0·2 to 0.3 per cent in the carbon dioxide content of the atmosphere, with a probable total rise of about 25 per cent by A.D. 2000. Insufficient information is available to say clearly what the effect will be except that it is more likely to be unfavourable and it would be wise to take steps, as far as possible, to ameliorate it.

The main source of carbon dioxide is the burning of fossil fuel. Fuels used today—coal, oil, petrol, natural gas, in fact all except nuclear fuels—contain carbon compounds produced originally by ancient and recent plants and animals. This carbon was at one time present in the atmosphere as carbon dioxide and burning these fuels returns carbon dioxide to the atmosphere. World production of fossil fuels has increased thirty times since the 1850s, and has doubled in the last twenty years. If the present rate of consumption is maintained, atmospheric carbon dioxide will be increased by 25 per cent approximately (1950 level) by A.D. 2000 but if the average annual rate of increase for the last twenty years continues the increase would be over 40 per cent. Human beings are another source of carbon dioxide pollution as each adult in breathing produces about 0·3 tonnes per year (i.e. the population of cities like Melbourne and Sydney each contribute nearly three quarters of a million tonnes a year). All animals, domestic and native, also add their contribution.

In terms of the geological time scale, about half of this (previously thought to be about 80 per cent) will eventually end up in the oceans, mainly as limestone, and dolomite rock. But this is a slow process, in terms of the human life scale, only about 15 per cent will be 'absorbed' by the oceans while the balance remains in the atmosphere. (Oceans contain only 0.01 per cent of carbon dioxide.) There still remains then a large volume of carbon dioxide in the atmosphere representing a problem in pollution which requires attention.

Control of carbon dioxide additions is beyond the scope of this publication but the production of such volumes of carbon dioxide represents a large, untapped, raw-material resource for plant growth. It is mixed throughout the whole of the atmosphere but even the quantity in the first 20 to 30 metres (the height of tree growth) is considerable, especially as the concentration to about 15 metres is usually about twice that of the atmosphere in general. Most of the emissions would originate in or near centres of population and it is likely that the effect there would be greater than the world-wide effect. It is at the local level that some amelioration could be expected.

As a rough rule, it takes 1.5 tonnes of carbon dioxide to produce one tonne of wood with one tonne of oxygen being released to the atmosphere in the process. An actively growing forest on a reasonable site will produce 20 tonnes of wood per hectare per year, using 30 tonnes of carbon dioxide and releasing 20 tonnes of oxygen in the process. Carrying out an exercise in figures: the people in cities, such as Melbourne or Sydney, produce enough carbon dioxide in the course of breathing to support about 25 000 hectares of forest.

To have any significant effect on the world-wide pattern, reforestation on a grand scale would be required. By coming down to city or local level some useful benefits could be obtained. Recreational and commercial forestry are tradition-ally ventures for the remoter parts of the country but there is a good case for introducing them to the urban areas, particularly where the green belt or wedge development concepts of town planning are in operation. Large forests of thousands of hectares could be developed. Cost would a major factor in the initial phase but, with a proper balance between commercial and recreational forestry, the project would become self-supporting after some years.

## NOISE

Noise is a form of air pollution which is becoming increasingly prevalent. It is not confined to city traffic and factory noise but can be a problem to people near main roads, railways, aerodromes, etc. It is a very subjective matter; a level or frequency annoying to one person may not affect another. Variation of this type makes it difficult to be precise about results and effects.

Frequency or pitch of a sound is measured in cycles per second, or hertz (abbreviated to Hz). The range of audibility for most people is from about a low of 20 hertz to a high-pitched sound of 15 000 hertz, with experience showing that people are more susceptible to noise in the 500 to 2000 hertz range. (International concert pitch of A is 439 hertz.) Traffic noises cover almost the full audible range but peak over a certain range; cars from about 80 to 2000 hertz, and heavier vehicles such as buses, trucks, semi-trailers, etc. from 100 to 1000 hertz. Aircraft cover a wider range, and often produce a disturbing rumble at low frequencies (50 to 100 hertz).

The 'loudness' of sound is measured in decibels over a range from 0 to 135. It is a logarithmic scale which really measures sound energy. A value of 0 is perfect quiet (never experienced under natural conditions), 25 decibels represent a quiet room or library, at 65 to 70 decibels discomfort occurs, 80 to 85 decibels represent a very noisy street intersection and 135 decibels are literally deafening. An increase of 10 decibels, although it represents a ten-fold increase in sound energy, to most people is a doubling of the noise or 'twice as loud'. A further standard commonly used is speech interference level; at 60 decibels normal conversation is possible over about 1 metre, but the voice has to be raised to permit conversation over 3 metres. At about 65 to 70 decibels most people consider noise is no longer tolerable and has become a nuisance.

Noise can be reduced by barriers of trees. They function by diffusing and absorbing the sound rather than by reflecting it like a solid barrier. They cannot eliminate noise but can reduce it. Their effectiveness depends on the height, width, and density of the barrier, and distance from the noise source. Leaf shape or size is of only minor significance. Ground conditions cause some reflection, so if grass or other low cover is planted underneath the reduction is greater than, for example, along a roadway where bitumen is laid to the edge of the trees. The object then is to grow a dense mass of foliage from ground level up and, provided this can be achieved, species of tree does not generally have a major effect. Conifers, which retain branches to ground level and have relatively uniform distribution of foliage, are the best; the taller eucalypts and others which shed their lower branches require under-planting, while deciduous trees are of very limited value during winter. Since density is important, planting distances within rows need to be the minimum suitable, but rows should be a minimum of 3 to 5 metres apart to avoid suppression of the lower branches. The length of the belt needs to be a minimum of twice the distance from the noise source to the edge of the belt, otherwise the eddying effect around the ends will reduce its effectiveness.

Practical considerations will determine the size of the planting. Best results are obtained with a belt 25 to 30 metres wide (seven to ten rows) with trees 12 to 15 metres minimum height in the centre and smaller ones in the outer rows. With this arrangement a reduction of 7 to 10 decibels can be expected. If the belt is dense, with grass at ground level, the reduction can be as high as 12 to 15 decibels. With narrower belts the effect will be less; in very general terms each 30 to 40 metres in width would give a reduction of about 10 decibels. The reduction behind the sound belt is uniform regardless of the distance (i.e. a 5 decibel reduction just behind will still give the same reduction 100 metres to the rear).

In siting a belt, it is important that it be placed close to the noise source rather than the hearer. With road noise, having the near edge about 10 to 15 metres from the traffic is better than further away. But where this is not practicable and the belt is some distance away, as for example, around the boundary of a recreational area or airfield reduction of 6 to 8 decibels per 30 metres in width can be expected on the lee side. A similar position exists within a timbered area as each 30 metres reduces the sound level by 8 to 10 decibels. In selecting a camping site near a noisy road a depth of 100 metres of dense foliage would reduce noise level from 80 decibels to about 55 decibels.

At Tullamarine Airport, Melbourne, some 3500 metres of sound barriers have been planted. About 1000 metres is 100 metres wide and 2500 metres is about 160 metres wide. A basic nine row pattern with rows 5 metres apart is repeated as necessary to the width required while the species used vary in height and shape to give maximum interference to the sound waves. Plantings are around the residential boundaries as it is obviously not feasible at an airport to plant close to the sound source (aircraft).

## PARKS

Despite a popular belief, most Australian communities and especially the suburbs of larger cities are not adequately provided with parks. Large areas are devoted to playing fields, golf courses, etc., and while these are necessary for active recreation especially amongst the younger age groups, further substantial areas solely for passive recreation are required.

Parks exhibit all the moderating effects on wind, temperature, shading, carbon dioxide use, noise, etc. referred to earlier. They provide a form of mental relaxation to most people who, irrespective of their age or physical capacity, need the more restful atmosphere of a good park at some time.

Parks of various sizes are necessary. Rest areas in or near local shopping centres may be only small, perhaps a quarter hectare or possibly the size of one or two typical suburban allotments. In the residential zones, neighbourhood or residential parks of about 15 to 20 hectares are needed at frequent intervals so that a park is within convenient walking distance for most of the residents. In the larger towns and cities, town or city parks of a minimum area of 50 or more hectares are required. In most built-up areas, these parks can be established along streams, as wide buffer zones around industry, along main highways, on prominent ridges or hilltops, in areas not well suited for residential or industrial purposes, or on poorly-drained areas (which usually lend themselves to the development of water and lagoon features).

The density and size of parks will be influenced by many factors, not the least of which is the distance to the nearest regional park of several hundred or thousand hectares. They need to be related to population density and, with the trend to higher concentrations of people, the area allocated should be generous. Numerous estimates of the number of hectares per 1000 people have been made over the years but it is apparent that the proportion is increasing with time. Forty or fifty years ago 1 to 2 hectares per 1000 people was commonly recommended but nowadays, when the population is becoming increasingly concerned with the quality of life, proportions of 3 to 4 hectares and up to 5 hectares per 1000 people is a more acceptable ratio.

Land acquisition would be a major initial cost and would be a serious limiting factor in built-up areas. In the smaller towns, or in areas being developed around the larger towns and cities, municipal authorities are empowered to ensure adequate reserves for future use. Maintenance cost can be high if garden beds, annuals, etc., are established. Although this is desirable if it can be provided, the type of park envisaged is one in which primarily trees and hardy shrubs are grown; maintenance involved would be little more than grass slashing with an occasional trimming of trees. With the mechanical equipment available today this is not expensive.

In past generations many fine reservations were made and, bearing in mind the size and population of the town at the time, were frequently as good if not better than the standards of 5 hectares per 1000 people referred to above. The last generation or two seems to have been content to live on the foresight of their predecessors and have done little to establish areas for the present and future generations, who will have to cope with the pollution problems of the way of life developed by the present generation. While much can be done to reduce the problems, it will be impractical to eliminate them completely. Amelioration will become an increasingly important approach and one of the most effective methods will be to take full advantage of the systems evolved by Nature. Extensive use of trees in newly-created parks and forests near and within municipal limits will need to be one of the major techniques used in the near future.

FARM FORESTS

Farm forests may be established for their aesthetic or amenity value, as a source of farm timber, or as an investment. They can be of any size convenient to the owner but to attain a forest-like formation a minimum area of 2 to 3 hectares is desirable. Either a natural stand may be developed or a new one created by planting.

Many properties have a natural patch of timber or a bush paddock which can be fairly easily developed to a farm forest. The aim should be to get a range of sizes in seedlings, young, medium and old trees. Old stag-headed, deformed, or faulty trees need to be removed. Gaps in the stand can be restocked by natural means or by planting. For natural regeneration, cultivate the ground during late spring to a miminum of 6 to 8 centimetres to expose and loosen the surface soil. A light fire will also expose sufficient soil, but cultivation gives a better seed bed. During the late summer, seed from the older trees re-seeds the bare patches and sufficient seeds will germinate during the following autumn or spring to restock the area. Some seed falls each year but good seed crops usually occur every four to six years; they can be predicted by observing the flowering during the previous summer as the seed matures and falls about a year later. If planting is preferred to natural seeding, species covering a range from small shrubs to trees should be used, particularly if the intention is to create an aesthetic area. Most of the native shrubs (acacias, melaleucas, callistemons, etc.) are ideally suited for this purpose as they provide a habitat for and attract birds and animals.

Scattered or even isolated trees may also be used as a nucleus. They seed an area two to three times the height of the parent tree and if an appropriate area is prepared and protected from domestic stock and rabbits an adequate crop of seedlings will be obtained.

If the intention is for the stand to serve also as a source of farm timbers, durable species may be planted. However, their growth rate is slow and it is much better to plant less durable but faster growing trees as, with present timber preservation techniques, non-durable species can be treated to last as long as the traditional durable ones.

Planting as an investment is a sound proposition but economics dictate that *Pinus radiata* is about the only species worth considering for southern Australia and then only if conditions are favourable. It is a capital asset appreciating from virtually nothing to one with a substantial value at maturity; economic studies have shown that in most cases it will give an average annual return over a rotation of forty years which is better than that from normal farming.

An investment stand of *P. radiata* requires a minimum annual rainfall of about 700 to 750 millimetres. Soil should be at least a metre deep and, for ease of working, topography should not be steep. A minimum width of 40 to 50 metres is preferred otherwise the edge trees, which are usually much branchier, occupy too large a proportion of the area. Cultivation to 12 to 15 centimetres or more is advisable to give the trees a better chance of establishing and to reduce the amount of grass and competing scrub which has to be removed later. A planting distance of 2.5 metres by 2·5 metres (1600 trees per hectare) is a good general stocking. On good sites it can be reduced to 2·5 metres by 2 metres (2000 trees per hectare) or even 2 metres by 2 metres (2500 trees per hectare). Rabbits need to be well controlled and the trees protected from domestic stock (by fencing) for the first four to five years; after this the area may be grazed. Fertilizing with superphosphate or a nitrogenous type at planting will generally give a response, but it is usually not sustained so the long term economics are doubtful. Aerial application some years after planting is likely to be more beneficial as well as being cheaper.

Economically, establishment costs need to be kept to a minimum otherwise interest charges over the life of the stand can seriously erode the profit. Rough clearing and preparation is adequate as most of the debris left after clearing a timbered area will decay within a few years. It is easy to spend a lot of money on 'polishing' a project but this does not help the stand and is really money wasted.

Yields commence at ten to twelve years and a thinning may be made every five to six years thereafter. Early yields are not large, about 40 per cent of the volume and 50 per cent of the financial return will be obtained at clear falling. Clear falling age at forty years is common; it may be earlier (thirty to thirty-five years) although the volume and financial yields will be less. Conversely the stand will continue to grow for much longer than forty years should this be preferred.

Two States, Victoria and New South Wales, have financial assistance schemes. In Victoria, up to $8000 may be obtained as a loan for planting; it is interest free for the first twelve years and repayable over twenty-five years. Further details can be obtained from the forestry authorities in each State.


## BIRDS

One of the more commendable aspects of tree planting today is the desire of many people to plant trees which attract birds. Normally they refer to the smaller and colourful insect- and nectar-eating species since it is not easy to establish quickly a habitat suitable for birds of prey, water and swamp birds, or for those which require hollow logs, branches, etc., for nesting.

Birds will be attracted provided their basic requirements of a food supply and suitable habitat are met. They are remarkably adaptable and provided the food or habitat is adequate they are not particular whether a species is native or exotic—cockatoos destroy large numbers of pine cones in their attempts to get the seed, while lyrebirds scratch over large areas of the pine litter in their search for insects. Some sources advise planting only native species on the grounds that birds are instinctively conditioned to them, but this is a very doubtful argument. Encouraging insectivorous birds can also have its side effects as birds do not normally distinguish between 'pest' or 'beneficial' insects within their food range, but eat any that are available.

Providing a food source is probably easier than establishing a suitable habitat. A single tree may suffice for food but a group would be preferred for shelter and nesting. A mixture of species and sizes can be used to attract a wider range, the larger birds tending to use the larger trees with the smaller ones using the bushier plants.

Almost every tree has some value for birds but some species because of the availability of food (insects and nectar), shelter, and nesting sites are more attractive. Amongst the natives many of the *Banksia, Callistemon, Calothamnus, Eucalyptus, Grevillea, Hakea, Leptospermum* and *Melaleuca* species are highly favoured. *Acacia* species are not very good for food but supply good shelter. Native and introduced shrubs such as *Pittosporum, Crataegus, Pyracantha,* etc. provide a good source of berries. Further species are listed in the reference tables at the end of this publication.

## AIR POLLUTION

Injury to and destruction of vegetation in the vicinity of certain types of industrial processes, for example, smelting, has been known for centuries, but it is only in the last two to three decades that the widespread occurrence of lesser injury has been recognized. Industrial damage becomes more common each year but fortunately so far it is only in specific localities, such as close to some ·industrial plants, that death of vegetation has occurred. In the great majority of cases the injury is only to the leaves and, while this may spoil the appearance of the plant for the time being, new leaves will eventually appear.

Leaves are usually the first part of the plant to appear affected. Those with smooth or waxy surfaces are less likely to be damaged than those with rough or hairy surfaces but, contrary to what might be thought, the surface of the leaf is rather resistant. Its surface layer or cuticle is fairly inert; the visible symptoms originate internally and indicate damaged or destroyed tissues since pollutants enter via the stomata and react with compounds within the leaf to produce substances which cause the injury.

The fact that pollutants enter via the stomata means that damage is more likely to occur under conditions which favour open stomata, such as moist atmosphere or high humidity, rather than on hot, dry days. Hence, depending on whether conditions favour open or closed stomata, the same concentration of pollutant may on one occasion cause damage and on another have little effect. The critical level above which damage occurs can cover a fairly wide range of concentrations rather than be a precise figure. Following from this it is apparent that duration of exposure is also important. With open stomata, an hour or so may be sufficient to cause injury but if the stomata are almost closed, several hours exposure at the same concentration could probably be tolerated by the plant without any ill-effect.

Many of the symptoms of pollution damage are similar to those of other injurious agencies, for example, lack of moisture, nutrient deficiencies, fungal

attack, etc., so that it is often impossible to determine if a pollutant was responsible. Some work has been done overseas and is being done in Melbourne on the exposure of plants to known concentrations of specific pollutants. While resulting damage patterns follow broad lines, they are not distinctive enough in themselves to indicate conclusively which pollutant, if any, caused the injury. When identification in the field is involved, it is usually a case of using experience and a subjective assessment of the local factors, for example, presence of industrial plants, etc., rather than a purely objective approach.

(a) *Sulphur Dioxide*  The increasing consumption of petroleum products by the larger consumers such as industry, office buildings, hospitals, etc. has led to an increase in sulphur dioxide damage to plants. Oil fuel with a high sulphur content may produce over 100 kilograms of sulphur dioxide per tonne compared with 20 to 25 kilograms for black coal and less than 0·5 kilogram for natural gas. The oil companies are well aware of the disadvantages of high sulphur content in regard to plants and for this and other reasons they take steps to reduce it to as low a level as is reasonably practical.

Injuries are generally localized and found close to the source compared with photochemical injury which is usually widespread. Symptoms normally are light brown, brown to reddish-brown markings between the veins on broadleaf trees or, with conifers, similar colours at the ends of the needles.

(b) *Photochemical Agencies*  These are due to substances produced by the action of sunlight on nitric oxide, nitrogen dioxide and hydrocarbons. The main source of these is vehicle exhausts (8 to 9 kilograms of nitrogen oxide is produced per 1000 litres of petrol used) and, while they can be harmful in themselves, under the action of sunlight they give rise to a number of other compounds. The two main ones are (i) ozone and (ii) peroxyacetyl nitrate. In America ozone and peroxyacetyl nitrate are believed to be the most widespread cause of general pollution damage.

(i) Ozone is normally present in very minute quantities in the atmosphere but is confined mainly to the upper levels. The quantity produced in the sunlight/nitrogen oxides/hydrocarbon series of reactions is small but then plants are damaged by concentrations of only 10 to 30 parts per hundred million; even less if substantial quantities of sulphur dioxide are also present. Symptoms are red-brown, light yellow, and purple, to almost black spots or larger bleached patches.

(ii) Peroxyacetyl nitrate. This compound was first identified as a cause of pollution injury about 1963. Very small concentrations of 5 to 10 parts per hundred million cause serious injury. Symptoms are usually a glazing or bronzing of the rapidly expanding leaves; mature leaves or very young leaves just emerging do not appear to be affected.

Since both ozone and PAN (peroxyacetyl nitrate) are dependent on sunlight for their formation, concentrations tend to be highest in the afternoon, decreasing to a low level at night. Atmospheric conditions restricting dispersion, for example an inversion layer, can aggravate the position. The initial reaction involved in the ozone cycle is when nitric oxide (NO), principally from vehicle exhausts, reacts with ozone ($O_3$) producing nitrogen dioxide ($N_2O_4$) and oxygen ($O_2$). This continues throughout the day and night but during the day, under the influence of the ultra-violet of sunlight, the reverse reaction also occurs in which ozone $(O_3)$ is formed. This leads to a build-up in concentration of ozone in the afternoon and a decrease at night. The sequence of reactions for the peroxyacetyl nitrate, however, has not (to the author's knowledge) been clearly defined.

(c) *Fluoride*   Fluoride injury has been known for a long time but has only come into prominence in the last decade or so. Fluorides are widespread and occur as minor impurities in many of the primary minerals used in industrial processes, for example, aluminium, steel, clay products, fertilizer, phosphates, and in virtually any industry which requires the heating of a mineral product. Damage then is more likely to be seen around these industries. The susceptibility of plants varies more widely than for other pollutants, but concentrations of 5 to 10 parts per hundred million will produce injury. The first symptom is a 'water soaked' discoloration turning in a few days to brown or yellow-brown patches as if the leaf has been scorched.

(d) *Ethylene*   Ethylene is one of the minor components of coal gas, and injury or death of plants from leaking mains was fairly common. However, natural gas does not contain it so is not a source of potential injury to plants. While ethylene has many sources, including normal plant metabolism, the risk of injury is highest in the vicinity of factories producing polyethylene film, or where exhaust fumes from cars, etc., can reach high concentrations. External concentrations of 10 parts per million are harmful. Symptoms are a yellowing of the leaves, death of the apical shoot and reduced growth. The effect is slow to show and often takes several days to a few weeks to become apparent. In very minute amounts ethylene acts as a growth regulator. After flowering, concentrations of 0.1 to 1.0 parts per million are produced which initiate fruit and seed ripening. Following some forms of insect or fungal attack, large local increases (ten to twenty times normal) may occur which probably trigger the formation of phenolic compounds which function as antibiotics.

(e) *Carbon Monoxide*   Although this is the main emission from vehicle exhausts (approximately 2500 kilograms per 1000 litres of petrol used) it is not a significant cause of injury to plants. Concentrations of the order of 1000 parts per million are required and, as these are rarely met with in practice, little work has been done on its effect.

(f) *Smog*   The term smog is commonly used to describe two broad forms of atmospheric 'fog'.

(i) Wet smog. A mixture of normal meteorological fog and smoke pollution which occurs in the early morning with low temperatures and moist air.

(ii) Dry smog. Primarily a photochemical fog which occurs in the afternoon with moderate to high temperatures and dry air.

Wet smog has little deleterious effect on plants, in fact it is more likely to be beneficial than harmful. Dry smog, however, has the opposite effect. (See (b) Photochemical Agencies, page 117.)

(g) *Dust*   Dust, such as that from quarrying operations, clay products, cement works and similar industries, may cause the death of plants by covering the leaves and blocking the stomata. Before this stage is reached, the dust is generally thick enough to cause inconvenience to people and remedial measures are usually taken in time to preserve vegetation.

(h) *Lead* The most common sources of lead pollution are motor vehicle exhausts and industrial fumes. In the Wollongong area of New South Wales, concentrations of over twenty times higher than in nearby farmland have been found in the surface soil close to industrial works. Plants can accumulate fairly large amounts of lead (up to several hundred parts per million) in their tissues, particularly in the roots, without adverse effects. The main interest is in its absorption by food plants used by man and animals. In America, horses are reported to have died from grazing on pasture containing 50 to 100 parts per million of lead. Lead can be absorbed both from the ground by the roots, and from the atmosphere by the leaves, with little exchange between them. High lead concentrations in the foliage can be expected where atmospheric concentration is high.

(i) *Cadmium* Cadmium has only come under suspicion as a pollutant in recent years. Its action is indirect in that it is chemically very similar to the essential element zinc and competes with it. If plentiful it is absorbed to excess, giving rise to zinc deficiency. Being comparatively volatile, atmospheric pollution occurs from metallurgical processes, industrial fumes, and attrition of vehicle tyres. Again in the Wollongong area of New South Wales, cadmium contents two to three times higher than in adjoining rural land have been recorded. High levels along main roads, arising from the attrition of tyres, have also been reported. It is also found as an impurity in most phosphatic fertilizers; superphosphate contains 40 to 50 parts per million. Plants absorb cadmium readily, the highest concentration being in the roots and the lowest in the seed. Although Australian soils have a long history of fertilization with superphosphate, the cadmium content of food plants is very low. It is believed that 3 milligrams per day is the safe limit for adults and to achieve this it would be necessary to eat 20 to 40 kilograms of vegetables daily.

(j) *Ammonia* The increasing use of anhydrous ammonia as a fertilizer increases the risk of damage to plants by vapour from spillage or other careless handling.

(k) *Miscellaneous* Included in this group are compounds which are injurious to plants in very small concentrations but where usage is small or confined to special cases so the risk of damage is slight. Amongst these is chlorine (used for purification in swimming pools), leakage of hydrochloric acid fumes from a few industrial plants, and of course negligent use of pesticides.

(l) *Susceptibility of Species* Little is known of the susceptibility of native plants to pollutants. Some work on exposing a wide range of native plants to known concentrations of pollutants is being undertaken at the University of Melbourne; preliminary results suggest that they are fairly sensitive, particularly the eucalypts, and should not be planted in risk areas. If further work confirms these trends, it appears that the tried and proven introduced species may have to be planted in the critical situations, or that planting be confined to the few natives which have proved fairly resistant, for example, *Eucalyptus botryoides*, *E. maculata* and *Melaleuca incana*. In general terms, most *Eucalyptus* species are fairly susceptible to sulphur dioxide damage while *Melaleuca* species and *Leptospermum* species are moderately susceptible. *Banksia* species are readily affected by ozone but *Eucalyptus*, *Melaleuca* and *Leptospermum* species seem to be more tolerant of ozone than of sulphur dioxide. *Casuarina*, *Lagunaria* and *Hakea* species appear to be fairly resistant to both sulphur dioxide and ozone damage.

The more sensitive species among introduced plants (as recorded in literature) are given below. Although this is based on overseas work (principally in America and Britain), local observations generally agree with the overall rating. Insufficient knowledge of Australian conditions is available to allow comment on the actual pollutant responsible.

## Susceptible Species

| Sulphur dioxide | Ozone | Peroxyacetyl nitrate | Fluoride |
|---|---|---|---|
| Australian Native Species | | | |
| most *Eucalyptus* sp., *Melaleuca* sp. and *Leptospermum* sp. | most *Banksia* sp. | not known | not known |
| Introduced Species | | | |
| *Betula* sp. | *Acer negundo* | not known | *Acer campestre* |
| *Catalpa* sp. | *Acer saccharinum* | | *Acer negundo* |
| *Larix* sp. | *Alnus* sp. | | *Acer saccharinum* |
| *Malus* sp. | *Catalpa* sp. | | *Fraxinus pennsylvanica* |
| *Pinus ponderosa* | *Gleditsia triacanthos* | | *Larix* sp. |
| *Pinus strobus* | *Malus* sp. | | *Picea pungens* |
| *Populus nigra* | *Pinus ponderosa* | | *Pinus contorta* |
| *Pyrus* sp. | *Pinus strobus* | | *Pinus ponderosa* |
| *Ulmus americana* | *Platanus occidentalis* | | *Pinus strobus* |
| | *Populus tremuloides* | | *Pinus sylvestris* |
| | *Salix babylonica* | | *Populus nigra* |
| | | | *Populus simonii* |
| | | | *Populus tremuloides* |
| | | | *Pseudotsuga taxifolia* |
| | | | *Thuja* sp. |
| | | | *Tilia cordata* |

# NATIVE PLANTS

For a long time many Australian native plants were not very highly regarded for their attractiveness or usefulness. Possibly this view was encouraged by a general lack of obvious seasonal changes in the bush compared with the forests and countryside of northern Europe, but even from the early days of settlement some people appreciated the uniqueness and colour of the local flora. Sir George Grey, during his exploration in the north west of the continent, recorded in his diary (8/2/1838): 'A sweet short herbage had been raised by the heavy rains, from the sandy soil, and amongst this the beauteous flowers, for which Australia is deservedly celebrated were so scattered and inter-mixed that they gave the country an enamelled appearance' (*The Discovery and Exploration of Australia* by E. H. J. Feekem, G. E. E. Feekem and O. H. K. Spate, 1970.) The reference to 'beauteous flowers, for which Australia is deservedly celebrated' seems to indicate that native plants were already widely appreciated for their beauty by 1838, even though to some later generations they do not appear to have had the same appeal.

What constitutes an Australian native plant is an interesting point. The continent of Australia is a single political identity, but it is far from being a single unit in the ecological sense. There are three main divisions in the flora, as referred to previously (see page 16), each so distinct from the others that only one could be called Australian. With the numerous subdivisions that exist it seems that the expression 'Australian native plants' refers more to the political entity than to the ecological one.

Growing native plants has become very popular in the last twenty to thirty years. However, it would be wrong to assume that this is a recent development as many species, especially eucalypts and acacias, have been grown by nurseries for eighty to one hundred years. But the period since World War II has seen a tremendous increase in the number of species raised, particularly those from the drier parts of the continent, many of which have been shown to be well adapted to southern Australia. Genera such as *Melaleuca*, *Banksia*, *Grevillea*, *Hakea* and *Leptospermum* are now well known to most people.

While they may have some attractive feature such as colourful flowers, unusual fruits, etc. and are well worth growing for these alone, frequently they are also well adapted to difficult sites like low rainfall regions; calcareous, lateritic, saline, poorly-drained or erodable soils; and exposed and adverse situations. Their introduction has widened considerably the range available so that there are now plants suitable for almost every situation. The big increase in planting in recent years for beautification around farm homesteads, community amenities in towns, and utility projects like windbreaks, shade trees, and farm forests, is due in some measure to these introductions.

However, it is still necessary to select species with discretion. A large number came from latitudes further north than south-eastern Australia and some have not been very successful in southern Victoria or Tasmania. Except for the northern Mallee, Victoria is further south than the southern-most point of Western Australia, while the goldfields area of Western Australia, where many of the better known species are found, is in the same latitude as northern New South Wales. In home gardens where each plant can be given individual attention, the success rate can be expected to be higher than in mass planting like windbreaks, parks, golf courses, roadsides, etc. In these projects more care needs to be exercised in the selection of species. The fact that one grows well in a suburban garden does not necessarily mean that it will be equally satisfactory under field conditions; many are but then many are not.

A disadvantage of many natives is their wide variation in form. To meet the demand seed still needs to be collected in many instances from natural stands with the inevitable result that some plants from the same collection will have good form and others not so good. The popular *Eucalyptus macrocarpa* (Mottelcah or Rose of the West) is a good example as the same collection of seed may produce bushy plants with a hemispherical shape with up to 3 metres spread, or ones consisting of a single upright but twisted stem of less than 2 centimetres diameter. This is a reflection of the genetic constitution and is common to all plants; little can be done immediately other than to collect seed from parents with the desired form. Some straggly specimens will still be obtained but the proportion will be less; this is of little consolation to those who plant several only to find that most of them turn out to be a poor shape. Reputable seed merchants and nurserymen give a lot of attention to seed origin and provenance as it is only by collecting from good specimens through successive generations that the preferred form can be obtained with certainty. The process will take many decades and generations of plants but is the only solution and will be justified in the end. Introduced or exotic plants have been going through this process of continued selection for specific features for up to several hundred years so that inconsistency of form has long since been eliminated and a certain shape, flower characteristic, etc. can be assured. Most Australian native plants, on the other hand, have only recently been selected for desirable features.

Reputable nurserymen have planted or know of good specimen trees and collect seed from only these trees to increase the proportion of acceptable plants. One organization, the Forests Commission, Victoria, has established some 15 hectares of seed gardens and is continuing planting for the express purpose of growing good specimens for seed production alone. In these gardens, a number of plants of each species are planted, the poorer specimens are then cut out so that pollination and seed production takes place between plants with the preferred characteristics. This will be a continuing process but in time it will help to raise the standard and consistency of the plants produced in its nurseries. It is a policy and an approach which is recommended to all who intend to remain in the business of growing native plants.

Vegetative propagation, such as cuttings, layering, grafting, etc., transmits the characteristics of the parent plant, since the chromosomes transmitted are exact replicas of the parent chromosome content. Some natives can be propagated vegetatively and this technique should be used wherever practicable.

Seeds, since they arise from the fusion of a pollen grain (male) and an ovule (female), exhibit some of the characteristics of each parent but they may still vary from each parent, even if the pollen and ovule came from the same individual plant. This is due to the dominance or recessiveness of each gene allele of the chromosomes and the form of the offspring is dependent on whether the dominant or recessive allele is transmitted in the particular pollen grain and ovule involved in fusion. These differences between vegetative and seed propagation are common to all plants, whether they be introduced or native.

Observations of the conditions under which a native plant (or any other plant for that matter) grows will give a good indication of the kind of site it can tolerate. However, these are not necessarily the optimum conditions. A plant may be growing there only because circumstances at some stage of its life prevented it from becoming established on more favourable sites. Survival is frequently determined by such factors as speed or ease of seed germination and early rate of the growth; the plants with the shortest germination period and/or the most vigorous early growth will suppress other species and occupy the site. The same reaction results in the more favourably-equipped plant ultimately becoming the dominant species over extensive areas with the less well-equipped plants relegated to the more unfavourable areas. If transplanted or planted under better conditions, species from adverse sites sometimes make better growth and produce a better form than in their natural environment.

This reaction is quite common with native plants under cultivation. Their size and form under natural conditions does not necessarily mean that these will be the same under cultivation, although it is often a good guide. Similarly, when a plant from a particular region and climate is transferred to another, growth may be better or worse. On the other hand, if in a natural locality the plant has evolved to adapt itself to this site, then the site conditions are probably the optimum. The correct conclusion cannot be determined except by planting a species under a wide range of conditions and finding out by experience which are the most suitable. These unknown factors give rise to some of the more stimulating and rewarding aspects (or disappointing and frustrating aspects if the plant fails) of growing native plants.

Plants growing naturally on sandy soils, particularly in the lower rainfall areas, can also give apparently contradictory results under cultivation. With the open texture of sandy soils, the roots may extend down for some metres. If the annual rainfall is only 250 to 300 millimetres, the individual plant then has access to a greater supply of water than the rainfall suggests. When planted in a heavier soil in higher rainfall regions (for example, 700 millimetres), the species sometimes does not thrive because of lack of water despite the much higher rainfall. The heavier soil restricts the root system to the surface horizon, and the summer reserve in this zone may be inadequate compared with the water stored and available to the plant in the sandy soil.

There are many other interesting and often apparently conflicting reactions when introducing a new plant. Much can be learned from its growth in its natural habitat and while this serves as a useful guide it is a serious mistake to assume that the natural conditions are the optimum. Evolution and ecology are complex sciences and nothing in them is straightforward and simple. Herein lies one of the reasons why 'new' plants are such a fascinating field of study and experiment. Probably one of the best examples of habitat contradiction, and one which is so commonplace that most people are not even aware of it, is silky oak (*Grevillea robusta*). Under natural conditions, it is a rain-forest tree of southern Queensland and northern New South Wales which is found growing in regions with rainfalls of 1000 to 2000 millimetres. It is a widely-planted tree which grows, without any attention or supplementary irrigation, in many parts of southern Australia—up to 1500 kilometres south of its natural habitat and in the warmer districts with rainfalls as low as 500 millimetres a year. A review of its natural habitat to determine growing conditions would obviously be misleading. Of course, this is the exception rather than the rule but the moral is that nothing is proved until demonstrated.

# PART II

## SPECIES DESCRIPTIONS

## HOW TO USE THIS SECTION

On the following pages species have been arranged in alphabetical order of botanical names. To economize on space and avoid repetition descriptions are given in the form of notes rather than narrative and, for ease of reference, these have been arranged in the following pattern:

*In the first lines*, reading from left to right, may be found: (a) Botanical name; (b) synonyms, if any, in brackets; (c) common name; (d) the letter E (indicating evergreen) or D (deciduous), followed by (e) the letter C (conifer or softwood) or H (hardwood); and (f) its native locality. Where more than one common name is given, the first is the preferred one; where none is given there is no generally-accepted common name.

*In the following lines*, reading from left to right: (a) the usual means of propagation; (b) height in metres; (c) minimum rainfall in millimetres; (d) notes on soil types (when appropriate); (e) approximate growth rate.

*In the next line*, its uses, and occasionally the words 'New introduction'.

*and then*, the description.

The first part of a botanical name refers to the genus. In an alphabetical list all members of a genus will occur together.

The origin of the name, notes on number of species and main areas of natural occurrence are given. Notes under the genus heading also apply to the descriptions of the individual species and should be read in conjunction with them. Where the feature is common to the whole genus, for example, *Acacia* and *Eucalyptus*, which are propagated by seed, the word 'Seed' is indicated under the genus heading only, showing that it applies to all the species listed and is not repeated for each one. Any information given under the species is additional. Characteristics given under this heading refer only to the species listed, not necessarily to other members of the genus which are not described.

*Propagation* refers to the method which is usually the most reliable or convenient but other methods are often used. The words 'seed', 'cutting', 'layers' or 'grafts' are usually used.

*Height*, in particular, can vary according to the site. The figures given are averages for suitable areas in south-eastern Australia and are heights which could be expected in a reasonable time. With long-lived species these may be considerably exceeded. They also refer to species under cultivation which may differ from those under natural conditions.

*Rainfall* is approximate only. The figures given being the minimum necessary for reasonable establishment and growth. Many plants will grow outside the areas indicated, particularly if irrigation, ground water, or garden watering is available. The figure 650 millimetres (26 inches) is commonly used as it represents the approximate division between the drier northern and moister southern parts of the continent.

*Soil Type* Trees generally grow best on good, well-drained loams but most of the species described will grow satisfactorily on a wide range of soils. Where no reference to soil type is given it can be assumed that the species will grow on most soils. The exceptions are indicated by the word 'prefers' e.g., 'prefers sandy soils' etc.; and, in the cases where they will grow on an adverse site, by using the words 'will grow', e.g., 'will grow on wet soils'. However, it should be noted that with the latter these will usually still grow better on a better soil. The term deep soil as used means a well structured loam or clay loam without a sharply defined clay subsoil.

*Growth Rate* is very approximate, it depends on so many factors that only broad indications are possible. Very generally: 'fast' means more than about 0·5 metres height growth per year, 'moderate' about 0·5 metres, and 'slow' well below 0·5 metres.

*Uses* given are the ones for which the species is most suitable but most can be used for other purposes, and all are ornamental to some degree. Most of the terms used are self-explanatory although the following have been used to denote certain uses:

Farm forests—suitable for farm and commercial timber production.

Parks—municipal gardens, recreational reserves, golf courses, school grounds and similar areas where space is not so limited.

Roadsides—suitable for roads where there is plenty of space for growth.

Streets—species which can be grown in towns and are less likely to cause trouble with overhead and underground installations.

Avenues—symmetrical or formal in shape.

Ornamental—have some distinctive feature of ornamental value.

*New Introduction* This means the species has only recently been introduced into cultivation and its responses are not well known. The details given refer more to its growth under natural conditions than under cultivation.

*Shape* The diagram on the left hand side indicates the shape at early maturity. Shapes have been grouped into nine basic forms.

*Descriptions* generally follow the pattern of: comments on the appearance of the tree, its silvicultural characteristics and some details to aid identification. 'Slight frost-resistance' means the species will stand temperatures to about $-1°C$, 'moderate frost-resistance' withstands about $-5°C$ and 'severe or frost-resistant' lower than $-5°C$. The information on identification is not botanically complete. It refers to distinctive features only and should serve to identify most of the species which the average person is likely to find in south-eastern Australia.

*Horticultural Varieties* These are mutants of the parent type. Usually they are distinguished for colour or shape of foliage, branching habit, unusual flowers or fruit, or some other distinctive feature which could have ornamental value.

*Date of Introduction* At the conclusion of some species descriptions the dates at which they were first introduced into cultivation are given. Normally this refers to its introduction to Britain, Europe, North America or Australia since many species from other parts of the world may have been under cultivation in their native habitats from earlier times. Where a year is not given the date of introduction to cultivation is not known with sufficient accuracy. The letters 'L.C.' mean 'long cultivated'.

*General*

The notes given in the descriptions are intended as a general guide only. There will always be notable exceptions and as species are tried in new localities it is inevitable that with time many will be found to be suitable beyond the range given in the description. Knowledge on plant growth is never static; but the information given represents a general summation of current experience and knowledge.

# DESCRIPTIONS

## ABIES—Firs E.C.

Seed                              Slow to moderate
Farm forests, windbreaks, shade

*Abies*—The old Greek name derived from *abeo* (to rise) alluding to the height of the tree. About forty species in Europe, Asia and North America. Pyramidal trees, open-grown specimens retaining branches to ground level. Shade-tolerant, frost-resistant. Prefer moist well-drained sites; more suitable for the cooler districts. Not tolerant of smog. Bark rather smooth with resin blisters; leaves (needles) flat, usually notched at the apex, leaving a circular disc-like scar on the branchlets after falling; cones upright on the branches, disintegrating during seed fall. *Abies* are the true firs. Not very common in southern Australia (trees frequently called or sold as firs are Douglas fir, *Pseudotsuga menziesii*). True firs are distinguished from other single-needle conifers by (a) the cones being upright on the branches, (b) the cones disintegrating on the tree as soon as the seeds have fallen (obtaining complete cones is therefore difficult), and (c) the single needles when pulled off the stem leaving a clean disc-like mark whereas the single needles of other conifers either pull away with a few shreds of bark attached to the needle or leave a prominent leaf stub.

*A. alba*—**European Silver Fir**    Central and S. Europe
25–30 m    1000 mm    Prefers heavier,      Moderate
                       deep loams

The common fir of Europe. Tends to shed its lower branches giving a clean trunk. Withstands dry summers once established. Requires good drainage. Withstands frosts and some snow. Tolerant of shade, useful for underplanting. Produces a clean straight trunk and good quality timber. Bark silver grey; leaves 2 mm wide, 1–3 cm long, notched, glossy green; cones 10–15 cm long, 2·5–5 cm wide.

*A. cephalonica*—**Grecian Fir**             Greece
20–25 m    600 mm                   Moderate

Attractive, broadly pyramidal tree with deep green foliage. Most heat- and drought-resistant of the firs. Needles 2–2·5 cm long, 1–2 mm wide with a horny point; cones 15 cm long, 4 cm diam., bracts protruding beyond the scales. 1824.

*A. concolor*—**White Fir**           Western USA
25–30 m    800 mm    Prefers deep,       Moderate
                      moist loam

Attractive, tall and straight tree reaching 50 m in its native habitat. Prefers cool moist mountain situations but will grow in warmer and drier districts. Not satisfactory on sands or heavy clays. Root system shallow. Very tolerant of shade, suitable for under-planting. Withstands snow. Probably the best of the firs for local planting. Leaves 5–8 cm long, 2 mm wide, silvery-blue; cones 8–12 cm long, 2–3 cm diam., purplish when immature. 1851.

*Abies alba*

*Abies nordmanniana*

**A. magnifica**—**Californian Red Fir**  Western USA
+30 m   1200 mm   Prefers moist,   Moderate then
cool climate   fast
Narrow-crowned tree with short horizontal branches. Over 70 m tall in its native habitat. Needles 2·5–4 cm long, 1 mm wide, bluish, upper surface ridged, tip rounded; cones large, 15–25 cm long, 7–12 cm wide, purplish when young. 1851.
Similar to *A. procera* but upper leaf surface of *A. procera* is grooved (not ridged).

**A. nordmanniana**—**Caucasian Fir**   S.E. Europe
20–30 m  800 mm   Well-drained loams   Moderate
Attractive, well-shaped, pyramidal tree. Planted more commonly in the past, still worth planting. Tends to shed its lower branches giving a clean trunk. Withstands dry summers once established. Leaves dark green, silvery underneath, notched, 2–4 cm long, 2 mm wide; cones cylindrical, 12–14 cm long, 4–5 cm wide, usually covered in resin, slightly protruding soft bracts. 1848.

**A. pindrow**—**West Himalayan Fir**   Himalayas
+30 m   1200 mm   Prefers moist,   Moderate
cool climate
Narrow crowned tree, up to 70 m tall in its native habitat. Tolerates cold and heavy snow. Bark smooth; needles bright green, up to 6 cm long, 1·5 mm wide, apex often divided into two horny tips; cones 10–15 cm long, 5–7 cm wide. 1834.

**A. pinsapo**—**Spanish Fir**   Spain
10–15 m   800 mm   Deep soils, will grow   Slow
on limestone
Attractive tree, retaining its horizontal branches to ground level. Grows to 30 m in its native habitat. Leaves short, 1–2 cm long, 2 mm wide, stiff, arranged around the shoot; cones 8–12 cm long, 2–4 cm diam. 1837.

**A. procera** (A. nobilis)—**Noble Fir**  North-western USA
+30 m   1200 mm   Prefers moist,   Moderate
cool climate   then fast
Attractive, slender, narrow, conical tree up to 80 m tall in its native habitat. Needles dense, bluish, 2·5–4 cm long, 1 mm wide, upper surface grooved, characteristically longest in the middle part of the shoot; cone large, 15–25 cm long, 7–10 cm wide, distinctive large brown bracts protruding between the scales and almost covering the cone ('feather-cone'). Similar to *A. magnifica*. 1830.

**A. sibirica**—**Siberian Fir**   Siberia
15–25 m   1000 mm   Prefers cold regions   Slow
Slender tree, one of the silver firs. Tolerates severe cold and prolonged snow. Only satisfactory in cold climates. Bark smooth; needles very soft, 2–4 cm long, 1 mm wide, greyish underneath; cones 5–8 cm long, 3–4 cm wide. 1820.

*Abies pinsapo*

130

# ACACIA E.H.

Seed

*Acacia* is derived from the Greek name *Akakia* which is believed to be derived from the Greek *ake* (point) alluding to the thorny acacias of northern Africa. The common name 'wattle' is an old Anglo-Saxon word meaning flexible twig—used for weaving old fashioned hurdles (portable fences) and in wattle and daub. Over 750 species occur, mainly in Australia but with representatives in the warmer regions of most continents. Most acacias are fast growing and so can provide temporary shelter while slower trees are becoming established. They are generally short lived (10–20 years), but notable exceptions are *A. aneura*, *A. melanoxylon* and *A. terminalis* which may live for more than a century. Having nitrifying properties growth can be rapid on relatively poor soils and it is only with a few species that soil type is likely to be limiting. Good pollen-producers for bee-keepers. Acacias generally have 'leaves' which are in fact flattened leaf-stalks, known as phyllodes, but which function as leaves; true leaves only occur in the juvenile stages or in the pinnate forms. In the descriptions given 'flowers' refers to the typical globular ball even though botanically this comprises a group of 4–60 flowers according to species. On old trees the fungus *Uromycladium* sp. may form unsightly reddish-brown galls up to 8–10 cm in diameter.

**A. acinacea—Gold Dust Acacia**            S.A., Vic., N.S.W.
2 m             350 mm      Will grow on heavy        Fast
                            sandy, wet or dry
                            soils
Low shelter, ornamental

Showy, spreading, rather sprawling, shrub with arching branches. More suitable for the hotter districts. Phyllodes 5–15 mm long, narrow and pointed. Yellow flowers, globular, on stalks as long as the phyllodes. Late winter to spring-flowering; pod curved, often coiled, 2–4 cm long, 3–5 mm wide. 1842.

**A. acuminata—Jamwood Wattle**            W.A.
3–6 m         250 mm      Will grow in wet sites      Slow
Farm forests, windbreaks, shade, parks

Attractive, small tree with pendulous branches and spreading crown. Wood very durable with a strong odour of raspberry jam. Not suitable for limestone soils. Phyllodes 8–25 cm long, very narrow; flowers rich yellow in spikes 2–3 cm long; late winter-flowering; pods 5–15 cm long, 2–6 mm wide, straight, thin, smooth.

**A. adunca** (A. accola)—**Wallangarra Wattle** or   N.S.W.
            **Golden Glory Wattle**         Qld
6 m           450 mm                         Moderate
Parks, roadsides, ornamental

Attractive, forming a shapely, dense, small tree. Hardy, will tolerate moderately adverse conditions. Can be trimmed; hard trimming temporarily gives juvenile re-growth. Phyllodes 5–10 cm long, very narrow, green; flowers golden, early spring.

*Acacia adunca—R. Elliot*

**A. amoena—Boomerang Wattle**            Vic., N.S.W.
2 m           500 mm                         Fast
Parks, low shelter

Small, shapely shrub. Suitable for dry positions. Often found growing in rocky situations. Hardy. Phyllodes 4–7 cm long, about 1 cm wide, dark green, sharp tip; flowers yellow, in racemes 3–5 cm long, spring-flowering; pods 5–7 cm long, 5 mm wide, flat, funicle encircling the seed about three times.

**A. aneura—Mulga**            W.A., S.A., N.S.W., Qld, N.T.
4–5 m         250 mm                         Slow
Windbreaks, shade, fodder, ornamental

Small tree with ascending branches and spreading crown. Branches erect. Grows best in the hot, dry inland; a low shrub under hard conditions. Will grow on a wide range of soils, including heavy clays. Foliage suitable as fodder. Timber hard, durable, rich brown. Phyllodes 4–7 cm long, usually only about 3 mm wide but can be wider, even almost round; flowers yellow, in spikes 1–2 cm long, spring-flowering or in dry areas after good rain; pods flat, broad, 2–4 cm long, 1–1.5 cm wide.

**A. baileyana—Cootamundra Wattle**            N.S.W.
5–6 m         450 mm      Will grow in heavy        Fast
                          or wet soils
Low windbreaks, parks, gardens, avenues, ornamental

Attractive, spreading, bushy, bluish foliage covered in masses of bright yellow flowers in late winter. Hardy, withstands drought and moderate coastal exposure. Tolerant of some shade. Suitable for most districts except where snow lies on the ground. Can be trimmed as a hedge; trimming after flowering prolongs life. Single trees also benefit from regular light pruning. Occasionally seen as a garden escape. Pinnate foliage, pinnae 2–4 cm long, 5–10 mm wide, leaflets 5 mm long, 2 mm wide; flowers globular, in racemes 5–8 cm long, winter-flowering; pod 4–7 cm long, 5–10 mm wide, flat, nearly straight. The cultivar var. *purpurea* has purplish new growth while the var. *aurea* has golden tips. 1873.

*Acacia baileyana*

*Acacia baileyana—flowers*

*Acacia baileyana—golden-tipped form*

*A. boormanii* (A. hunteriana)—**Snowy River**   Vic.·
                          **Wattle**      N.S.W.
3–4 m     650 mm                  Fast
Low shelter, roadsides, parks, ornamental

Slender-branched, tall shrub or small tree. One of the most decorative of the wattles, deserves to be more widely grown but hard to obtain. Prefers moist well-drained soils. Suckers profusely. Branchlets numerous, angled; foliage soft, grey-green, very narrow, 5–8 cm long, 2–3 mm wide; rich yellow globular flowers in spikes, flowers abundantly early spring; pods 6–8 cm long, 5 mm wide.

*A. botrycephala* (A. discolor)—**Sunshine**   Tas., Vic.,
                          **Wattle**   N.S.W., Qld
3–5 m     650 mm               Moderate
Roadsides, parks, ornamental

Attractive, spreading shrub or small tree suitable for most soils in the cooler districts. Foliage green, bi-pinnate, 2–6 pairs pinnae, widely spaced, 10–15 pairs leaflets, oval, 1 cm long; flowers creamy-yellow in 5–8 cm racemes appearing in late autumn and winter, long flowering period; pods 3–8 cm long, up to 1 cm broad. 1788.

*A. brachybotrya*—**Grey Mulga**    S.A., Vic., N.S.W.
2–3 m    250 mm   Will grow on wet   Moderate
                        soils
Low shelter, ornamental

Rounded shrub with stiff branches. Hardy. Foliage silver-grey, oval, thick, 1–2 cm long; flowers globular, bright yellow, on stalks, 1–5 per stalk amongst the leaves, spring-flowering; pod straight, 2–3 cm long, 5 mm wide.

*A. buxifolia*—**Box Leaf Wattle**    Vic., N.S.W., Qld
2–3 m    550 mm                  Fast
Streets, parks, ornamental

Large compact bush, erect branches. Can be trimmed, occasional pruning necessary to improve shape. Foliage like English box, oval, thick, 2–3 cm long, 5–10 mm wide; flowers profuse, golden-yellow, globular, 4–10 per raceme of 3–5 cm length amongst the foliage, late winter-flowering, (some plants sold as this species have terminal racemes); pods flat, 6–10 cm long, 5–8 mm wide.

*A. calamifolia*—**Wallowa**    S.A., Vic., N.S.W., Qld
2–3 m    350 mm   Will grow on wet sites   Fast
Low shelter, parks, ornamental

Attractive, fine-leaved, rounded shrub with upright branches, suitable for warm districts. Prefers lighter soils. Withstands dry conditions. Foliage very slender, 5–15 cm long, less than 3 mm wide, not dense; flowers rich yellow, solitary or in pairs on stalks amongst the foliage, winter- to spring-flowering; seed pods very long, 8–20 cm, narrow, wrinkled, brittle. 1820.

*A. cambagei*—**Gidgee**                     S.A., N.S.W., Qld
5–7 m          300 mm          Prefers sandy soils          Moderate
Shade

Spreading small tree, similar to *A. pendula*. Ascending branches. Will grow on heavy soils, around swamps, and along watercourses. Hardy, wind-firm, frost- and drought-resistant. Long-lived. Timber very durable. Exudes an unpleasant odour when wet—not suitable for planting near houses, in parks, or along roadsides. Phyllodes 2–10 cm long, 5–10 mm wide, parallel veins; flowers globular, in groups of 2–6 in the leaf axils, flowering after heavy rain; pods 6–8 cm long, 5–10 mm wide, flat, straight.

*A. cardiophylla*—**Wyalong Wattle**                     N.S.W.
3–4 m          450 mm          Prefers lighter soils          Fast
Low shelter, parks, roadsides, ornamental

Very attractive, free-flowering, hemispherical shrub, often as wide as it is high, outer branches arching and pendulous. Foliage soft green, fern-like, phyllodes pinnate, 15–20 pinnae, 8–12 leaflets, very small, 2 mm long, oval, hairy; flowers yellow, produced in quantity, almost covering the bush in spring.

*A. cognata*—**River Wattle**                     Vic., N.S.W.
3–5 m          650 mm          Prefers deep loams,          Moderate
                              but not sands
Low shelter, parks, streets, ornamental

Very attractive tree with distinctly pendulous foliage and stems. Should be more widely grown. Phyllodes long and narrow, 8–12 cm long, 1 cm or less wide with 2 prominent parallel veins; flowers pale yellow, solitary balls on short stems, spring-flowering; pod 3–10 cm long, 2 mm wide. This species is often confused with *A. subporosa*.

*A. colletioides*—**Wait-a-while**     W.A., N.T., S.A., Vic.,
2–3 m          300 mm          Prefers warm dry          N.S.W.
                              sites          Moderate
Ground cover, low shelter, erosion

Compact, rounded, rigid shrub spreading to more than its height. Requires well drained soil. Hardy. Long-lived. Forms a good stock-proof hedge. Very showy when in flower. Phyllodes very stiff, 1–3 cm long, about 2 mm wide, needle-like; flowers yellow, prolific, single or in small clusters, spring-flowering; pods flat, curved, 2–4 cm long, 5 mm wide.

*A. conferta*—**Crowded Leaf Wattle**          N.S.W., Qld
2–3 m          400 mm          Prefers moist soils          Fast
Low windbreaks, parks, streets, ornamental

Attractive, compact, large shrub or small tree. Suitable for hot districts. Free-flowering. Hardy, drought-resistant. Phyllodes 5–10 mm long, narrow, thin; flowers yellow, solitary, spring-flowering; pod 5–7 cm long, 5–10 mm wide.

*A. cultriformis*—**Knife Edge Wattle**          N.S.W., Qld
2–3 m          500 mm          Most soils, except          Moderate
                              alkaline ones
Low shelter, parks, ornamental

Spreading shrub which flowers profusely in spring. Can be trimmed to form a hedge. Phyllodes small, triangular, silver-grey, 1–2 cm, set closely to the stem; flowers yellow, extending beyond the foliage in sprays; pod rich brown, 4–7 cm long, 5 mm wide. 1820.

*A. cyanophylla*—**Orange Wattle**                     W.A.
3–5 m          400 mm          Grows on sandy and          Very fast
                              calcareous loams
Low shelter, parks, ornamental, sand-drift

Spreading, bushy, colourful, large shrub with angular drooping branches; very good for temporary low shelter. Tends to be short-lived. Coppices freely. Drought-resistant. Withstands moderate coastal exposure. Very good for stabilizing drifting sand. Tolerates saline soils. Phyllodes usually 10–20 cm,

*Acacia cyanophylla—flowers*

*Acacia cyanophylla*

*Acacia dealbata along a stream*

occasionally up to 30 cm long, 1–2 cm wide, often slightly bluish; flowers in large balls, yellow to orange-yellow, flowers late winter; pods flat, narrow, 8–12 cm long, 5 mm wide. 1838.

*A. cyclops*—**W.A. Coastal Wattle**                        W.A.
2–3 m       400 mm       Good on sandy soils       Fast
Low shelter, sand-drift

Compact, low, spreading, rounded shrub. Older plants may become straggly but can be regenerated by pruning. Slightly frost-tender. Drought-resistant. Wind-firm. Withstands severe coastal exposure, useful for sand binding. Tolerates saline soils. Coppices freely. Phyllodes 5–10 cm long, 5–10 mm wide; flowers yellow, in clusters of 2–3, summer-flowering; pods slightly curved, 5–8 cm long, 5–10 mm wide.

*A. dealbata*—**Silver Wattle**          Tas., Vic., N.S.W.
5–30 m       650 mm       Best on deep soils       Fast
Farm forests, windbreaks, gully erosion

Known as mimosa in Europe. Height very variable, from 5 m in poorer sites to well over 30 m in high rainfall areas. Upright, producing straight timber when in forest formation. Suckers freely from the roots so is useful for control of gully erosion but not for garden planting. Subject to fire blight disease. Trunks grey with silvery patches; foliage grey-green, pinnate, 3–10 pairs pinnae, 10–20 pairs leaflets, 2–4 mm long; flowers yellow in racemes, 5–10 cm long; pods 7–10 cm long, 5–10 mm wide, often slightly bluish. Flowers profusely in late winter. 1792.

*A. deanei*—**Deane Wattle**               Vic., N.S.W., Qld
5–6 m       500 mm       Prefers sandy soils       Fast
Windbreaks,

Upright, small tree with a moderately dense crown. Moderately drought- and frost-resistant. Leaves bi-pinnate, pinnae 4–8 pairs, 15–25 pairs leaflets each 3–5 mm long, young shoots often yellowish; flowers yellow.

*A. decora*—**Graceful Wattle**            Vic., N.S.W., Qld
2–3 m       450 mm       Will grow on stony       Moderate
                                    and sandy soils
Low shelter, streets, parks, ornamental

Compact, graceful shrub, can be trimmed as a hedge. Grows up to 5 m under natural conditions. Phyllodes narrow, thick, 2–3 cm long; flowers deep yellow in racemes extending beyond the foliage, spring-flowering; pod thin, flat, 6–10 cm long.

*A. decurrens* (now includes A. normalis)—
          **Early Black Wattle**      Vic., N.S.W., Qld
5–10 m       650 mm       Well-drained soils       Fast
Farm forests, windbreaks, shade

Upright tree with dense, dark green, bi-pinnate foliage. Produces clean straight stems when close grown. Wind-firm. Bark dark; branchlets winged; foliage 5–15 pairs of pinnae, widely spaced, 20–30 pairs of leaflets, very narrow, 5–7 mm long; flowers creamy-yellow, globular, in racemes 8–10 cm long, spring-flowering; pods 8–10 cm long, 5 m wide. Before 1850.

*A. doratoxylon*—**Currawong**    Vic., N.S.W., Qld
4–8 m    250 mm    Moderate
Windbreaks, parks, roadsides

Attractive, small to medium tree, more suitable for the warmer districts. Requires well-drained soils but otherwise will grow on most types, including stony soils and dry ridges. Hardy. Frost- and drought-resistant. Phyllodes 8–20 cm long, 2–5 mm wide, thick, slightly curved; flowers yellow, in cylindrical spikes arranged in racemes, spring-flowering; pod 7–10 cm long, 2–4 mm wide, round.

*A. drummondii*—**Drummond Wattle**    W.A.
1–3 m    550 mm    Moderate
Low shelter, parks, ornamental

Bushy, low growing shrub 1–1·5 m across. One of the best shrubby types which should be more widely planted. Sometimes difficult to raise and establish. Good seed can be hard to obtain. Moderately frost-sensitive. Leaves small, pinnate, 4–8 pinnae, each with 4–12 leaflets, less than 1 cm long; lemon-yellow flowers in cylindrical spikes 3–5 cm long, drooping, on a stalk 2 cm or more long, flowers freely from the second year in late winter to spring; pods small, 2–3 cm long, 5 mm wide, flat.

*A. falciformis*—**Hickory Wattle**    Vic., N.S.W., Qld
6–12 m    800 mm    Prefers moist good soils    Fast
Windbreaks, farm forests

Small upright tree with a straight bole. Suitable for the cool mountain districts. Hardy and frost-resistant. Phyllodes 10–15 cm long, 2–3 cm wide, narrowed at the base, usually curved to almost a sickle shape; flowers pale yellow in short clusters, spring-flowering; pod flat, 8–12 cm long, 5–10 mm wide.

*Acacia decurrens*

*Acacia falciformis—flowers and foliage*

*A. farinosa*—**Mealy Wattle**    S.A., Vic., N.S.W.
2 m    400 mm    Moderate
Low shelter, ornamental

Spreading low shrub; leaves and stem often mealy. Phyllodes 4–5 cm long, narrow, broader towards the tip, thick sometimes almost fleshy; flowers solitary on a short stalk, spring-flowering; pod curved, twisted, 3–5 cm long, 2–3 mm wide.

*A. farnesiana*—**Perfume Wattle**
    Old World tropics,
2–3 m    400 mm    N.S.W.
Low shelter, parks, hedges, ornamental    Fast

Thorny shrub or small tree, rather straggly on poor sites. Unique as it is the only *Acacia* native on all continents. Will not stand severe frosts. Grown extensively in France for the perfumery industry. Foliage bi-pinnate, leaflets 4–6 mm long; flowers deep yellow, almost orange, strongly perfumed, spring-flowering; thorns 2 cm or more long; pods 4–8 cm long, curved, thick, woody, when inflated will float on water for days without injury to the seed.

*A. fimbriata*—**Fringe Wattle**    N.S.W., Qld
3–5 m    750 mm    Prefers good,    Moderate
    moist soils
Parks, streets, ornamental

Attractive shrub or small tree with pendulous foliage. Best growth on moist alluvial soils in warm districts. Phyllodes 3–5 cm long, 5 mm wide, bright green; flowers yellow, abundant, in clusters 3–5 cm long, spring-flowering; pod 4–7 cm long, flat, straight, bluish. Very similar to *A. prominens*. 1828.

*A. floribunda*—**Catkin Acacia**    Vic., N.S.W.
3–8 m    650 mm    Will grow on wet sites  Very fast
Low shelter, sand-drift, ornamental

Dense, bushy, attractive, very fast-growing shrub or small tree. Best growth on lighter soils. Withstands severe coastal exposure. Phyllodes narrow, 7–10 cm long, 5–10 mm wide, pointed; flowers yellow, produced abundantly in 2–5 cm long spikes, spring-flowering; pods flat, rigid, 4–10 cm long, 3–5 mm wide.

*A. gladiiformis*—**Sword Leaf Wattle**          N.S.W.
2–3 m          350 mm          Will grow on sandy          Moderate
                              and rocky sites
Low shelter, ornamental

Bushy, multi-stemmed shrub with phyllodes 10–15 cm long, 1 cm wide, dark green; flowers pale yellow, in racemes shorter than the phyllodes; pods 6–12 cm long, 5–10 mm wide, narrow between the seeds. Similar to *A. notabilis* but foliage of this species is greyish.

*A. glandulicarpa*—**Hairy Pod Wattle** or          Vic.
                **Gland Wattle**          Moderate
1–2 m          250 mm
Ground cover, ornamental

Dense, hardy, hemispherical shrub growing close to the ground and giving very good low shelter. Attractive appearance. Phyllodes oval, 5–10 mm long, with wavy edges, fresh yellow-green colour; flowers globular, solitary, shorter than the foliage; pods straight, hairy.

*A. glaucescens*—**Coastal Myall**          N.S.W., Qld
6–10 m          650 mm          Fast
Parks, roadsides, ornamental

Attractive, small, shapely tree. Hardy, growth best in coastal areas, particularly along streams. Withstands moderate coastal exposure. Not recommended for planting where stock can gain access to the foliage— foliage is regarded as poisonous, the only acacia in Australia that is so. Bark fibrous; phyllodes 10–15 cm long, 1–2 cm wide, curved, greyish; flowers in catkins 5–8 cm long, spring-flowering; pod straight, 5–8 cm long, 2–3 mm wide.

*A. gracifolia*          S.A.
2–3 m          250 mm          Moderate
Parks

Rather open, graceful shrub. Prefers a warm, sunny position. Hardy, frost- and drought-resistant. Branchlets angular; phyllodes 5–8 cm long, very thin and slender, sometimes resinous; flowers yellow, in short cylindrical almost globular spikes, spring-flowering; pods very narrow, 5–6 cm long, 2–3 mm wide.

*A. hakeoides*—**Hakea Wattle**          W.A., S.A., Vic.,
                                        N.S.W., Qld
3–4 m          250 mm          Moderate
Low shelter, parks, gully erosion, ornamental

Dense, bushy, showy, hardy shrub giving good ground cover. Should be planted more extensively. Suckers freely from the roots. Flowers profusely in spring giving a very attractive display. Phyllodes hakea-like, 5–12 cm long, 2–3 mm wide, broader towards the tip; flowers yellow, in racemes as long as the foliage, during late winter and spring; pod curved, 7–10 cm long, 5–7 mm wide.

*A. harpophylla*—**Brigalow**          N.S.W., Qld
5–10 m          400 mm          Will grow on heavy          Fast
                              clay soils
Low shelter, gully erosion, subsistence fodder

Small tree, shapely with a straight and upright trunk when given sufficient space, otherwise can be spindly and straggly. Generally similar to *A. salicina*. Hardy, tolerant of adverse conditions. Wind-firm. Sprouts from the roots. Foliage can cause 'Brigalow itch'. Phyllodes 10–20 cm long, 1–2 cm wide, thick, curved, parallel veins, with a characteristic blue-grey sheen; flowers globular, in clusters of 6–10 on slender stalks, amongst the foliage, winter to spring-flowering; pod narrow, distinctive, 5–12 cm long with longitudinal stripes.

*A. homalophylla*—**Yarran**          S.A., Vic., N.S.W., Qld
5–10 m          250 mm          Moderate
Shade, windbreaks, subsistence fodder

Small tree with a straight trunk and spreading crown when open grown; otherwise irregular in shape. Suitable for most soils. Hardy, frost- and drought-resistant. Wind-firm. Long-lived. Tolerates limited trimming. Phyllodes very narrow, 3–8 cm long, up to 5 mm wide, thick, stiff, greyish; flowers globular, in pairs or groups of 3–6 amongst the foliage, spring-flowering; pods thin, slightly curved, 3–8 cm long, 5 mm wide, bluish.

*A. howittii*—**Sticky Wattle**          Vic.
5–8 m          650 mm          Medium to heavier soils          Fast
                              but adaptable to most
                              sites
Windbreaks, parks, streets, ornamental

Attractive, graceful, dense, nearly conical tree. One of the better-shaped acacias; well worth planting. Wind-firm. Can be pruned. Phyllodes dark green, oval, 1–2 cm long, slightly sticky; flowers pale yellow, solitary, amongst the foliage, flowers early spring; pods 4–6 cm long, 5 mm wide, thin. 1893.

*A. implexa*—**Lightwood**          Vic., N.S.W.
6–12 m          550 mm          Fast

Very similar in appearance, uses, etc. to *A. melanoxylon* and so is rarely grown. Phyllodes are slightly longer (12 cm), and sickle shape. It can only be definitely distinguished by the seed in that the funicle (thread attaching the seed to the pod) is cream in colour and folded under the seed whereas in *A. melanoxylon* it is red and completely encircles the seed.

*A. inophloia*          W.A.
2–3 m          300 mm          New introduction
Low shelter

Upright shrub, showing promise for the warmer districts. Hardy, drought-resistant. Phyllodes needle-like, fine, soft; flowers yellow, in cylindrical spikes, spring-flowering.

*Acacia iteaphylla*

*Acacia ligulata*

**A. iteaphylla—Gawler Range Wattle**       S.A.
3 m       250 mm       Fast
Low shelter, parks, hedges, ornamental

Dense, bushy shrub; attractive because of its drooping branches. Phyllodes narrow, often silver-grey, 5–10 cm long, 5–7 mm wide; flowers globular, on short stalks, winter-flowering; pods flat, 6–12 cm long, 10–12 mm wide.

**A. jonesii—Jones Wattle**       N.S.W.
2–4 m       400 mm       Very fast
Low shelter, ornamental

Bushy, spreading shrub with fine foliage. Suitable for gardens under cooler conditions. Foliage fresh green colour, pinnate, 8–10 pairs pinnae 1 cm long, leaflets numerous, 2–3 mm long; flowers rich yellow, abundant, globular, late winter-flowering; pods 4–8 cm long, 5–7 mm wide, flat, slightly curved, more or less glaucous.

**A. leprosa—Leper Wattle**       Vic., N.S.W.
3–5 m       500 mm       Fast
Low windbreaks, ornamental

Willowy habit with weeping pendulous branches when young; more erect, small tree later. Isolated specimens in exposed situations may require staking. Hardy, frost-resistant, and shade-tolerant. Phyllodes 5–8 cm long, 1 cm wide, spotted and rough; flowers globular, pale yellow, in pairs, with a cinnamon fragrance, flowers early spring; pod slightly curved. 1817.

**A. ligulata—Small Cooba**       W.A., S.A., Vic.,
2–3 m       250 mm       N.S.W., Qld, N.T.
Low shelter       Fast

Hardy, compact, spreading and rounded shrub, suitable for low shelter on a wide range of soils, particularly sandy types. Sometimes tends to be straggly, hence pruning for shape is required. Withstands slight coastal exposure. Tolerates saline and alkaline soils. Phyllodes thick, narrow, 5–8 cm long, 5–7 mm wide with a sharp tip; flowers single or in clusters within the foliage, bright yellow to yellow-orange, winter to early spring-flowering; pods hard, brittle, 5–10 cm long, 5–10 mm wide, constricted between the seeds.

**A. linifolia—Flax Wattle**       N.S.W., Qld
3–5 m       650 mm       Prefers moist, well-       Fast
              drained soil
Low shelter

Compact shrub on drier sites, small tree on good sites. Spreading to give good ground protection. Withstands moderate coastal exposure. Phyllodes 2–4 cm long, 3 mm wide, slender, straight; flowers pale yellow, in clusters same length as the phyllodes, winter-flowering; pod very flat, 5–10 cm long, 1 cm wide. 1828.

**A. loderi**

Very similar to *A. sowdeni* in most characteristics, except that the phyllodes are narrower and longer.

*Acacia jonesii*

*Acacia longifolia*

**A. longifolia—Sallow Acacia**     Vic., N.S.W., Qld
3–5 m     650 mm     Will grow on wet sites   Very fast
Low shelter, parks, sand-drift, ornamental

Dense, bushy, round shrub; very fast-growing but tends to be short-lived. Flowers and foliage attractive. Wind-firm. Withstands severe coastal exposure. Can be lightly trimmed. Similar to *A. floribunda* in appearance. Phyllodes 8–15 cm long, 1–2 cm wide, prominent veins; flowers bright yellow, in spikes 5–8 cm long, spring-flowering; pods flat, rigid, leathery, 4–10 cm long, 7–10 mm wide. 1792.

**A. longifolia var. *sophorae*—Coastal**     S.A., Tas., Vic.,
                                  **Wattle**           N.S.W., Qld
2–4 m     450 mm     Prefers sandy soils     Very fast
Low shelter, ornamental, gully erosion, sand-drift

Spreading, hemispherical, compact and dense bush, excellent for control of coastal sand drift. Wind-firm. Withstands severe coastal exposure. Tolerates saline soils. Retains branches to ground level and on sandy soils in particular these may root at the nodes. Phyllodes dark green, 5–12 cm long, 2–5 cm wide, blunt, several prominent parallel veins; flowers bright yellow, in spikes 5–6 cm long, flowering profusely in spring; pods flat, leathery, 5–10 cm long, 5–10 mm wide.

**A. mearnsii (A. mollissima)—Black**     S.A., Tas., Vic.
                                **Wattle**         N.S.W., Qld
6–20 m     550 mm                           Very fast
Shade, windbreaks, farm forests

Usually small tree but can reach 25–35 m in cool mountain areas. Will grow under hard conditions. Rather short-lived. Subject to fire blight disease (attack by *Paropsis* insects). Phyllodes grey-green, pinnate, pinnae 8–25 pairs, leaflets very small, 2–3 mm long, 20–50 pairs; flowers creamy-yellow, globular, in racemes 5–8 cm long, late spring-flowering; pods 7–10 cm long, 5–7 mm broad. 1810.

**A. melanoxylon—Blackwood**     S.A., Vic., Tas.,
                                   N.S.W., Qld
10–20 m     650 mm     Prefers deep, moist soil     Fast
Windbreaks, shade, parks, farm forests, gully erosion

Attractive, dense, upright tree which under good conditions can grow to 40 m. Best growth is obtained in high-rainfall mountain gullies but it will grow on much poorer sites including sandy loams. Unlike most acacias it is long-lived, growing for a century or more. Produces the blackwood timber of commerce. Retains branches to ground level. Suckers freely from stumps and exposed roots; has a vigorous and spreading root system. Wind-firm. Withstands slight coastal exposure and will grow on wet sites better than most wattles. Not drought-resistant. Juvenile leaves pinnate, 8–14 pairs leaflets, each up to 5 mm long, phyllodes 7–10 cm long, 1–2.5 cm wide, parallel veins; flowers creamy-white, globular, in racemes shorter than the phyllodes, spring-flowering; pods rather flat, slightly twisted, 5–10 cm long, 5–10 mm wide. 1858.

*Acacia mearnsii*

*Acacia mearnsii—flowers*

*Acacia melanoxylon—young tree*

**A. microbotrya—Manna Wattle**                    W.A.

3–5 m      300 mm      Not suitable for light soils      Fast
Low shelter, ornamental

Bushy shrub or small tree. Does not grow so well on light soils. Phyllodes 7–12 cm long, 1 cm wide; flowers yellow, globular, small, in short racemes. Spring-flowering.

**A. montana—Mallee Wattle**          S.A., Vic., N.S.W.

2–3 m      350 mm      Not suitable for light soils      Fast
Low shelter, ground cover, ornamental

Dense shrub, providing good ground shelter. Attractive green colour. Phyllodes small, narrow, 2–3 cm long; flowers solitary or in pairs, late winter-flowering; pod 2–5 cm long, 2–4 mm wide, hairy.

**A. myrtifolia—Myrtle Wattle**          W.A., S.A., Vic., Tas., N.S.W., Qld

1–2 m      500 mm                                  Fast
Low shelter, parks, ornamental

Bushy shrub, suitable for low shelter or gardens. Requires occasional trimming to prevent straggly growth. Stems often reddish. Phyllodes oval, pointed, 2–4 cm long, 1 cm wide; flowers pale yellow, in groups of 2–5, winter-flowering; pod 4–10 cm long, 7–9 mm wide, curved.

**A. notabilis—Flinders Wattle**                    S.A.

2–3 m      250 mm      Prefers heavier soils      Moderate
Low shelter, streets, ornamental

Attractive, bushy wattle. Best in the hot and dry inland districts. Phyllodes 5–15 cm long, 1–3 cm wide, thick, grey; flowers globular, pale yellow, in short racemes, flowers early spring; pod straight, broad, 3–7 cm long, 1 cm wide.

**A. obliquinervia—Mountain Hickory Wattle**      Vic., N.S.W., Qld

5–8 m      750 mm      Prefers good soils      Fast
Windbreaks, farm forests

Upright tree with a straight bole and angular branches. Hardy, frost-resistant, suitable for sub-alpine conditions. Withstands light snow. Phyllodes 5–12 cm long, 2–4 cm wide, margins near stalk not opposite each other, mid-rib offset (not in the centre of the leaf); flowers pale yellow, in short clusters, spring-flowering; seed pod distinctive, very long for an acacia, 15–20 cm long, 1 cm wide.

**A. oswaldii—Umbrella Wattle**      W.A., S.A., Vic., N.S.W., Qld

2–4 m      300 mm      Prefers the lighter soils      Moderate
Shade, low shelter, subsistence fodder

Spreading, large bushy shrub or small tree with a moderately dense crown. Hardy, drought-resistant. One of the few long-living acacias. Similar in appearance to a small *A. homalophylla* but distinguished by its hard long pods. Pods reputedly eaten by stock. Phyllodes 2–7 cm long, 2–10 mm wide, thick, rigid, with a fine tip, often silvery when young; flowers yellow, globular, sessile, single or in pairs amongst the foliage, summer-flowering; pods distinctive, 5–25 cm long, 5–8 mm wide, curly and woody, hard, produced in large quantities.

*Acacia melanoxylon—windbreak*

139

*Acacia pendula*

*A. pendula*—**Weeping Myall** or **Boree**          Vic., N.S.W., Qld

5–10 m          300 mm          Prefers heavier soils          Slow to moderate

Shade, fodder, ornamental

Small tree with grey-green pendulous foliage, almost like a weeping willow. One of the most attractive of the acacias. Hardy. Withstands severe hot, dry conditions and periodic flooding. Phyllodes narrow, 5–8 cm long, 5–8 mm wide, silver-grey; flowers in short racemes, during late winter; pods flat, slightly curved, with a small wing, 2–7 cm long, 10–20 mm wide, slightly woody.

*A. penninervis*—**Hickory Wattle**          Vic., N.S.W., Qld

5–10 m          900 mm          Fast

Shelter, windbreaks, farm forests

Upright, small tree, with a dense crown when open grown or a long straight trunk and small crown when close grown. Good tree for mountain districts. Prefers better soils but will grow on the shallower ridge sites. Frost-resistant. Tolerates snow. Suitable for sub-alpine areas. Phyllodes 7–12 cm long, usually 1–2.5 cm wide, but often larger, tapered at both ends, mid-rib central; flowers creamy-yellow, globular, in clusters; pod 10–12 cm long, 1 cm wide, flat. 1824.

*A. podalyriifolia*—**Queensland Silver Wattle**          N.S.W., or **Mt Morgan Wattle**          Qld

3–5 m          650 mm          Will grow on heavy soils          Fast

Ornamental

Handsome, bluish, small tree. Withstands moderate frosts. Very good garden specimen. Phyllodes bluish, oval, 2–5 cm long; flowers profusely in winter, deep yellow, 15–20 large globular heads, in spikes 5–8 cm long, during winter to early spring; pod flat, broad, 4–8 cm long, 1–2 cm wide. 1824.

*A. pravissima*—**Ovens Acacia**          Vic., N.S.W.

3–8 m          600 mm          Fast

Low shelter, parks, streets, ornamental

Hardy, upright, bushy, rounded shrub. Showy, branchlets slightly pendulous, good garden specimen. Grows well in lighter soils, may be only 3–4 m tall in heavy types. Can be trimmed; occasional trimming desirable to prevent straggly growth. Withstands frosts and slight coastal exposure. Branches angular; phyllodes triangular, 1–2 cm broad, 2–3 prominent veins; flowers bright yellow, showy, in racemes 7–10 cm long, flowers late spring; pods flat, 5–10 cm long.

*A. prominens*—**Golden Rain Wattle**          N.S.W., Qld

5–8 m          600 mm          Fast

Low shelter, parks, ornamental

Small tree; flowers perfumed, in long pendulous sprays. One of the most attractive wattles. Best in the warmer lowland districts. Withstands moderate coastal exposure. Phyllodes 2–5 cm long, 5–10 mm wide, grey-green; flowers globular, in slender racemes, flowers early spring; pods straight, flat, 3–8 cm long, slightly bluish. Very similar to *A. fimbriata*.

*A. pruinosa*—**Frosty Wattle**          N.S.W.

5–8 m          650 mm          Prefers deeper soils          Very fast

Parks, ornamental

Usually a small tree but can be straggly. Silvery leaves. Best in the warmer lowland areas. Withstands slight coastal exposure. Foliage bi-pinnate, 10–20 pairs of 1 cm leaflets; flowers yellow, perfumed, in loose terminal sprays, summer-flowering; pods usually straight, 6–10 cm long, narrowed between the seeds.

*A. pubescens*—**Hairy Wattle**          N.S.W.

3–5 m          600 mm          Fast

Windbreaks, parks, ornamental

Attractive, graceful, dense shrub or small tree. Uncommon in Australia (both naturally or planted) but common overseas, particularly as an indoor or glasshouse plant. Young stems and foliage covered with long, soft, silvery hairs. Foliage bi-pinnate, 3–5 cm long, pinnae 6–12 pairs, each with up to 20 oblong leaflets, 2–5 mm long; flowers prolific, rich yellow, globular, 5–7 mm diam., in clusters up to 8 cm long, spring-flowering. 1790.

*Acacia pravissima*

**A. pycnantha—Golden Wattle**    S.A., Vic., N.S.W.
4–5 m    450 mm    Fast
Low shelter, ornamental

Small hardy tree, or a compact bush when open grown. Young plants can be straggly but this disappears in time. On older trees outer branchlets become pendulous. Grows on a wide range of soils, including heavy, shallow, and dry. Useful for conservation works. Withstands moderate coastal exposure. Widely used as under-shelter to sugar gum windbreaks. Phyllodes broad, 7–12 cm long, 1–2 cm wide, sickle-shape, young leaves up to 5 cm wide, glossy green; flowers globular, large, rich yellow, in racemes about 8 cm long, spring-flowering; pod straight, 5–10 cm long, 5–10 mm wide.

**A. retinodes—Wirilda**    S.A., Vic., Tas.
5–10 m    500 mm    Will grow on wet    Very fast
sites
Low shelter, ornamental

Small tree, variable, but usually shapely and compact when open grown. Good for mass under-planting. Withstands moderate coastal exposure, some flooding and slightly saline areas. Suckers freely from cut stems. Stems reddish; phyllodes long and slender, 7–20 cm long, less than 1 cm wide; flowers pale, creamy, globular, 4–8 in short racemes, not very conspicuous but some present almost any time of the year; pods straight, flat, 3–10 cm long, 5–10 mm wide. 1871.

**A. riceana—Rice Wattle**    Tas.
2–3 m    750 mm    Fast
Low shelter, streets, ornamental

Very attractive, shapely, rich green shrub, with drooping branches and soft foliage. Prefers cooler conditions than most wattles; should be planted more widely. Usually not recognized as a wattle until it flowers. Phyllodes almost needle-like, pointed but soft, 3–5 cm long, bronze when young, arranged in groups; rich yellow flowers, in spikes 4–6 cm long, scattered through the foliage, during early spring; pod very narrow, 5–8 cm long.

**A. rigens—Nealie**    S.A., Vic., N.S.W.
2–3 m    300 mm    Prefers lighter soils    Fast
Low shelter, parks, ornamental

Compact, grey-green, rounded shrub, good for dry areas. Phyllodes needle-like, fairly stiff, 3–10 cm long; flowers bright yellow, globular, solitary, or in twos and threes, winter-flowering; pod curved, twisted, hairy, 5–12 cm long, about 3 mm wide.

**A. rotundifolia** (A. obliqua)—**Spoon Acacia**    Vic.
2–3 m    400 mm    Fast
Low shelter, parks

Much-branched, rounded shrub. Hardy. Suitable for hot districts. Phyllodes variable, almost round, 6–12 mm long, mid-rib to one side; flowers yellow, small, under 6 mm diam., solitary, on stalk 1 cm long, scattered along the stem; pod 2–4 cm long, 2–3 mm wide, usually coiled. This species is regarded as only a variety of A. acinacea by some authorities.

**A. rubida—Red Stem Wattle**  S.A., Vic., N.S.W., Qld
3–10 m    650 mm    Very fast
Low shelter, parks, ornamental

Large shrub or small tree, stems turning reddish in winter. Will grow on adverse sites. Somewhat like A. pycnantha in appearance; should be more widely planted. Phyllodes sickle-shape, 7–10 cm long, 5–15 mm wide, usually with some bi-pinnate leaves remaining for many years; flowers yellow, globular, in short racemes of 3–5, flowers profusely during spring; pod straight, 5–6 cm long, 5–7 mm wide.

*Acacia prominens*

*A. salicina*—**Willow Acacia**        All States except Tas.
5–6 m        300 mm                    Fast
Low shelter, parks, hedges, gully erosion, ornamental

Attractive, large shrub or small tree with a round, bushy crown and drooping foliage, sometimes to ground level. Best growth on river soils in hot areas or where irrigation water is available. Long-lived. Wind-firm. Drought-resistant. Withstands slight coastal exposure. Tolerates saline and clay soils. Some value as fodder. Suckers freely. Phyllodes willow-like, 5–15 cm long, 6–12 mm wide; flowers globular, solitary or in twos and threes, usually bright golden yellow, winter to early spring-flowering; pods straight, 5–12 cm long, 1 cm wide, thick, woody.

*A. saligna*—**Golden Wreath Wattle**        W.A.
3–8 m        400 mm        Will grow on        Very fast
                              heavy soil
Low shelter, parks, roadsides, ornamental

Small tree, bushy, spreading crown with large rich yellow or sometimes orange-yellow flowers in spring and early summer. Can be grown in southern districts but better in the warmer parts. Hardy. Wind-firm. Withstands moderate coastal exposure. Tolerates smog. Occasionally found as a garden escape. Phyllodes slender, very variable in length, mostly 7–15 cm but up to 30 cm long particularly on young plants, 6–20 mm wide; flowers prolific, globular, usually in racemes 7–10 cm long, sometimes up to 20 cm, spring-flowering; pods flat, narrow, 7–12 cm long, 5–7 mm wide. (Occurs as a garden escape in the Melbourne area. Such specimens are often not typical and can be very variable— phyllodes up to 25 cm long and less than 1 cm wide may be found.)

Often difficult to distinguish from *A. cyanophylla*, although the flowers of the latter usually have a slight orange tinge. 1818.

*A. sclerosperma*        W.A.
5–6 m        250 mm                    Moderate
Windbreaks, roadsides

Small tree. Wind-firm. Suitable for planting on coastal sand dunes; withstands severe coastal exposure. Often only 1 m in exposed situations. Phyllodes 5–10 cm long, narrow, rounded; flowers yellow, winter to spring-flowering; pods woody, constricted between the seeds.

*A. silvestris*—**Red Wattle**        E. Vic., N.S.W.
15–20 m        650 mm                    Fast
Farm forests, windbreaks

Attractive, upright wattle with a clean, straight stem when close planted. May be bushier when open grown. Similar to *A. dealbata*. Hardy, frost-resistant. Will grow on dry ridges. Re-shoots vigorously. Trunk grey and smooth; foliage pinnate, similar to *A. dealbata* but much larger, pinnae 10–25 pairs, leaflets 30–65 pairs; flowers profusely in spring. 1957.

*A. sowdenii*—**Western Myall**        S.A.
4–6 m        300 mm        Will grow on        Slow
                              calcareous soils
Windbreaks, shade, parks, roadsides

Compact, neat, small tree with drooping branches and silvery foliage. Not suitable for wet sites. Long-lived. Frost- and drought-resistant. Good tree for dry areas. Sheep are fond of the foliage. Phyllodes linear (needle-like), 7–10 cm long, silver-grey; flowers yellow, globular, flowering irregular, but prolific in good seasons; pod curved, 7–10 cm long, 5–7 mm wide.

*A. spectabilis*—**Mudgee Wattle**        N.S.W., Qld
3–5 m        560 mm                    Fast
Low shelter, streets, ornamental

Showy shrub or small tree. One of the most attractive of the wattles, suitable for garden and street planting. Bark powdery-white; foliage sparse, phyllodes bi-pinnate, fern-like, 2–5 pairs pinnae, leaflets oval, 8–10 mm long, shiny, purplish when young; flowers rich yellow, in racemes 8–15 cm long, beyond the foliage, flowers late winter; pod purplish-grey, 5–10 cm long, 1 cm wide. 1820.

*A. stenophylla*—**Eumong**        S.A., Vic., N.S.W.,
                                      Qld., N.T.
5–8 m        250 mm        Prefers heavy soils        Moderate
Low shelter, streets, ornamental

Small tree with pendulous silver-grey foliage, suitable for the warmer districts. Tolerates saline soils and some flooding. Phyllodes very long and slender, 15–45 cm long, 5–10 mm wide; flowers globular, in short racemes of 3–6 heads, flowering irregular; pod leathery, 12–20 cm long, 1 cm wide, very narrow between the seeds, often breaking into one seed particles.

*A. stricta*—**Hop Wattle**        Vic., S.A., Tas.,
                                    N.S.W., Qld
1–5 m        650 mm                    Fast
Windbreaks

Variable shrub or small tree. Often only 1–2 m tall, but under suitable conditions forms a small tree 3–5 m tall. Frost-resistant. Tolerates some poor drainage. Phyllodes usually 5–10 cm long and about 1 cm in width, grey-green, with a prominent main vein; flowers yellow, solitary, winter to spring-flowering; pod flat, 3–5 cm long, 5–10 mm wide. 1790.

*A. suaveolens*—**Sweet Wattle**        Tas., Vic., N.S.W., Qld
2–3 m        650 mm        Prefers lighter soils        Very fast
Low shelter, ornamental

Well shaped, slender, bluish-green shrub; erect and narrow if space restricted. Useful on lighter and poorer soils. Phyllodes thick, 6–12 cm long, 1–2 cm wide; flowers globular, pale yellow, in short racemes 3–5 cm long, autumn to winter-flowering; pod flat, 2–4 cm long, 1–2 cm wide, leathery.

*Acacia terminalis*

*Acacia terminalis—flowers*

**A. terminalis** (A. elata)—**Cedar Wattle**         N.S.W.
12–20 m    650 mm                                    Fast
Windbreaks, shade, parks, ornamental

Attractive, large, spreading tree with branches to ground level. One of the largest of the wattles, often spreading to 10–12 m across. Long-lived. Slightly frost-tender when young. Withstands slight coastal exposure. Foliage attractive, ash-like, bi-pinnate, compound leaves, 20–30 cm long, pinnae 4–6 pairs, 10–15 pairs leaflets, each leaflet 3–5 cm long and 1 cm wide; flowers pale yellow to cream, in panicles 20–30 cm long, summer-flowering; pods straight, 5–10 cm long, 1 cm wide.

**A. trineura**—**Hindmarsh Wattle**    S.A., Vic., N.S.W.
2–5 m    400 mm                                    Fast
Low shelter, ornamental

Dense, rounded shrub giving good ground cover. Phyllodes grey-green, 2–5 cm long, 1 cm wide, 3 veins; flowers dark yellow, in short racemes during late winter; pod narrow, 5–6 cm long, 5–7 mm wide, slightly curved.

**A. triptera**—**Spur Wing Wattle**    Vic., N.S.W., Qld
3–5 m    500 mm                                    Fast
Low shelter, ornamental           New introduction

Bushy shrub with branches to ground level; often as wide as it is high. Rare species in Victoria, restricted to one part of the Warby Ranges, and only introduced into cultivation in 1969. Growth rate and soil requirements not yet determined, but occurs naturally amongst granitic rocks in a 550 mm rainfall area. Phyllodes stiff, curved but flat, running down the stem, 2–4 cm long; flowers globular, bright yellow, in loose spikes amongst but longer than the foliage, flowers late spring; pod slightly curved, 5–6 cm long, 3–6 mm wide.

**A. subporosa**                                    N.S.W.
6–10 m    650 mm    Prefers deep loams    Moderate
Low shelter, parks, ornamental

Attractive, small tree, branches slightly pendulous. Young shoots with a distinctive and conspicuous red tinge. Phyllodes long and narrow, 7–12 cm long, 3–6 mm wide, with 3–4 prominent parallel veins; flowers pale yellow, solitary, on short stems, spring-flowering; pod 4–5 cm long, 2 mm wide. This species is very similar to *A. cognata* but does not occur naturally in Victoria as previously believed. The Victorian type is *A. cognata*.

**A. subulata**—**Awl Leaf Wattle**             N.S.W.
2–3 m    400 mm    Prefers lighter soils    Fast
Low shelter, ornamental           New introduction

Attractive shrub, usually with several stems. Relatively new introduction which is proving promising. Seems to prefer the lighter alluvial soils. Phyllodes long and narrow, 5–15 cm long, 2–4 mm wide; flowers pale yellow, in short racemes during late winter to early spring; pod straight, narrow, 7–15 cm long, hard.

*Acacia triptera—flowers*

*Acacia verticillata*

*Acacia verticillata—flowers*

*A. undulifolia*—**Wavy Leaf Wattle**     N.S.W., Qld
2–3 m     500 mm     Will grow on rocky sites     Fast
Low shelter, ornamental

Shrub, sometimes straggly, with long pendulous branches. Better in warm areas but
will grow in the southern parts of Australia.
Phyllodes greyish-green, almost round, with
wavy margins, 2–3 cm diam.; flowers
solitary, on slender stalks, late spring-flowering; pods
broad, flat, 3–5 cm long, 1–2 cm wide, woody.

*A. verniciflua*—**Varnish Wattle**     S.A., Vic., Tas.,
                                        N.S.W., Qld
3–6 m     450 mm     Will grow on stony,     Fast
                     dry soils
Low shelter, ornamental

Variable in height, but open grown specimens form a neat, compact tree up to 6 m
tall in high rainfall areas, about 3 m in drier
areas. Phyllodes bright green, 5–10 cm
long, 5–20 mm wide, surface on fresh
specimens often sticky or with a varnished appearance;
flowers deep yellow, globular, solitary, flowering profusely in spring; pod straight, 6–10 cm long, 5 mm wide.

Right   *Acacia vestita*

*A. verticillata*—**Prickly Moses**     S.A., Vic., Tas., N.S.W.
2–5 m     700 mm     Will grow on poorly-     Fast
                     drained soils
Low shelter, ornamental

Variable, rich green shrub or small tree,
more usually only 2–3 m tall. Suitable for
sites where vertical drainage is not good.
Foliage prickly, tends to discourage stock.
Wind-firm. Phyllodes almost needle-like,
1–2 cm long, arranged in whorls around the stem;
flowers yellow, in spikes 2–4 cm long, spring-flowering;
pods flat, straight, narrow, 2–4 cm long, 1–2 mm wide.
1780.

*A. vestita*—**Hairy Wattle**     N.S.W.
2–3 m     450 mm     Fast
Low shelter, ornamental

Large, attractive shrub, suitable for most
districts. Branches hairy, phyllodes broadly
oval, 1 cm long; flowers golden yellow in
long showy sprays, spring-flowering; pod
flat, straight, bluish, 4–7 cm long, 5–10 mm
wide.

*A. victoriae*—**Bramble Wattle**     All States except Tas.
4–5 m     300 mm     Will grow on alkaline     Fast
                     and stony soils
Low shelter

Large shrub, tending to be straggly and
much-branched. Numerous spines 5–10 mm
long. Well suited for mass, hedge-like
plantings in the warmer districts. Phyllodes
narrow, 2–5 cm long, 3–6 mm wide;
flowers globular, pale yellow, in racemes 5–8 cm long,
occasionally frequent enough to form sprays, flowering
irregular but generally spring; pods flat, straight,
3–8 cm long, 5–10 mm wide.

*Acer palmatum—autumn foliage*

*A. wattsiana*—**Gladstone Wattle**                    S.A.

2–3 m          500 mm          Will grow on          Moderate
                               limestone soils

Low shelter

Small neat shrub. Phyllodes narrow, 4–8 cm long, 3–6 mm wide; flowers yellow, in short racemes, spring-flowering; pods narrow, 6–12 cm long, 5–7 mm wide.

# ACER—Maples   D.H.

Seed          Moist, well-drained soils
Windbreaks, summer shade, parks, roadsides, autumn foliage

*Acer* is the old Latin name. Over one hundred species of deciduous trees from Europe, North America, China and Japan are known. Planted mainly for their colourful autumn foliage; the most satisfactory colours are obtained in the cooler mountain areas with over 900 mm rainfall and early autumn frosts. Timber from natural stands is used extensively for furniture. Trees range in height from a few metres to over 30 metres. The bark is rough and grey, and furrowed in older trees. Leaves are usually orbicular palmate with three to fifteen lobes but usually with only three to five. Flowers are inconspicuous and white or greenish-white. The seeds have wings 1–5 cm long. Seeds have a life of a few months only and should either be sown as soon as they fall or stored in moist sand. Root systems are spreading and generally shallow.

*A. buergerianum* (A. trifidium)—**Trident**     China,
                                   **Maple**       Japan

2–5 m          750 mm                               Fast

Attractive, well-shaped small tree. Frost-resistant. Young shoots coppery. Foliage turns to a brilliant red in autumn. Leaves 5–10 cm long, 3 lobes each pointing forward, trident-like; seed wings parallel. 1890.

*A. campestre*—**Common Maple** or          Europe,
               **English Maple**            W. Asia

10–12 m        750 mm                       Moderate

Compact tree, branchlets often corky. Leaves 7–12 cm diam., with 3–5 large lobes and several minor ones. Seeds in pairs, wings opposite each other. Sap milky. Several horticultural forms available. L.C.

*A. cappadocicum*          Asia Minor to India

10–15 m        900 mm                       Fast

Round-headed densely-crowned tree, similar to *A. platanoides* but young foliage rich red and colourful. (Local specimens are believed to be one of the horticultural varieties and not the parent type.) Leaves 7–15 cm wide, 5–7 lobes on a stalk 10–20 cm long; seed wings 4–5 cm long, diverging at a wide angle (almost straight).

*A. ginnala*—**Amur Maple**              China, Japan

6–8 m          800 mm                       Moderate

Graceful, spreading tree preferring the cooler districts. Hardy, withstands severe cold—good substitute for *A. palmatum* in alpine areas. Colours bright red in autumn, with leaves persisting for some time. Leaves 3 lobes, 7–10 cm long, 5–8 cm across, rich red in autumn, doubly-serrated; seeds freely, seeds 2·5 cm long, wings nearly parallel. 1860.

*A. grosseri*                               China

3–6 m          800 mm                       Fast

Small tree, tending to be shrubby. Colours well. Hardy, prefers cool climates. Bark striped with white; leaves 5–8 cm long, 4–7 cm wide, 3 lobes or sometimes without lobes, turning rich red in autumn; seeds 2–3 cm long, wings diverging at a very obtuse angle (nearly opposite). 1927.

*A. negundo*—**Box Elder Maple**         North America

10–12 m        600 mm                       Fast

Large, bushy, spreading, hardy tree. Will grow on poorer soils and in drier climates than most other maples. Withstands moderate winds and severe cold. Tolerates smog. Can be trimmed. Roots freely from cuttings taken in the spring. Leaves are not like the typical maple but are compound, 15–25 cm long with 3–7 leaflets. Seeds in pairs, wings 4 cm long forming a distinct narrow 'V'. Numerous silver and gold horticultural varieties available. 1688.

*A. palmatum* (A. polymorphum)—**Japanese**     Japan
                                **Maple**        Fast

3–10 m         650 mm

Grown mainly as a garden specimen for its brightly-coloured foliage (red, brown and yellow) in autumn. Taller heights only obtained in the cooler higher rainfall areas. Best results are obtained under cool, moist conditions; not recommended for hot, dry districts. Leaves 7–10 cm across, 5–9 lobes; seed wings curved. Numerous horticultural varieties of leaf dissection and of colour are available. Varieties are raised from cuttings and grafts. L.C.

145

*Acer rubrum—autumn leaves*

**A. platanoides—Norway Maple**          N. Europe
12–20 m    900 mm                    Fast then moderate

Spreading, round-headed tree, not as large as some other maples but requires a large area for good development. Hardy, frost-resistant. Withstands moderate coastal exposure. Tolerates snow and smog. Transplants readily. Yellow in autumn. Leaves large, 15–20 cm long, 12–18 cm across, 5 lobes, tips of lobes pointed; seed with wings forming a very broad 'V'. Sap milky. Several horticultural varieties available.  L.C.

**A. pseudoplatanus—Sycamore**          Europe, W. Asia
15–20 m    800 mm                              Fast

Large, spreading, vigorous tree. Suitable for a wide range of conditions in cool mountain areas. On deep soils can spread rapidly by seed; control measures may be necessary. Suckers freely. Suitable for exposed situations. Tolerant of shade and smog. Withstands slight coastal exposure. Prolific crop of seed (and seedlings) in most years. Large leaves, 12–20 cm across, 5 lobes; wings of seed more or less at a right angle. Several horticultural forms available; one with purple underside of the leaf instead of the normal ashy green occurs commonly.  L.C.

**A. rubrum—Red Maple** or          E. North America
**Canadian Maple**                              Fast
12–20 m    900 mm

Large, spreading tree; very colourful (red and orange) in autumn. Suitable for a wide range of soils; will stand wet sites better than most maples. Not suitable for exposed positions. Moderately tolerant of shade. Coppices prolifically. Tolerates smog and slight coastal exposure. Leaves about 10 cm diam., 5 lobes (2 are small), white underneath; seed wings short, 15–20 mm long, forming a narrow 'V'. 1656.

**A. saccharinum—Silver Maple**          E. North America
20–25 m    900 mm                              Fast

Large spreading tree. Prefers deep, moist soils but will also grow on heavier types. Very similar to *A. rubrum* in appearance and characteristics. Shade-tolerant. Should not be planted near underground drains. Leaves 12–20 cm across, 5 lobes divided into smaller lobes, silver or white underneath; seed wings 4–6 cm long (largest of all maples) and forming a right angle. 1725.

**A. saccharum—Sugar Maple**          E. North America
20–25 m    1000 mm                         Moderate

Large, oval, compact crown when open grown, narrow when close grown. Brilliant colours (red and orange) in autumn, one of the most colourful of the maples. Long-lived. Wind-firm, frost-resistant. Shade-tolerant, responds well to thinning. Withstands snow. Coppices freely and suckers from the roots. Sap clear and sweet, the source of maple syrup. Leaves 8–12 cm across, 5 lobes (2 are small), pale green underneath; seed wings 2–3 cm long, 'U' shaped. The variety *pyramidalis* is narrow-crowned, even when open grown, and is more suitable for garden planting. 1753.

**A. tegmentosum—Manchurian**          Manchuria,
          **Maple**                            Korea
5–6 m    900 mm                                Fast

Attractive, upright, slightly spreading, small tree. Hardy, frost-resistant, withstands light snow. Moderately shade-tolerant. Bark with white stripes but confined to the branches on older trees. Leaves 7–15 cm long, 5–12 cm wide, 3 lobes, green underneath, colouring to bright orange-red in the autumn; seed and wings 2–3 cm long, wings nearly opposite each other. 1892.

# ACTINOSTROBUS

*Aktin* is derived from the Greek word for ray and *strobus* from the Greek for cone which alludes to the ray-like arrangement of the cone scales. There are two species from Western Australia.

**A. pyramidalis—Swan River Cypress**  E.C.    W.A.
Seed
2–3 m    500 mm                            Moderate
Low shelter, parks, streets

Dense, upright conifer very similar to *Callitris*, with branches divided into fine sprays. Retains its branches close to the ground. Not suitable for cool districts. Satisfactory on wet sites. Withstands moderate coastal exposure and saline soil conditions. Can be trimmed to form a hedge. Leaves very small, 2–3 mm long in 3 rows closely pressed to the branchlet; cones oval, 1 cm long, 6 triangular scales.

*Aesculus hippocastanum*

*Aesculus hippocastanum—flowers*

# AESCULUS

This is the ancient Latin name. There are about twenty-five species from Europe, Asia and North America.

*A. hippocastanum*—**Horse Chestnut**   D.H.   S.E. Europe
Seed
9–12 m   700 mm   Deep soils   Moderate
Shade, parks, avenues

Handsome, well-shaped, stately tree, with a dense crown. Moderately hardy. Conical when young, spreading when older. One of the last trees in leaf and one of the earliest to fall. Withstands wind, snow and moderate coastal exposure. Leaves palmate, the group about 20 cm long, comprising usually 5–6 large toothed leaflets; flowers in very conspicuous clusters, large, upright, white; fruit large, shining brown nut, 2–4 cm diam. in a prickly husk, not edible. 1576.

*A. x carnea*

A hybrid between *A. hippocastanum* and the North American *A. pavia*. Similar to *A. hippocastanum* except that it is more frost-resistant (to about –10°C) and the flowers are a bright reddish-pink to red. Before 1818.

*A. indica*—**Indian Horse Chestnut**   D.H.   Himalayas
Seed
12–20 m   900 mm   Deep loams   Moderate
Parks, ornamental

Attractive, upright tree. Uncommon but deserving of more attention. Similar to *A. hippocastanum* but 5–9 leaflets; flowers in larger and narrow panicles, up to 40 cm long, individual flowers slightly larger, with yellow or red markings on the petals; seed husk is not prickly. 1858.

# AGATHIS

*Agathis* is the Greek word for a ball of thread which refers to the catkin. There are about twenty species from Australia, New Zealand, Fiji, New Caledonia, Malaya and Phillipines.

*A. australis*—**Kauri Pine**   E.C.   N.Z.
Seed
10–25 m   700 mm   Prefers good soils   Moderate
Parks, roadsides

Upright, narrowly conical tree when young, old trees flat-topped. Grows to over 30 m in New Zealand. Produces good timber. Will grow on a wide range of soils including slightly swampy and poor sands. Spreading

*Aesculus × carnea and flowers*

*Agathis australis*

# AGONIS

*Agon* means a collection and alludes to the number of seeds. About twenty species occur and are all Australian.

*A. flexuosa*—**Willow Myrtle**   E.H.                    W.A.
Seed

| | | | |
|---|---|---|---|
| 5–6 m | 600 mm | Will grow on heavy and wet soils | Moderate |

Streets, parks

Handsome, spreading, hemispherical, eucalypt-like tree with pendulous foliage 1–2 m long. Slightly frost-tender when young. Best in mild to warmer areas. Drought-resistant when established. Can be trimmed. Grows well on limestone. Withstands moderate coastal exposure. Leaves willow-like, 7–10 cm long, 1 cm wide; flowers small, white, in clusters of 8–10; fruit woody, fused into a globular cluster, 4–7 mm diam.

*A. juniperina*—**Juniper Myrtle**   E.H.                    W.A.
Seed

| | | | |
|---|---|---|---|
| 5–6 m | 500 mm | | Fast |

Streets, parks

Attractive, small tree with soft green foliage suitable for a wide range of sites. Does not stand severe frosts. Withstands moderate coastal exposure. Needs regular trimming to retain shape. Leaves juniper-like, dark green, 1 cm long, pointed; white, heath-like flowers in clusters at the end of twigs, good cut flower; fruit globular, woody, 4–7 mm diam., in clusters.

# AILANTHUS

*Ailanta* is a native name. About nine species occur in south-eastern Asia and northern Australia.

*A. altissima* (A. glandulosa)—**Tree of Heaven**   D.H.
Seed, root cuttings                    China

| | | | |
|---|---|---|---|
| 10–20 m | 550 mm | | Fast |

Shade, parks, gully erosion

Tall, vigorous, upright tree with a distinctive appearance. Suckers readily from root cuttings; spreads easily in light or open soils. Should not be planted where root suckers could be a problem. Will grow on light clay soils, adverse sites, and eroded areas. Frost-tender when young. Can be pollarded. Tolerates smog. Timber brittle. Male plant not recommended for planting because of its strong disagreeable odour. Leaves large, borne umbrella fashion on thick stems, pinnate, 60–90 cm long, leaflets 5–20 pairs, 7–10 cm long; seeds double-winged, carried in large reddish clusters. 1751.
In 1965 this species was declared a noxious weed in Victoria (except in the metropolitan area) and should not be planted in country districts of that State.

root system. Frost tender when young, later hardy. Prefers shading for the first few years but then is intolerant of shade. Withstands moderately dry conditions once established. Leaves 3–5 cm long, 5–10 mm wide, thick, almost sessile and opposite, veins indistinct; cones woody, broadly globular, 5–8 cm diam. 1823.

*A. robusta* (A. brownii)—**Queensland Kauri**   E.C. Qld
Seed

| | | | |
|---|---|---|---|
| 10–15 m | 650 mm | Prefers good soil | Moderate to slow |

Parks, avenues

Upright, compact, and neat tree; growing to over 30 m in Queensland and producing excellent timber. Good specimen tree. Fairly hardy, but may be damaged by frost until about 2–3 m tall. Leaves 5–12 cm long, 2–4 cm wide, thick, parallel veins; cones woody, almost globular, 8–12 cm long.

# ALBIZIA

Named after Albizzi, an Italian naturalist. There are about thirty species from the warmer regions of Europe, Asia, Australia and Mexico.

*A. julibrissin*—**Silk Tree**   D.H.          Persia to Japan
Seed
6–8 m       550 mm      Will grow on light,      Fast
                        sandy soils
Parks, avenues

Shapely tree with jacaranda-like foliage, suitable for the warmer districts. Hardy. Moderately frost-resistant. Leaves fern-like, 10–15 cm long, bi-pinnate, leaflets oval, 1 cm long, mid-rib to one side, leaflets close together at night; flowers yellow or pink in large fluffy clusters; fruit a flat legume, 15 cm long, 1–2 cm wide. 1745.

*A. lophantha*—**Cape Wattle**   E.H.          W.A.
Seed
4–8 m       600 mm                          Very fast
Low shelter, parks

Large shrub or small tree resembling a wattle. Can be pruned. Very fast-growing. May spread from seed on light soils. Withstands moderate coastal exposure. Foliage pinnate, 20–30 cm long, 20–30 pairs leaflets; flowers yellow, like an acacia, arranged in dense spikes 5–8 cm long; fruit a flat legume, 7–10 cm long, 1–2 cm wide. 1803.

*Below   Ailanthus altissima*

# ALNUS—Alders

Seed                    Will grow in wet sites
Windbreaks, parks, roadsides, gully erosion, stream banks
*Alnus* is the old Latin name. There are about thirty species from Europe, Asia, North America and southwards to Peru. The trees are slender and attractive, and similar to silver birch. They will grow in poorly-drained and wet areas and are good for stream bank protection and swampy sites. They are hardy and frost-resistant with a vigorous, surface, fibrous root system. Large specimens can be transplanted. Cut stumps will coppice well. They prefer a sunny position and are not very tolerant of shade. Roots have nitrogen-fixing properties.

*A. cordata* (A. cordifolia)—**Italian Alder**   D.H.
                                        S. Europe
9–12 m      600 mm                          Fast

Well-shaped tree with a long trunk and oblong crown. Can be over 20 m with access to water. Branchlets angled. Withstands cold and light snow. Can be heavily trimmed. Bark grey; leaves oval, delta shape, 10 cm long, finely toothed; flowers in catkins, 5–8 cm long; fruit woody, cone-like, 2–3 cm long, usually in threes; seed with corky wings, float long distances on water. Purple-leaved varieties are available. 1820.

*A. glutinosa*—**Common Alder**   D.H.      Europe, Asia
10–12 m     650 mm                          Fast

Slender tree; rarely upright, usually leaning. Withstands cold and light snow. Good for erosion control in wet gullies. Can be heavily trimmed. Bark grey-white, birch-like; leaves rounded, 5–10 cm long, notched at the apex and tapering towards the stalk; fruit a small woody cone-like structure, 1–2 cm long, remaining on the tree for several years.   L.C.

*A. jorullensis*—**Evergreen Alder**   E.H. Central America,
                                        Mexico
6–10 m      650 mm                      Moderate

Relatively recent introduction. Slender tree with a grey-white trunk. In cooler climates leaves usually remain on the tree during winter. Bark grey-white, papery; leaves 5–8 cm long, 2–4 cm wide, serrated, pointed at the apex; fruit like a woody cone, about 2–3 cm long.

# ALYXIA

*Alyxia* is the native Indian name. There are about fifty species from tropical Asia, Indonesia and Australia.

149

*Alnus jorallensis*

*A. buxifolia*—**Sea Box**   E.H.                   W.A., S.A., Vic.,
Seed, cuttings                                     Tas., N.S.W.
1–2 m        600 mm                                Fast
Low shelter, hedges

Small, compact, easily-grown shrub. Frost-resistant. Withstands severe coastal exposure; will grow readily on beach sand. One of the best plants for low shelter in exposed coastal areas. Can be trimmed to form a dense hedge. Leaves oval, 1–2 cm, glossy; flowers small, white, not conspicuous; fruit orange berry, 5–7 mm diam.

## AMELANCHIER

This is the French name for a European species, *A. ovalis*. There are about twenty-five species which are mainly North American, but also occur in Europe and Northern Asia.

*A. canadensis*—**Shadbush**   D.H.            North America
Seed
3–6 m        750 mm                                Fast
Low shelter, parks, roadsides, ornamental, birds

Erect, compact, and attractive small tree more suitable for the cooler districts. Old trees may exceed 10 m in height. Not widely grown but worth more attention; one of the finer autumn foliage plants. Fibrous surface root system. Hardy. Withstands dry but not exposed positions. Can be heavily pruned. Large specimens can be transplanted when leafless. Liable to attack by the pear slug. Leaves oval, 5–8 cm long, woolly underneath (young leaves woolly both sides), usually serrated, golden yellow to orange and red in autumn; flowers white, 2–3 cm, prolific, in clusters, appear before the leaves in spring; fruit blackberry-like, purple-black, edible. Seed very short-lived. 1623.

*Right    Angophora costata and bark*

## ANGOPHORA

Derived from *aggos* the Greek for a vessel and *phero* the Greek verb to bear, it refers to the shape of the fruit. There are about ten species from Australia.

*A. costata* (A. lanceolata)—**Gum Myrtle**   E.H.   N.S.W.,
Seed                                                        Qld
15–20 m      650 mm      Will grow on poor,        Fast
                         rocky, and heavy soils
Windbreaks, parks, roadsides, honey

Eucalypt-like tree, pyramidal when young, spreading when older. Fast growing on good soils, slow on poor rocky sites. Wind-firm. Withstands moderate coastal exposure. Around Sydney known as red gum. Bark orange-reddish, smooth, deciduous, very attractive particularly in summer; leaves 7–15 cm long, 2–4 cm wide, gum-like, generally opposite; flowers yellow-white, terminal clusters of 4–8 flowers, 5 cm across; fruit 1 cm long, oval, ribbed, similar to eucalypts. 1816.

*Araucaria araucana—young tree*

*A. araucana* (A. imbricata)—**Monkey Puzzle Pine**

Chile

20–30 m    750 mm    Well-drained soils    Moderate

Roadsides, parks, farm forests

Large, distinctive, spreading tree with a rather open crown. Suitable for well-drained soils in cool, moist climates. Will grow on poorer sites but can be ragged in appearance. Hardy and moderately frost-resistant. Very popular during the late 1800s but now out of fashion. Leaves flat, 5 cm long, 1–2 cm wide, crowded and over-lapping, fine-pointed; male catkins erect, 7–12 cm long and 5 cm wide; cones globular, 10–15 cm; seeds large 2–4 cm long, 1 cm thick. 1795.

*A. bidwillii*—**Bunya Bunya Pine**    Qld

20–25 m    750 mm    Deep soils    Moderate

Windbreaks, parks, roadsides, farm forests

Large spreading tree, dome-shaped crown to ground level when young; reduced to a wide tuft at the top when old. Slow-growing when young. Withstands moderate coastal exposure. Produces good timber. Leaves flat, 5 cm long, 6–12 mm wide, with a stiff point; male catkins pendulous, conspicuous, 10–20 cm long, 1 cm wide, at the ends of the branches; cones very large, up to 30 cm long, 20 cm wide, and up to 4 kg weight; seeds very large, edible. 1843.

*Araucaria bidwillii—young tree*

*A. floribunda* (A. intermedia)—**Rough Bark**    Vic.,
**Gum Myrtle**    N.S.W.,
E.H.    Qld

Seed

10–12 m    650 mm    Fast

Windbreaks, parks, roadsides, honey

Tree with a spreading crown, branches frequently crooked. Will grow on sandy soils. Withstands moderate coastal exposure. Can be trimmed. Bark rough, otherwise similar to *A. costata* above.

# ARAUCARIA    E.C.

Seed

*Araucaria* is taken from *Araucanos* its Chilean name. There are about eighteen species from Australia, south-western Pacific islands, New Zealand and South America. They are evergreen conifers of tree dimensions, which are usually tropical or sub-tropical. Branches are horizontal and in whorls. The trees are generally frost-sensitive and slow-growing when young, but once established growth and drought-resistance are good. Provided adequate attention is given when young and some irrigation water is available, they can be grown in the warmer districts. They may coppice slightly from cut branches and stumps. Cones are large and disintegrate when the seeds ripen.

# ARBUTUS

*Arbutus* is the old Latin name. About twenty species occur from Europe and North America.

*A. menziesii*—**Madrone**   E.H.          W. North America
Seed

10–12 m     650 mm                          Moderate
Windbreaks, parks

Attractive, often multi-stemmed tree; narrow crown when young, broader later. Grows to 25 m in its native habitat. Not very common but worth much more attention. Frost-resistant, tolerant of shade, resists dry conditions once established. Re-shoots vigorously from the roots. Hard to transplant, planting-out losses can be high. Bark attractive, bright reddish-brown to orange, peeling off in large flakes; leaves thick, oval, 7–12 cm long, 4–6 cm wide, shining green, silver underneath; flowers in upright clusters, 15–20 cm high, individual flowers pitcher shape; fruit a berry, bright orange, globular, 1 cm diam. 1827.

Left   *Araucaria cunninghamiana*

Below   *Araucaria heterophylla*

*A. cunninghamii*—**Hoop Pine**          N.S.W., Qld
25–35 m    600 mm    Prefers deep soils    Moderate
Parks, roadsides, farm forests

Fairly narrow tree, foliage at the ends of branches giving the tree a tufted appearance. Prefers deep moist soils but will grow on poorer soils. Frost-sensitive. Withstands slight coastal exposure. Leaves lance-shaped, 5–10 mm long, overlapping and crowded on the stem; male catkins 5–8 cm long; cones oval, 10 cm long, 7 cm wide.

*A. heterophylla* (A. excelsa)—**Norfolk Island**   Norfolk
                              **Pine**              Island
20–30 m    600 mm    Will grow on sandy    Moderate
                            soils
Parks, roadsides, farm forests, ornamental

Very symmetrical and open crown, branches widely spaced in horizontal, regular whorls. Essentially ornamental. Withstands moderate coastal exposure and is widely planted along beaches. Frost-sensitive. Wind-firm. Can be planted in the drier areas if irrigation water is available. Slow-growing for the first few years. Long-lived, over 100 years. Foliage soft, bright green, leaves 5–10 mm long, spreading on the younger shoots, overlapping on older; male catkins 2–4 cm long; cones globular, 7–10 cm diam. sometimes wider than long. 1793.

*Arbutus unedo*

*A. unedo*—**Irish Strawberry Tree**  E.H.      S. Europe,
Seed                                              Ireland
3–6 m      650 mm                                    Fast
Low shelter, parks

Bushy crown, compact shrub or small tree with persistent branches. Mainly planted as a garden shrub, but useful for low wind-breaks, or under-planting. In good sites can be taller than 5 m. Hardy, growing on most soils including calcareous types. Long-lived. Withstands slight to moderate coastal exposure. Can be trimmed to form a hedge. Leaves deep green, oval, serrated, 7–10 cm long; flowers similar to lily-of-the-valley; fruit globular 'strawberry', 1–2 cm diameter, red, rough, edible, but narcotic in quantity.   L.C.

## ATRIPLEX

*Atriplex* is derived from an old Greek name. Over one hundred species occur in temperate and sub-tropical regions of all continents.

*A. nummularia*—**Old Man Saltbush**  E.H.    S.A., Vic.,
Seed                                          N.S.W., Qld
2 m      300 mm      Prefers lighter soils           Fast
Low shelter, fodder, hedges

Large, compact shrub. Grows under hot dry conditions and on saline soils. Frost- and drought-resistant. Can be trimmed as a hedge. Good fodder plant. Leaves broad to round, greyish, semi-succulent, 1–3 cm diameter; fruit fan-shaped, sessile, about 5 mm long.

## AZARA

A genus named after J. N. Azara, a Spanish patron of science. There are about twelve species from Chile.

*A. microphylla*—**Box Leaf Azara**  E.H.         Chile
Seed, cuttings
3–6 m      900 mm      Prefers good soils      Moderate
Parks, streets, roadsides, ornamental

Attractive, compact, erect, small tree, rather like the native myrtle beech (*Notho-fagus*). More suitable for the cool mountain districts. Will not tolerate poorly-drained sites. Hardy, frost-resistant. Shade-tolerant. Can be pruned. Large specimens can be transplanted. Branchlets slender, brown felted; leaves broadly oval, 1–2 cm long, in pairs, one smaller than the other, deep shining green; flowers very small, without petals, yellow stamens, strongly vanilla scented, scent obvious some distance away; fruit a small mauve to dark reddish globular berry. 1861.

## BAECKEA

A genus named after A. Baeck, a Swedish physician and friend of Linnaeus. There are about sixty species which are mainly Australian.

*B. virgata*—**Tall Baeckea**  E.H.       Vic., N.S.W., Qld,
Seed                                  N.T., New Caledonia
2–3 m      800 mm                     Moderate to fast
Low shelter, parks, ornamental

Moderately compact shrub, suitable for a wide range of soils. Hardy, easily grown. Leaves 1–3 cm long, narrow, opposite; flowers white, small, in clusters of 2–8, summer-flowering. 1806.

## BANKSIA  E.H.

Seed              Require lime-free soils
*Banksia* was named after Sir Joseph Banks, the naturalist on Cook's first voyage to Australia. About sixty species, all Australian, are known. They are mainly shrubs and small trees with bottlebrush-like flowers and woody cones with distinctive woody valves. They are hardy and can withstand some coastal exposure and shade. A wide range of soils are suitable but lighter well-drained types are preferred. Leaves are deep green and usually silver or whitish underneath. They are good honey trees and attract birds.

*B. ashbyii*                                          W.A.
2–4 m      250 mm      Prefers light soils      Moderate
Parks, roadsides

Small tree or dense shrub, more suitable for the warmer districts. Frost- and drought-resistant. Some forms may be only 1 m tall. Leaves long and narrow, 15–20 cm long, 2–3 cm wide, lobed more than half way to the midrib; flowers bright yellow to orange, spikes 10–15 cm long, 7–10 cm diam.

*B. attenuata*                                    W.A.
3–6 m     650 mm     Prefers sandy soils     Moderate
Low shelter, parks                        New introduction

Upright, small tree, developing a spreading crown when older. Relatively new introduction, indications are that it is hardy and suitable for a wide range of soils. Leaves 8–15 cm long, narrow, thick, coarsely serrated, white underneath; flowers 15–20 cm long, bright yellow to yellow-orange, 'untidy' appearance compared with other banksias.

*B. ericifolia*—**Golden Banksia**              N.S.W.
2–3 m     500 mm     Will grow on          Moderate
                     moist soils
Windbreaks, low shelter

Shapely, dense bush, or small tree. Easy to grow. Prefers lighter soils. Wind-resistant. Withstands moderate coastal exposure. Can be pruned. Leaves linear, 1–4 cm long, 1 mm wide, closely packed; flowers very conspicuous, yellow to orange, large, 15–20 cm long, 5–8 cm diameter, scattered through the foliage. 1788.

*B. grandis*—**Bull Banksia**                   W.A.
5–6 m     560 mm     Will grow on          Moderate
                     deep sands
Parks, roadsides, streets

Single-stemmed, slightly spreading, small tree. Foliage confined to the outer branches in mature trees. Not suitable for heavy soils, requires good drainage. Hardy. Tolerates some shade. Withstands moderate coastal exposure but is then usually a smaller tree. Leaves distinctive, 15–40 cm long, 3–8 cm wide, divided to the midrib into triangular lobes 5 cm wide, 2–4 cm deep, underside hairy; flowers large, 20–30 cm long, sometimes 40 cm, 7–10 cm wide, yellow, dense; fruit, 20–40 cm long, 7–10 cm diam. 1794.

*B. ilicifolia*—**Holly Leaf Banksia**          W.A.
5–6 m     750 mm     Will grow on          Moderate
                     deep sands
Windbreaks, shade

Variable, small tree, usually spreading but upright forms occur. May be only a shrub on unfavourable sites. Hardy. Wind-firm. Withstands moderate coastal exposure. Requires good drainage. Leaves 5–8 cm long, dark green, glossy, holly-like; flowers usually yellow to orange when mature, arranged in a globular head, 4–5 cm diam. (not a cylindrical spike as in other banksias); fruit small, only 2–5 cm long.

*B. integrifolia*—**Coast Banksia**       Vic., N.S.W., Qld
8–10 m    600 mm     Will grow on              Fast
                     coastal sands
Windbreaks, sand drift

Hardy, compact tree when open grown. Withstands severe coastal exposure. Prefers well-drained soils and will grow on coastal sands. Leaves 10–15 cm long, 2–3 cm wide, smooth margin; flowers greenish-yellow, large, 15–20 cm long. 1788.

*B. littoralis*—**Swamp Banksia**               W.A.
3–10 m    550 mm     Will grow in             Fast
                     swampy soils
Windbreaks, parks, roadsides

Small, low-branching tree, often spreading. Tends to become straggly in unfavourable sites. Not suitable for heavy soils but will survive in swamps or sandy areas for 2–3 months during winter. Leaves 10–20 cm long, narrow, deeply serrated, end square, white underneath; flowers yellow, 15–20 cm long, in slender spikes, autumn and winter-flowering. 1822.

*B. marginata*—**Silver Banksia**          S.A., Tas.,
                                           Vic., N.S.W.
5–6 m     500 mm     Will grow on sandy soils     Fast
Windbreaks, parks, sand drift

Hardy, compact, small tree. Wind-firm. Withstands severe coastal exposure. Very suitable for planting on exposed coastal sands. Leaves 5 cm long, 1–2 cm wide, silver underneath, square (cut-off) tip, margins may be entire or serrated; flowers greenish-yellow, 5–8 cm long, 2–4 cm diameter. 1822.

*B. menziesii*—**Firewood Banksia**             W.A.
3–8 m     500 mm     Will grow in          Moderate
                     sandy soils
Windbreaks, parks, roadsides

Small, bushy tree with woolly branches. Tends to become straggly without shaping occasionally. Withstands moderate coastal exposure. Leaves 15–20 cm long, up to 3 cm wide, narrow, serrated, blunt end, woolly, dark green; flowers bright red to reddish-yellow, 10–15 cm long, broad, conspicuous, autumn-flowering.

*B. ornata*—**Desert Banksia**             S.A., Vic.
1–2 m     350 mm                          Moderate
Low shelter, parks, birds

Hardy, bushy shrub, very suitable as a garden shrub for hot dry conditions. Requires trimming occasionally. Leaves 7–12 cm long, rigid, broadly serrated; flowers globular to oval, 7–15 cm diameter, pale yellow.

*B. prionotes*—**Orange Banksia**               W.A.
5–6 m     500 mm     Will grow on          Moderate
                     heavier soils
Shelter, birds, ornamental

Bushy, stout-stemmed, small tree, branching from near ground level. May spread to 3–5 m across. One of the showiest of the banksias with its bright orange flowers. Prefers lighter soils and warm conditions. Leaves 20–30 cm long, divided into triangular lobes; flowers large, 15–20 cm long, 5–8 cm diam., very conspicuous.

*Banksia prionotes—flowers*

*B. serrata*—**Saw Banksia**                    Tas., Vic., N.S.W.
5–8 m          650 mm          Will grow in poor,          Moderate
                                sandy soil
Windbreaks, parks

Erect, rather massive, small tree. Strong-growing, wind-resistant. Withstands severe coastal exposure. Leaves 10–15 cm long, 1–2 cm wide, leathery, white underneath, deeply serrated; flowers greyish-yellow to reddish, large, 8–15 cm long, 7–10 cm diam.; fruit cylindrical, silvery-grey.

*B. spinulosa* (B. collina)—**Hairpin Banksia** or          Vic.,
                             **Hill Banksia**          N.S.W.
1–3 m          500 mm                            Moderate
Low shelter

Neat, rather upright and compact bush. Will grow on poorly-drained soils. Leaves very narrow, serrated, flat, 5–8 cm long, 3–6 mm wide; flowers yellow-amber, 12–15 cm long with hooked, purplish stamens, 5–8 cm long. 1788.

# BETULA—Birch    D.H.

Seed

Parks, roadsides, ornamental, autumn foliage

*Betula* is the old Latin name. About sixty species occur in northern Europe, Asia, and North America. The trees are slender, graceful and upright with pendulous branchlets. They will grow on poorly-drained soils but not as well as the closely-related and similar alders (*Alnus* sp.). Hardy and wind-firm, they will withstand snow and cold and are suitable for sub-alpine planting. They are tolerant to sun, some shade and frost. The shallow, fibrous root systems allows them to be grown closer to underground drains than most trees. Large specimens can be transplanted in winter. Being suitable for impoverished areas, in native habitats they are important pioneer species. The bark is papery and is furrowed towards the base on older trees. Leaves

are broadly oval to triangular, pointed and serrated, and have relatively long stalks. Flowers are inconspicuous and in catkins 2–5 cm long. The fruit is catkin or cone-like and the seeds are minute nuts (about 1000 per gm) with two broad, papery wings. They are very short-lived and require to be sown within a few days of falling. Of the four species described, Silver Birch (*B. pendula*) is the one usually planted, the other three species are uncommon. A leaf rust fungus, *Melampsoridium betulinum*, has recently appeared in Victoria but it can be easily controlled by dusting with sulphur.

*B. lutea*—**Yellow Birch**                    North America
12–20 m          750 mm          Prefers moist sites          Moderate
Farm forests

Upright, single-stemmed tree, reaching 30–40 m in its native habitat. Largest of the birches. Produces good timber. Grows on a wide range of soils, including sandy loams but prefers moist situations. Hardy, frost-resistant, more suitable for the colder districts. Bark brownish-yellow, flaky; leaves large, oval, 7–12 cm long, 4–5 cm wide, yellow-green underneath. 1800.

*B. maximowiczii*—**Regel**                    Japan
10–15 m          750 mm                            Fast

Probably the most attractive of the birches because of its bark and large leaves. Bark conspicuous yellow-orange; leaves large, 7–12 cm long, 5–10 cm wide, deeply indented at the base; fruit a slender, 'nodding' cone. 1800.

*B. papyrifera*—**Canoe Birch**                    North America
8–12 m          750 mm                            Fast

Pyramidal when young, often developing a crooked trunk. Rather open crown. Bark very white, waterproof (used for canoes); leaves 5–8 cm long, 4–5 cm wide, pale green underneath; fruit catkin-like. 1750.

*B. pendula* (B. alba, B. verrucosa)          Europe,
          —**Silver Birch**          North America
5–10 m          650 mm

Attractive tree, with graceful foliage. Widely planted in south-eastern Australia. Will tolerate wet sites better than most birches. Leaves triangular, 3–6 cm long; fruit catkin-like. Numerous horticultural varieties are available. L.C.

# BRACHYCHITON

Seed

*Brachychiton* is derived from the Greek *brachy* meaning short and *chiton* meaning a covering which refers to the short hairs covering the pod, etc. There are about eleven

*Brachychiton populneus—young tree*

Below    *Brachychiton populneus—mature tree*

species all from Australia. The trees are upright with attractive foliage and are more suited to the warmer inland areas. While frost-tender when young, they later become moderately hardy.

*B. acerifolius* (Sterculia acerifolia)—**Flame**        N.S.W.,
                                    **Tree**   D.H.       Qld
6–10 m        650 mm       Will grow on heavy soils     Slow
Shade, parks, avenues, ornamental

Compact, pyramidal tree. Flowers very showy, best displays occur in hot, dry seasons and about every 6–7 years. Leaves maple-like, 10–20 cm across on long stalks up to 30 cm long, very variable, usually divided into 3 lobes, most dropping before flowering; flowers bell-shaped, 1–3 cm across, vivid orange-red, in terminal sprays 20–30 cm long, trees sometimes 20-years-old before flowering; fruit boat-shaped pod, 8–12 cm long; seed, oval, 1 cm long.

*B. discolor*—**Queensland Lacebark**   D.H.      N.S.W.,
                                                  Qld, N.T.
5–15 m     600 mm      Will grow on heavy        Slow to
                        soils                     moderate
Parks, avenues

Compact, slightly rounded tree. Particularly beautiful when in flower. Moderately hardy but withstands light frosts only. Leaves 15–20 cm across, usually dissected into 5 lobes; flowers large, 5 cm long, pinkish, brown felt behind, falling soon after opening; fruit boat-shaped pod, 8–12 cm long.

*B. populneus* (Sterculia diversifolia)—**Kurrajong**   Vic.,
                              E.H.   N.S.W., Qld,
                                              N.T.
10–12 m     350 mm                              Slow
Shade, avenues, parks, fodder

Symmetrical, densely-crowned tree with a strongly tapering bole, considerably swollen at the base. Suitable for lower rainfall zones if irrigation water is available in the first few years; very drought-resistant once established. Will grow on limestone and sandy soils. Very versatile. Deep-rooting. Crops can be grown close to the trunk. Valuable fodder tree, can be lopped and trimmed. Leaves very variable, from simple to deeply lobed, 5–15 cm long, 2–12 cm across depending on dissection into lobes; flowers bell-shaped, white, spotted reddish brown, 1–2 cm long with five 'petal' points; seeds yellow, in brown boat-shaped pods, 5–8 cm long.

# BURSARIA

The name is derived from *bursa* the Latin word for a pouch, which refers to the shape of the fruit. There are two species—one Australian and one from the Phillipines.

*Bursaria spinosa—flowers*

*Callistemon citrinus—flowers*

*Callistemon citrinus—young spring foliage*

*B. spinosa*—**Sweet Bursaria**   E.H.   Vic., N.S.W.
Seed
3–6 m       550 mm                           Fast
Low shelter, hedgerows, honey

Compact, thorny shrub or small tree with a short, often crooked trunk. Leaves, frequency of thorns, and height very variable. Often straggly in nature but compact under cultivation. Hardy, frost-resistant. Not widely planted but worthy of more extensive planting for hedgerows and low shelter. Can be trimmed, forms a good hedge. Leaves very variable, usually oval, 3–4 cm long, stalkless; flowers white, small, abundant, scented; fruit a flat capsule, purse-like, 5–8 mm long, produced in quantity. 1793.

# CALLISTEMON—Bottlebrushes   E.H.

Seed               Tolerate poor soils            Fast
Low shelter, parks, streets, ornamental, birds
*Callistemon* is derived from the Greek *kallos* meaning beauty and *stemon* meaning stamen. There are about forty species occurring in Australia. They are hardy shrubs or small trees and grow in a wide range of soils and climates. They tolerate poor soils and hot conditions and will stand some degree of coastal exposure and shade. They can be trimmed and become straggly and unattractive if not shaped every few years. They do not transplant readily and the honey is not particularly good. Leaves are usually crowded and linear. Flowers are spikes like a bottlebrush and are conspicuous and attractive. Seed capsules are sessile, globular (3–7 mm diam.), more or less crowded on the stem and may not open for several years. The seed is very small and almost like dust.

*C. acuminatus*—**Coast Bottlebrush**            N.S.W.
2–3 m       750 mm

Low, round-headed, not widely planted, tree. Wind-resistant. Best in the warmer coastal areas. Leaves 7–10 cm long, 1–2 cm wide, smooth, wavy margins; flowers crimson, 6–8 cm across, with conspicuous leafy bracts in each brush.

*Callistemon citrinus—Gawler hybrid*

*C. citrinus* (*C. lanceolatus*)—**Crimson**   Vic., N.S.W.,
                                       **Bottlebrush**       Qld
2–3 m       550 mm      Will grow on heavy soils

Hardy, best known of the bottlebrushes. Occasionally a small tree to 5 m. Withstands moderate coastal exposure. Grows on almost any soil. Flowers twice a year under good conditions. Requires trimming every 2–3 years otherwise it becomes straggly and untidy. Can be trimmed to form a hedge. Leaves 5–10 cm long, 5–15 mm wide, reddish when young, strong lemon odour when crushed; flowers rich red, 7–10 cm long. Several horticultural forms available. 1788.

*Callistemon pallidus—flowers*

*Callistemon rigidus—flowers*

*C. lilacinus* (C. violaceus)—**Lilac Bottlebrush**   N.S.W.
2–3 m      450 mm

Erect shrub. Withstands moderate coastal exposure. Leaves 5–10 cm long, 1–2 cm wide, smooth, thin; flowers small, unusual, stamens purplish, tipped with gold.

*C. linearis*—**Narrow Leaf Bottlebrush**   N.S.W.
2–3 m      500 mm

Erect, bushy shrub. Very narrow leaves, 8–12 cm long, 2–3 mm wide, stiff; flowers rich crimson, 8–12 cm long, 5 cm wide.

*C. macropunctatus* (C. rugulosus, C. coccineus) S.A., Vic.,
—**Scarlet Bottlebrush**   N.S.W.
3–4 m      450 mm      Will grow on wet soils

Attractive, hardy shrub; best on light soils. Drooping branches. Good street tree. Leaves 3–6 cm long, 3–6 mm wide, coppery-red when young; flowers crimson-red, 7–10 cm long.

*C. pallidus*—**Lemon Bottlebrush**   Tas., Vic., N.S.W.
3–5 m      650 mm

Small, upright tree with slightly pendulous branchlets. Leaves hairy and silvery when young, smooth when older, 3–6 cm long, 1–2 cm wide; flowers 5–8 cm long, pale yellow.

*C. paludosus*—**River Bottlebrush**   S.A., Tas., Vic.,
N.S.W.
3–5 m      600 mm

Tall, rather straggly shrub. Grows well on river banks but also suitable for wet and poorly-drained soils. Best of the bottle-brushes for these sites. Only about 1 m tall under very adverse conditions. Leaves narrow, 5–10 cm long, 5–6 mm wide; flowers creamy-yellow, 5–8 cm long, 2–3 cm wide.

*C. phoeniceus*—**Fiery Bottlebrush**   W.A.
3–4 m      500 mm

Handsome, showy shrub; worth wider planting. Leaves narrow, 8–10 cm long, 5–7 mm wide; flowers rich scarlet, conspicuous, 10–12 cm long, 5–7 cm wide, contrasting with the greyish foliage.

*C. pinifolius*—**Green Bottlebrush**   N.S.W.
2–4 m      500 mm

Attractive, fine-foliaged, small tree. Hardy. Tolerates light shade. Leaves very narrow, stiff, up to 10 cm long, upper surface grooved; flowers green.

*C. rigidus*—**Stiff Bottlebrush**   N.S.W., Qld
2–3 m      300 mm

Erect, stiff shrub. Suitable for hot dry areas. Leaves rigid, flat, pointed, narrow, 7–12 cm long, 5–9 mm wide; flowers deep red, tinged with green, 8–10 cm long, 6–8 cm wide. 1815.

*C. salignus*—**Willow Bottlebrush**   S.A., N.S.W., Qld
5–7 m      500 mm

Small tree with a slender trunk and many upright branches. Will grow on brackish sites. Good street tree. Branchlets semi-pendulous; bark whitish, papery; leaves willow-like, coppery when young, grey-green when mature, 5–8 cm long, 1 cm wide, with a prominent midrib; flowers pale yellow, occasionally red, 5–8 cm long, 2–3 cm wide. Pink-flowering and red-flowering horticultural varieties are available. 1788.

*C. shiressii*   N.S.W.
3–5 m      700 mm

Dense, compact, spreading plant. Prefers some summer moisture, otherwise hardy. Pruning after flowering is an advantage. Leaves 10–15 cm long, 1–2 cm wide, broadly serrated, young leaves bright red; flowers cream, terminal, 5–10 cm long, spring-flowering but often in autumn as well.

*C. sieberi*—**Alpine Bottlebrush**          Vic., N.S.W.
2–3 m          650 mm          New introduction

Recent introduction, growth under cultivation still to be found out. Withstands cold conditions. At present not recommended for warm areas. Leaves almost needle-like, 1–3 cm long; flowers pale yellow, 2–5 cm long.

*C. speciosus*—**Albany Bottlebrush**          W.A.
2–3 m          600 mm

Stiff, erect shrub, suitable for wet and swampy conditions. Largest and most conspicuous flower of the bottlebrushes in general cultivation, deep scarlet, 8–15 cm long, 5–8 cm diam. Leaves, thick, narrow, 7–10 cm long, 5–10 mm wide, prominent rib and margin, stiff and crowded on the branches. 1823.

*C. teretifolius*—**Flinders Range Bottlebrush**          S.A.
2–3 m          300 mm

Small, compact, attractive, desert shrub, suitable for hot dry conditions. Leaves needle-like, 5–10 cm long; flowers crimson, 5–8 cm long, 2–3 cm diam.

*C. viminalis*—**Weeping Bottlebrush**          N.S.W.
5–10 m          550 mm

Large shrub or small neat tree, with a somewhat open crown. Largest of the bottlebrushes. Very attractive, and moderately hardy. Branches upright when young becoming semi-pendulous when older. Usually grows along streams but will tolerate adverse sites. Bark dark grey, furrowed; leaves 4–8 cm long, very narrow, 5–7 mm wide, bronze-red and downy when young; flowers crimson, 5–10 cm long, prolific.

A compact dwarf form reaching 1–1·5 m named Captain Cook was introduced as the Shrub of the Year 1970–71 by the Australian Selection Committee.

# CALLITRIS—Native Pines          E.C.

|  |  | All States |
|---|---|---|
| Seed | Prefer light, well-drained soils | Moderate |

*Callitris* is from *kallos* the Greek for beauty, referring to the tree. There are about twenty species from Australia. These neat, compact trees have attractive green or bluish foliage and semi-pendulous outer branchlets. They are not suitable for heavy clays. Being fibrous rooted, relatively large specimens can be transplanted. Should not be trimmed and are moderately frost-resistant. While often slow for the first 2–3 years. the growth is then moderate to fast. The timber is fragrant, durable and resistant to white ants. Leaves are very small (5–10 mm long) and reduced to scales in whorls of two to four closely pressed around the fine branchlets giving a 'jointed' appearance. Fruit is a cone, globular to oval, with the cone scales opening outwards from a common point at the base. There are usually six scales with a large and a small scale alternating. Seeds are short-lived; viable for about one year only. The botanical names of *Callitris* species have been confused. The synonyms listed are believed to be correct, but if not strictly correct the differences are not significant for the purposes of this publication.

*C. columellaris* (C. glauca, C. arenosa,          W.A., S.A., Vic.,
          C. intratropica, C. robusta)          N.S.W.,
          —**Murray Pine**          Qld
10–20 m          300 mm          Will grow on calcareous soils
Windbreaks, shade, farm forests, parks, roadsides

Compact, upright, pyramidal, rather formal tree with fine, bluish foliage. Hardy and attractive. Rich, dark brown bark. Will grow in exposed positions. Withstands moderate coastal exposure. Drought-resistant. Branchlets end in a fine spray; leaves deep green, very small; cones 1–2 cm diam.

*C. endlicheri* (C. calcarata)—**Black Cypress Pine**          Vic.,
          N.S.W., Qld
12–15 m          350 mm
Windbreaks, shade, parks, roadsides, farm forests

Attractive, upright tree with bright green foliage. Prefers warm conditions and will grow in sandy (particularly granitic) soils. Leaves small; cones oval, 1 cm long.

Below          *Callitris columellaris*

*Callitris endlicheri*

*C. oblonga*—**Tasmanian Cypress Pine**                Tas.

3–5 m      500 mm      Prefers well-drained soils

Low shelter, avenues, ornamental

Conical shape, bushy, dense, decorative; good centre piece for a garden or as a background plant. Needs to be planted closely to give shelter. Foliage grey-green, branchlets divided into fine sprays; leaves very small; cones oval, 1–2 cm long.

Below    *Callitris preissii*

*C. preissii* (C. robusta, C. gracilis, C. hugelii,        W.A. C. propinqua)—**Slender Cypress**  S.A., Vic., **Pine**          N.S.W.

20–25 m      550 mm      Will grow on limestone soils

Windbreaks, shade, roadsides, farm forests

Similar in form to *C. columellaris*. Foliage bronze to light green rather than bluish. Leaves very small; cones large, 2–4 cm long.

*C. rhomboidea* (C. tasmanica, C. cupressiformis)        S.A., —**Oyster Bay** or        Tas., Vic., N.S.W., **Port Jackson Cypress Pine**        Qld

6–8 m      450 mm      Requires well-drained soils

Low shelter, parks, roadsides

Attractive, upright, blue-green tree. Growth slow at first but improves later. Good ornamental. Withstands moderate coastal exposure. Can be trimmed. Branchlets comparatively coarse and angular. Foliage brownish red or bronze in winter. Leaves small, fine; cones 1 cm long, clustered, prominent point on each of the 6 valves.

*C. verrucosa*—**Mallee Pine, Scrub Pine** or        S.A., Vic., **Turpentine Pine**        N.S.W.

5–7 m      250 mm

Low shelter, hedges

Spreading, bushy tree suitable for low windbreaks. Drought-resistant, Can be trimmed. Foliage bright green, leaves very small; cones 2–3 cm diam. very rough and warty.

# CALOCEDRUS

There are about ten species from North and South America, New Zealand, south-western Pacific islands, China and Taiwan; although some authorities believe they should be assigned to the genus *Libocedrus*.

Below    *Calothamnus gilesii*

*C. decurrens* (Libocedrus decurrens)
——**Incense Cedar**   E.C.          Western USA
Seed

| 15–25 m | 900 mm | Will grow on heavy soils | Moderate to slow |

Parks, roadsides, avenues, ornamental

Tall, compact, pyramidal to columnar tree with a strongly tapering trunk. Grows to 70 m in its native habitat. Requires fertile, moist well-drained soil for good growth. Produces high quality, durable and aromatic timber. (Growth is too slow locally to be worth planting in farm forests.) More suitable for the cooler mountain districts. Very hardy. Frost-resistant. Cannot be trimmed. Withstands heavy snow but not smog. Bark fibrous, reddish-brown; foliage in flattened sprays, leaves dark green, 5–10 mm long, scale-like, pressed closely to the stem and arranged in fours; cones pendulous, 2 cm long, 1 cm wide, 6 scales, spreading wide when open.

# CALODENDRON

The name is derived from *kalos* the Greek for beautiful and *dendron* meaning a tree. There is one species from South Africa.

*C. capense*——**Cape Chestnut**   E.H.          S. Africa
Seed

| 10–12 m | 600 mm | Well-drained soils, but not sands | Slow |

Ornamental, parks, gardens, avenues, roadsides

Attractive tree with a dense, compact and rounded, symmetrical crown. Clean straight trunk, lower branches are shed. Prefers the warmer areas, but many good specimens can be seen in Melbourne. Can be grown where irrigation water is available. Frost-tender when young. Withstands slight coastal exposure. One of the showiest of all trees, growth slow when young but still worth planting. Leaves simple, glossy, 10–15 cm long, 5–7 cm wide; flowers whitish-pink, in large showy, terminal sprays, 15–20 cm high and 10–15 cm across, covering almost the entire crown; fruit a nut, 2 cm diam. 1789.

# CALOTHAMNUS—Net Bushes   E.H.

Seed                                         W.A.

| 2–3 m | 500 mm | Prefer lighter soils but will grow under poor conditions and on dry soils | Fast |

Low shelter, gardens, ornamental, birds

*Calothamnus* is from the Greek *kalos* meaning beautiful and *thamnus* meaning a shrub. There are about twenty-five species from Western Australia. They are colourful, low-growing shrubs which tend to be straggly.

*Calodendron capense*

*Calodendron capense—flowers*

Drought-resistant, can withstand hot dry conditions and moderate coastal exposure, but are not very satisfactory in the colder districts. Wind-firm. Annual trimming improves them and they can be trimmed as a hedge. While similar to callistemons, they are in many ways more attractive. Flowers are usually red with stamens in bundles forming a flat, one-sided group. Fruit are globular and woody. Species are very similar to each other so there is not much to choose between them.

*C. asper*——**Rough Net Bush**
Leaves variable, usually narrow, 1–2 cm long, rough, rigid; flowers drooping, brush-like.

*C. chrysantherus*——**Golden Net Bush**
Leaves needle-like, fine, 5–10 cm long; stems tend to be corky.

*C. gilesii*——**Giles Net Bush**
Leaves needle-like, 5–7 cm long; heavy branches.

*C. quadrifidus*——**Common Net Bush**
Probably the best, planted more frequently than the other species. Multi-branched, dome-shaped shrub. Will grow on heavy well-drained soil. Leaves greyish, very narrow and flat, almost needle-like, 2–4 cm long, 2 mm wide; flowers like a one-sided bottlebrush, 3–10 cm long; fruit 5 mm diam. 1803.

*C. rupestris*—**Cliff Net Bush**
Leaves needle-like, curved, 1–2 cm long; flowers crimson with green calyx.

*C. sanguineus*—**Blood Red Net Bush**
Leaves silky, needle-like, 1–4 cm long; flowers appear to be split into two halves, tassel-like.

*C. schaueri*—**Schauer Net Bush**
Leaves very fine, 10–20 cm long.

*C. torulosa*
Spreading, dome-shaped. Leaves needle-like, 4–5 cm long; flowers large, 10–15 cm long.

*C. villosus*
Very attractive bush, branches corky. Leaves narrow, soft, woolly, grey, 3–6 cm long; flowers deep red, 7–10 cm long. 1803.

# CAPPARIS

*Capparis* is the old Greek name. There are about 250 species from the warmer regions of Europe, Asia, Africa and Australia.

*C. mitchellii*—**Tree Caper** or     S.A., Vic., N.S.W.,
        **Native Orange**   E.H.     Qld, N.T.
Seed
3–8 m    250 mm    Prefers heavier    Slow to
                 soils        moderate
Shade, windbreaks, parks, fodder

Dense, compact, sturdy, small tree when mature, but can be straggly when young. Retains branches to ground level and may spread more than its height. Not widely planted but good in the warm to hot districts. Hardy, easily grown, will grow in heavy and wet soils. Leaves 5 cm long, 1–2 cm wide, rough, woolly; flowers white to cream, 4–6 cm diam., 4 petals, prolific, showy but short-lived, numerous stamens; fruit globular, woody, 3–4 cm diam., deep green, containing 2 large flat seeds.

# CARPINUS

This is the old Latin name. There are about thirty-five species from Europe, Asia and North America.

*C. betulus*—**Hornbeam**   D.H.     Central and
Seed                              S. Europe
15–20 m    900 mm    Prefers the better    Moderate
                   soils
Windbreaks, ornamental, parks, hedges

Attractive tree with a dense crown and beech-like foliage. Not widely planted. Wind-firm. Shade-tolerant. Suckers freely; can be pollarded or trimmed to form a hedge. Trunk fluted, grey; leaves oval, 5–10 cm long, serrated, turning rich brown in autumn; flowers inconspicuous, in catkins, 5–10 cm long; fruit a small nut with 3 prominent leafy outgrowths, distinctive, hanging in clusters. L.C.

# CARYA

*Carya* is the Greek name for a walnut. There are about twenty species occurring in North America and eastern Asia.

*C. illinoensis*—**Pecan**   D.H.      E. North America

Similar to *C. ovata* under local conditions. Produces the pecan nut of commerce. Distinguished from *C. ovata* by having 9–17 leaflets instead of 5. 1766.

*C. ovata*—**Hickory**   D.H.      E. North America
Seed
10–20 m    750 mm    Prefers deep soils     Slow
Shade, windbreaks, ornamental

Tree with a straight cylindrical bole and small crown when close grown; but with a large spreading crown when open grown. Growth in early years slow. Deep soils are preferable as it develops a long taproot. Suckers from stumps and roots. Moderately frost-resistant. Produces large quantities of nuts, useful as pig food. Bark very fibrous and shaggy; leaves pinnate, 25–35 cm long, leaflets usually 5, sessile, oval, large, terminal ones 12–20 cm long, 5–8 cm wide; fruit 2–5 cm diam., fleshy husk containing a reddish-brown, globular nut, 1–2 cm diam. 1629.

# CASSIA—Cassia or Buttercup Tree
     E.H.

Seed           Prefer light soil      Fast
Low shelter, parks, ornamental
*Cassia* is the old Greek name. There are about 500–600 species from the warmer regions of all continents. These compact shrubs have masses of bright yellow flowers and grow on a wide range of soils but prefer sandy loams. They are best in sunny, well-drained positions but withstand moderate coastal exposure. Can be lightly trimmed but are difficult to transplant. The leaves are pinnate. Flowers are yellow in terminal sprays and flowering over a long period each year. Fruit is a pod.

*C. artemisioides*—**Silver Cassia**     W.A., S.A., Vic.,
                             N.S.W., Qld
2–3 m    250 mm    Will grow on heavy soils

Bushy shrub. Leaves narrow silvery, leaflets, 4–6 pairs, 2–5 cm long; fruit a pod, 5–8 cm long.

*C. coluteoides* (also incorrectly known as *C. candolleana*)
                                  hybrid
2–3 m    400 mm

Hardy, rounded, fast-growing, nearly as wide as high. Leaves and flowers like *C. corymbosa*.

*Cassia nemophila* var. *coriacea—flowers*

**C. corymbosa** (C. floribunda)  South America
2–4 m  600 mm

Moderately hardy, rather spreading, much-branched shrub. One of the showiest. Autumn-flowering. Leaves 7–10 cm long, 2–3 pairs leaflets; flowers in clusters, 5–15 cm across. 1796.

**C. nemophila** (C. eremophila)—**Desert Cassia**  W.A.,
S.A., Vic.,
N.S.W., Qld
2–3 m  250 mm

Compact, hardy shrub. Good garden plant for hot, dry conditions. Leaves with 2–3 pairs leaflets, needle-like, 2–3 cm long.

**C. nemophila** var. *coriacea* (C. sturtii)—  W.A., S.A., Vic.,
**Dense Cassia**  N.S.W., Qld,
N.T.
2–3 m  250 mm

Showy, low-growing, dense shrub, Easily grown. Best in dry, hot districts. Leaves fine, greyish, leaflets narrow, 1–2 cm long.

# CASTANEA

*Castanea* is the old Latin name. There are about twelve species occurring in Europe, Asia and North America.

**C. dentata**—**American Chestnut**  D.H.  Eastern USA
Seed
20–25 m  800 mm  Prefers well-  Fast
drained soils
Shade, parks, windbreaks, nuts, autumn foliage

Large, spreading, long-lived tree, uncommon in Australia. Produces edible nuts and durable timber. Will grow on sandy loams as well as on heavier soils. Develops a strong taproot. Hardy, frost-resistant. Coppices vigorously from cut stumps. Leaves elliptical,

12–20 cm long, 4–5 cm wide, broadest half way to the tip, coarsely serrated; fruit 2–3 nuts in a spiny husk, 5–8 cm diam., each nut distinctly flat on one side.

Formerly very widespread in eastern USA but now few old trees remain. It has almost disappeared to the point of extinction because of chestnut blight, a bark disease caused by the fungus *Endothia parasitica* introduced into America from eastern Asia in the late 1890s. Arrangements are in progress to raise plants from selected seed in Victoria as the chestnut blight is not known in this country. 1800.

**C. sativa**—**Sweet** or **Spanish Chestnut**  D.H.
Seed  Mediterranean
12–15 m  900 mm  Deep, fertile soil  Moderate
Shade, parks, ornamental, nuts, autumn foliage

Large, long-lived tree, spreading, almost as wide as high, giving good summer shade. Requires a deep, fertile, well-drained soil with good rainfall. Frost-resistant. Coppices vigorously from cut stumps. Edible nuts (for best results grafted stock should be planted). Leaves elliptical, 15–20 cm long, 4–5 cm wide, coarsely serrated, broadest towards the tip; fruit 1–3 rounded nuts in a fleshy, spiny husk, 3–5 cm diam. L.C.

# CASTANOSPERMUM

The name is derived from *castanea* the Latin for chestnut and *sperma* meaning seed. There is only one species and that is from Australia.

**C. australe**—**Black Bean**  E.H.  N.S.W., Qld
Seed
8–15 m  800 mm  Prefers good, moist  Moderate
soil
Windbreaks, shade, parks, roadsides, ornamental

Attractive, densely-crowned, upright tree, more suitable for the alluvial soils in the warmer moist districts. Tends to be multi-stemmed in southern areas. Frost-sensitive. Leaves distinctive and conspicuous, compound, 30–45 cm long with 8–17 oval leaflets each 5–12 cm long, 1–2 cm wide; flowers pea-shaped, 2–4 cm long, yellow to orange, in sprays 10–15 cm long, usually not abundant and often hidden by the foliage; fruit is a legume, 15–25 cm long, thick, seeds like chestnuts. Green pods are reputed to be poisonous to stock. 1828.

# CASUARINA  E.H.

Australia
Seed  Will grow on adverse
sites. (Dry, heavy,
stony, etc. soils.)

*Casuarina* is named after *Casuarinus* (cassowary), alluding to the drooping branches being like the drooping feathers of the cassowary. There are about forty-five species which are mainly Australian but also occur in East Africa, India

*Casuarina cunninghamiana—young tree*

and the south Pacific Islands. These adaptable trees or large shrubs are suitable for a wide range of sites and are wind-firm. Foliage consists of fine branchlets, furrowed or grooved, drooping or pendulous, and has subsistence fodder value. Leaves are very small, rarely longer than 2 mm, and arranged in collars or whorls at regular intervals giving the branchlet a jointed appearance; leaves are referred to as leaf teeth. Trees are male or female; male trees have a rusty appearance in spring because of the pollen flowers. Flowers are inconspicuous. Fruit is a woody, cone-like structure.

*C. cristata* (C. lepidophloia)—**Belar**     S.A., Vic., N.S.W.
10–12 m    300 mm                                    Fast
Windbreaks, parks, shade, roadsides, farm forests

Attractive, with semi-erect branches. Best growth in warm areas on sandy soils but will also grow well on heavier soils. Salt-tolerant. Frost- and drought-resistant. Withstands moderate coastal exposure. Sometimes suckers freely from exposed roots. Leaf teeth 9–10; fruit cylindrical, 1–3 cm diam.

*C. cunninghamiana*—**River She-oak**     N.S.W., Qld, N.T.
10–20 m    450 mm                                    Fast
Windbreaks, farm forests, roadsides, parks, stream banks, gully erosion

Pyramidal, dense, attractive tree with persistent foliage to ground level. Spire-like when young. Largest of the casuarinas. Best growth on stream banks or riverain sites. Will grow in irrigation areas. Rapid early growth. Frost-resistant. Withstands moderate coastal exposure. Suckers freely from the roots. Useful subsistence fodder tree. Leaf teeth 6–8; fruit small, 5–10 mm diam.

*C. glauca*—**Grey Buloke**     Vic., N.S.W., Qld
10–12 m    400 mm                                    Fast
Windbreaks, shade, parks, roadsides, farm forests, gully erosion

Tree with bluish-green, drooping, spreading crown. Very hardy, suitable for most soils; will grow on marshy or saline soils. Frost- and drought-resistant. Withstands moderate coastal exposure, can be grown close to high tide zone in bays and estuaries. Can be lightly trimmed to form an open hedge. Suckers freely. Leaf teeth 10–15; fruit barrel-shaped, approx. 1 cm diam.

*C. leuhmannii*—**Buloke**     S.A., Vic., N.S.W., Qld

Similar to *C. cristata* in characteristics, etc., but able to withstand longer inundation. Foliage bitter, of little attraction to stock. Leaf teeth 10–12; fruit small, squat, 1 cm long, 1–2 cm diam.

*C. littoralis* (C. suberosa)—**Black She-oak**     Tas., Vic., N.S.W., Qld
7–10 m    500 mm    Will grow on light soils    Fast
Windbreaks, parks, roadsides, farm forests

Small, upright tree. Withstands moderate coastal exposure. Grows well in coastal areas and close to the shore-line. Other characteristics similar to *C. stricta*. Leaflets 6–8; fruit cylindrical, 2–3 cm long.

*Casuarina glauca*

164

*Casuarina stricta*

**C. muellerana—Slaty She-oak**                              Vic.
2–3 m     250 mm     Prefers sandy soils          Fast
Low shelter

Small shrub; slender, erect, greyish-green branchlets. Grows well in warm dry areas. Leaf teeth 5–7; fruit variable in shape, 2–3 cm long, 1–2 cm diam.

**C. obesa—Swamp She-oak**                              W.A.
8–10 m     500 mm                              Moderate
Shelter, ornamental

Little-known and not widely planted in the eastern States so that its full potential has not yet been assessed. Withstands brackish and salt water; suitable for planting close to the water-line in tidal bays, estuaries, etc.

**C. paludosa—Scrub She-oak**          S.A., Tas., Vic.,
                                                      N.S.W.
2–4 m     400 mm     Prefers moist sites          Fast
Low windbreaks, roadsides, parks

Moderately compact shrub. Suitable for wet and swampy sites in the warmer districts. Branchlets downy; leaflets 6–8; cones 2–3 cm long, 1–2 cm diam.

Right   *Catalpa bignonioides*

**C. stricta** (C. quadrivalis)—**Drooping She-oak**     S.A.,
                                          Tas., Vic., N.S.W.
6–10 m     400 mm                              Fast
Windbreaks, shade, parks, roadsides, farm forests

Attractive, shapely tree with dense crown of pendulous greyish branchlets, often on a clear trunk, 3–4 m tall. Hardy. One of the most versatile trees available. Suitable for all districts, except cold, mountain areas. Frost- and drought-resistant. Wind-firm. Withstands severe coastal exposure. Grows on limestone or saline soils, wet or brackish sites, dry sandy or stony ridges, and poor coastal sands. Provides useful subsistence fodder. Worth more extensive planting. Leaf teeth 9–12; fruit large, oval, 4–6 cm long and 4–5 cm diam. 1812.

**C. torulosa—Rose She-oak** or **Forest Oak**   N.S.W., Qld
8–15 m     550 mm                              Fast
Windbreaks, parks, roadsides, farm forests

Shapely, erect, pyramidal tree, with slender, drooping branchlets. One of the most attractive of the casuarinas. Will grow on light soils but more suited to the better types. Branchlets sometimes pinkish; leaf teeth 4; cones oval, 3–4 cm long, 1–2 cm diam.

# CATALPA

*Catalpa* is an American Indian name. There are about twelve species from North and South America and eastern Asia.

**C. bignonioides—Indian Bean**   D.H.   South-eastern USA
Seed
8–15 m     750 mm     Deep soils          Moderate
Summer shade, parks, streets, roadsides, autumn foliage

Broad-crowned, attractive and showy tree. Best in cool mountain districts. Frost-tender. Should be planted in an open and sunny but not exposed position. Not tolerant of shade. Large specimens can be transplanted in winter. Tolerates smog. Leaves large and distinctive, heart-shaped, 20–25 cm diam., unpleasant odour when crushed; flowers bell-shaped, 3–5 cm wide, frilled, white with purple markings, arranged in large conspicuous terminal clusters, 15–25 cm high; fruit a brown, bean-like pod, 15–30 cm long. 1726.

*Catalpa bignonioides—flowers*

*Catalpa bignonoides—fruit*

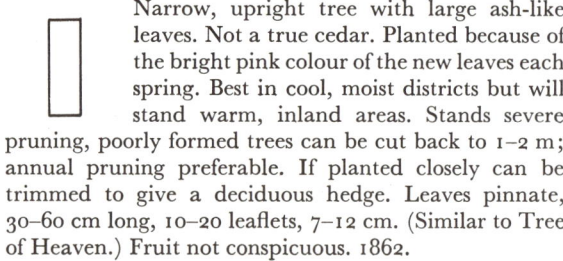

Narrow, upright tree with large ash-like leaves. Not a true cedar. Planted because of the bright pink colour of the new leaves each spring. Best in cool, moist districts but will stand warm, inland areas. Stands severe pruning, poorly formed trees can be cut back to 1–2 m; annual pruning preferable. If planted closely can be trimmed to give a deciduous hedge. Leaves pinnate, 30–60 cm long, 10–20 leaflets, 7–12 cm. (Similar to Tree of Heaven.) Fruit not conspicuous. 1862.

## CEDRUS—Cedars E.C.

Seed

| | | | |
|---|---|---|---|
| 10–25 m | 650 mm | Prefers deep, moist soils | Moderate to slow |

Windbreaks, parks, avenues, roadsides

*Cedrus* is the old Latin name. There are three species from the Mediterranean region and Himalayas. The trees are broadly-pyramidal, densely-crowned, shapely and attractive conifers with horizontal spreading branches. They are widely planted, mainly for decorative purposes but produce high quality timber in their native habitats. This genera comprises the true cedars; three species only are known, although some authorities consider these to be merely geographical variations of the one type. They will grow on a wide range of soils but best growth is on

### C. speciosa—**Western Catalpa** D.H.  USA
Seed

| | | | |
|---|---|---|---|
| 20–25 m | 900 mm | Prefers moist, well-drained loams | Fast |

Parks, ornamental

Attractive, large, spreading tree, pyramidal when young. Not widely planted, but hardier and more vigorous than *C. bignonioides*. Frost-sensitive. Intolerant of shade. Leaves large 20–30 cm long, oval; flowers white, large, occasionally with purple spots, in upright clusters 12–15 cm tall; fruit a brown bean, 25–40 cm long, 1–2 cm thick. 1756.

## CEDRELA

The name is a diminutive of *Cedrus*. There are about twenty species from warmer regions of Australia, southeast Asia and America.

### C. sinensis—**Chinese Cedar** D.H.  China
Root cuttings

| | | |
|---|---|---|
| 10–20 m | 750 mm | Good soils |

Shade, hedges, ornamental  Fast

*Cedrela sinensis—spring foliage*

*Cedrus atlantica*

*Cedrus deodara*

good loamy soils above 750 mm rainfall. They are frost-resistant, wind-firm, and cannot be pruned. Large trees should not be transplanted. Needles are arranged in clusters. Cones are upright and disintegrate on the tree during seed fall.

*C. atlantica*—**Atlas Cedar**      Atlas Mountains

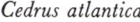 Stiff branches. Blue and golden forms are available, the former being particularly attractive. Will grow on poorer sites and calcareous soils. Needles 1–2 cm long; cones oval, 6–8 cm long, 4–5 cm wide. Recognized by its short needles. Before 1840.

*C. deodara*—**Himalayan Cedar** or **Deodar**   N.W. India

Probably the most attractive of the cedars. Blue and golden varieties available—both very attractive. Branches with pendulous tips; needles 3–4 cm long; cones barrel-shaped, flat-topped, 7–12 cm long, 4–5 cm wide. Recognized by pendulous branchlets. 1831.

*C. libanii*—**Cedar of Lebanon**      Lebanon

Stiff branchlets, ends do not droop. Blue variety available. Needles stiff, 2–4 cm long; cones 8–10 cm long, barrel-shaped. Distinguished from *C. deodara* by its stiff branchlets. 1638.

*Cedrus deodara* var. *aurea*

*Cedrus libani*

## CELTIS

*Celtis* is the old Greek name. There are about seventy species from Europe, Asia and North America.

*C. australis*—**Nettle Tree**　D.H.　　　Mediterranean
Seed, cuttings
8–12 m　　500 mm　　　　　　　　　　Moderate
Shade, streets, ornamental

Attractive, shapely tree with a spreading crown and flexible branches; shedding its lower branches to give a long, clean and straight bole. Light green foliage, smooth grey bark. Hardy, very similar to an elm but endures hotter and drier conditions. Will grow on slightly wet sites. Does not produce root suckers but coppices freely. Large specimens can be transplanted when leafless. Leaves elm-like, but longer and narrower, 7–10 cm long, 2–4 cm wide, long tapering point; fruit a small berry, reddish brown, edible. 1796.

*C. occidentalis*—**Hackberry**　D.H.　　Eastern USA,
Seed　　　　　　　　　　　　　　　　　Canada
12–15 m　　450 mm　　　　　　　　　　Moderate
Windbreaks, shade, roadsides, parks

Attractive, upright tree with a rounded bushy crown and slender, sometimes pendulous, branches. Lower trunk often clear of branches for one-third to half tree height. Occasionally 30 m or more in its natural habitat. Elm-like appearance. Will grow on limestone but not wet or poorly-drained sites. Frost-hardy. Withstands long, dry periods and severe cold and snow; good for cold windswept plains (but deciduous). Can be pollarded and trimmed, but does not sucker freely from the roots. Bark

dark brown, thick, hard; leaves oval, 5–12 cm long, 4–5 cm wide, with a tapering tip, serrated; fruit orange-red, then purple to black, globular, 6–10 mm diam. on a long stem. 1636.

## CERATONIA

*Ceratonia* is derived from *keraton* meaning a pod. There is only one species which occurs in the eastern Mediterranean countries.

*C. siliqua*—**Carob**　E.H.　　　　　Mediterranean
Seed
6–15 m　　400 mm　　　　　　　　　　Moderate
Windbreaks, hedges, shade, parks, roadsides, fodder

Attractive, branching, compact tree with glossy green leaves, and crooked branches. Will grow on sandy and calcareous soils. Suited to hot, dry, inland areas particularly where irrigation water is available. Frost- and drought-resistant. Can be trimmed to a hedge. Tolerates smog. Seed pods are good stock fodder (rich in sugar and protein, contain a sweetish pulp), borne on female trees only. Leaves pinnate, 20–30 cm long, leaflets 6–10, broadly oval to round, 3–5 cm long; flowers in 12–15 cm terminal clusters; fruit a large leathery, legume-like pod, 12–20 cm long. L.C.

## CERATOPETALUM

The name is derived from the Greek *keras* meaning horn and *petalum* meaning a petal, referring to the shape of the petals. There are five species from Australia and New Guinea.

*C. apetalum*—**Coachwood**　E.H.　　　N.S.W., Qld
Seed
6–10 m　　750 mm　　Prefers good,　　　Moderate
　　　　　　　　　　　moist soils
Parks, roadsides, ornamental

Attractive, upright tree preferring moist warm sheltered sites. Not very satisfactory as a windbreak tree. Will grow on heavy soils. Frost-sensitive. Leaves 7–12 cm long, elliptical, slightly serrated; flowers white, then deep red, 2–4 cm diam.

*Ceratonia siliqua*

*C. gummiferum*—**New South Wales**         N.S.W.
        **Christmas Bush**  E.H.
Seed, cuttings
5–10 m    650 mm    Prefers moist,    Moderate
                    fertile soil
Ornamental, avenues, roadsides, parks

Semi-hardy, attractive tree planted for its conspicuous 'flowers' which may cover the entire tree. Not very wind-firm. Withstands slight coastal exposure. Leaves light green, small, 4–6 cm, arranged in threes. True flowers small, white petals drop off, 3–4 bracts enlarge to form the showy 'flower', pink then red, last 2–3 months on the tree. 1823.

# CERCIDIPHYLLUM

*Cercis* is taken from the likeness of the leaf to *Cercis*, and *phyllon* from the Greek for leaf. The genus comprises one species from eastern Asia.

*C. japonicum*—**Katsura Tree**  D.H.    China, Japan
Seed, cuttings
6–10 m    900 mm    Prefers deep loams    Fast
Shade, parks, roadsides, ornamental

Compact, dense and neat tree, one of the most attractive available but not widely grown. Upright with ascending branches when young, spreading to an oval-shaped crown when older. Sometimes produces several stems. Grows to 30 m in its native habitat. Suitable for the cool mountain districts on good, rich soils. Hardy, frost-resistant. Easily-grown. Tolerates some trimming. Leaves round, heart-shaped, like *Cercis*, opposite, 5–10 cm diam., purplish when young, yellow and red in autumn; flowers inconspicuous; fruit a pod, 1–2 cm long, containing many seeds. 1865.

*C. japonicum* var. *sinense*  D.H.    W. China

Similar to *C. japonicum* but has a single stem only and grows taller. Relatively new introduction. Sheds the lower branches, may be free of branches for about half its height. 1907.

# CERCIS

The name is from the Greek name *Kerkis*. There are seven species from Europe, China, and North America.

*C. siliquastrum*—**Judas Tree**  D.H.    E. Mediterranean
Seed
5–8 m    500 mm                Moderate
Windbreaks, parks, ornamental

Showy, small, spreading tree, branching close to the ground. Hardy. Drought- and heat-resistant, suitable for the warmer districts. Growth slow at first, then moderate. Does not transplant easily. Withstands slight coastal exposure. Leaves heart-shaped, 7–10 cm across;

*Chamaecyparis lawsoniana*

flowers purplish-red, pea-shaped, 1–2 cm, appearing in masses on the branches (even larger ones) in spring before the leaves appear, very showy; fruit a flat reddish pod, 12–15 cm long, 1–2 cm wide. Before 1600.

# CHAMAECYPARIS  E.C.

Seed
Windbreaks, parks, streets, ornamental
*Chamaecyparis* is derived from *chamai* meaning dwarf and *kuparissos* meaning cypress. The significance of *chamai* (dwarf) is obscure. About fifteen species occur in North America, Japan and Taiwan. They are attractive, dense, compact, neat trees, which are generally pyramidal, retaining branches to ground level. Grow to over 35 m in their native habitats. Very suitable for ornamental plantings and produce high-quality reddish timber. Prefer good conditions, particularly high rainfall and well-drained loams and are not very satisfactory on alkaline soils or where heavy smog is common. More suitable for the cooler districts as they can withstand cold, and frost, but are not suitable for coastal areas. Can be lightly trimmed and are wind-firm. Bark is red-brown, shed in fibrous strips. Branchlets are plate-like. Leaves are similar to cypress—very small, scale-like and closely pressed to the stem. Cones are globular, under 1 cm

*Chamaecyparis lawsoniana—golden form*

diam. and ripen in the first year. Most *Chamaecyparis* sp. are very similar in appearance. They can be propagated by cuttings, although it is not easy to obtain a good strike. Hundreds of horticultural varieties are known, most of them being dwarf forms, ones which retain the juvenile foliage, or are mutants with colourful or distinctive foliage. The first species *C. lawsoniana* is common in southern Australia but the others are very uncommon in the adult or type form. The horticultural forms, however, are very widely planted.

*C. lawsoniana*—**Lawson Cypress**    North-western USA
10–15 m    750 mm    Deep, loamy soils    Moderate

Very compact, shapely, conical tree retaining its form for many years. Best-known and most compact species, fairly widely planted. Leaves 2 mm long; cones bluish, 5–8 mm diam., scales 8. Sometimes still known as *Cupressus lawsoniana* even though this name has been invalid for over 100 years. 1854.

*C. nootkatensis*—**Alaska Cedar**    Alaska, British
                                                              Columbia
15–20 m    1000 mm    Moderate

Branches tend to be slightly drooping on older trees. Will grow on slightly alkaline or shallow stony soils in the cooler districts. Hardy, vigorous on good sites. Leaves dull green, 3 mm long; cones 6–10 mm diam., scales 4–6. Similar to *C. lawsoniana* but the foliage is coarser and the male flowers are yellow compared with pink to crimson on *C. lawsoniana*. 1853.

*C. obtusa*—**Hinaki**    Japan
15–20 m    900 mm    Moderate

Similar to the previous 2 species but more 'rectangular' in shape. Leaves of 2 sizes, 2 rows, 5–6 mm long, and 2 rows 2–3 mm long, blunt tips; cones 7–10 mm diam., orange seed, scales 7–8. 1861.

*C. pisifera*—**Sawara Cypress**    Japan
20–25 m    900 mm    Fast

Very similar in appearance, characteristics, and foliage to *C. lawsoniana*. Leaves 2–3 mm long; cones globular, 5–7 mm diam., dark brown, scales 10. This species has been cultivated for hundreds of years. Numerous varieties have been developed and were previously known as *Retinospora*. L.C.

*C. thyoides*—**White Cypress**    North America
10–20 m    900 mm    Moderate

Also very similar to *C. lawsoniana*. Leaves 2 mm long; cones 5–7 mm diam., bluish-purple, scales 6. Of interest mainly because a number of small horticultural varieties are widely grown. L.C.

# CHAMAECYTISUS

There are about twenty species from Europe and eastern Atlantic islands.

*C. proliferus* (Cytisus proliferus)—    Canary Islands
        **Tree Lucerne**    E.H.
Seed
5–6 m    450 mm    Fast
Low windbreaks, hedges, fodder

Small, rounded tree, with several branches and no clearly defined trunk. Hardy and vigorous but will not stand excessive cold. Widely-planted. Can be trimmed to form a hedge. Withstands moderate coastal exposure. Good fodder tree. Leaves simple, 2–4 cm long, 1 cm wide; flowers white, pea-shaped; fruit a pod, 4–6 cm long.

# CHAMAELAUCIUM

The origin of the name is not known. There are about twelve species all from Australia.

*C. uncinatum*—**Geraldton Waxflower**    E.H.    W.A.
Seed, cuttings
2–3 m    450 mm    Well-drained soils    Fast
Low shelter, ornamental

Hardy shrub. Requires well-drained site (fails in wet soils) and a sunny position. Responds to cutting, can be trimmed as a hedge, annual trimming for shape desirable. Drought-resistant once established. Leaves needle-like, soft, 2–3 cm long; flowers in clusters, waxy, tea-tree-like, flowering for a long period in spring and summer, usually pink to pinkish-purple.

*Cinnamomum camphora*

# CINNAMOMUM

*Cinnamomum* is derived from the Greek name *Kimmamomom*. There are about 250 species from south-eastern Asia, and Indonesia.

*C. camphora*—**Camphor Laurel**  E.H.    China, Japan
Seed
6–12 m    500 mm    Moist, well-    Fast
drained soils
Streets, avenues, large hedges, ornamental

Handsome, shapely tree with yellow-green foliage. Frost-tender when young. Withstands hot, dry conditions once established. Suitable for the warmer areas, particularly if irrigation water is available. Leaves soft, simple, oval, 7–12 cm long, emitting a strong camphor smell when crushed; flowers greenish-white, inconspicuous; fruit a small berry. 1727.

# CODONOCARPUS

*Condonocarpus* is from *kodon* meaning a bell and *carpos* meaning fruit, alluding to the bell-shaped fruit. There are about three species all occurring in Australia.

*C. cotinifolius*—**Bell Fruit Tree,**  All States, except Tas.
**Australian Poplar** or
**Native Poplar**  E.H.
Seed (?)
5–7 m    250 mm    Deep, light soils    Moderate
Parks, roadsides, ornamental

Slender, pyramidal, small tree with a rather open crown. One of the most attractive of the native trees although not well known. Prolific crops of seedlings occur naturally but it is difficult to propagate in nurseries so plants are not readily available. Very drought-resistant. Has a deep and strong root system—not suitable for soils with a strong root barrier (clay layer, etc.). Bark pinkish, smooth; leaves grey-green, 2–5 cm long, broad to almost round (poplar-like), stalk 2–4 cm long; flowers not conspicuous; fruit bell-shaped, soft, pear-like, 1 cm long.

# COMBRETUM

The origin of the name is obscure. About 250 species occur in tropical and subtropical areas of Africa, Asia and America.

*C. caffrum* (C. salicifolium)—**Bush Willow**  D.H.
Cuttings    S. Africa
5–8 m    600 mm    River banks    Fast
Windbreaks, parks, roadsides

Shapely, handsome, rather spreading, densely-crowned tree, similar in appearance and requirements to willows. Trunk usually crooked. Rare. Prefers moist but well-drained sites, such as river banks. Generally semi-deciduous in southern Australia, old leaves remaining over winter. Roots are not troublesome. Slightly frost-sensitive. Can be heavily pruned. Colours well in autumn to a bright red, more attactive than willows in this regard. Bark light-coloured; leaves 6–15 cm long, 1 cm wide, opposite, slender, willow-like but not serrated, bright red-brown in autumn; flowers small, inconspicuous, borne in round balls; fruit 4-winged, 2 cm long, 1 cm wide, produced in large numbers, giving the tree a reddish-brown appearance in autumn.

# COPROSMA

The name is from *kapros* meaning dung and *osma* meaning a smell, referring to the unpleasant odour of the leaves of some species when bruised. There are about ninety species from Australia, New Zealand, Malaysia and Chile.

*C. repens* (C. baueri, C. retusa)—**Mirror Plant**  E.H.
Seed, cuttings    N.Z.
3–5 m    600 mm    Fast
Low shelter

Hardy, easily-grown, compact shrub. Withstands drought and severe coastal exposure. Can be trimmed heavily to form a dense hedge; more often seen in this form than as a free-growing plant. Leaves very shiny, glossy green, leathery, round, 3–5 cm diam.; flowers white, not very conspicuous; fruit a yellow berry, borne in clusters, 5–10 mm diam. 1876.

# CORNUS

*Cornus* is an old Latin name. About forty species occur in temperate Europe, Asia and America.

*C. capitata* (Benthamia fragifera)    N. India,
—**Evergreen Dogwood**  E.H.    Himalayas,
China
Seed
5–12 m    800 mm    Deep soils    Fast
Parks, avenues, windbreaks, ornamental

Compact tree, very showy when in flower. More suited to cool mountain areas on good soils. Withstands severe frosts. Can be lightly trimmed. Leaves oval, 7–12 cm long; flowers small, 1 cm but subtended by 4–5

*Cornus capitata*

*Cornus capitata—flowers*

large, yellow bracts, broad, 5–10 cm long, forming a 'flower' 10–12 cm across, very conspicuous and attractive; fruit globular, strawberry-like, 2–3 cm diam., dull red. 1825.

*C. florida*—**Flowering Dogwood**   D.H.   Eastern USA
Seed
3–6 m      900 mm      Prefers good soils      Fast
Parks, roadsides, ornamental

Small tree with conspicuous flowers and autumn foliage (orange and scarlet). More suitable for the cool mountain districts. Hardy, frost-resistant, and easily grown. Sensitive to dry conditions. Tolerates semi-shade. Coppices freely. Responds to pruning for shaping. Large specimens can be transplanted when leafless. Easily killed by earth fills. Leaves oval, 7–12 cm long, 4–6 cm wide; 'flowers' white, 8–12 cm across, 4 'petals' at right angles—actual flower is very small, several form a cluster of about 1 cm diam. in the centre of the 'flower'—apparent 'petals' are enlarged bracts, oval, 5–8 cm long,

notched at the apex, spring-flowering; fruit a berry, 5–10 mm diam.; seed often takes 12 months to germinate. 1731.

*C. kousa*—**Chinese Dogwood**   D.H.      China, Japan
Layering
5–7 m      750 mm                                    Fast
Parks, roadsides, autumn foliage, ornamental

Attractive, slightly-spreading, small tree with arching branches. One of the best of the dogwoods (*Cornus* sp.). Hardy. May require pruning for shape. Leaves 5–8 cm long, oval, pointed, wavy margin, richly-coloured in autumn; 'flowers' creamy, 5–10 cm across, erect, comprising 4 large but narrow bracts each about 2.5–5 cm long; fruit bright red. 1875.

*C. kousa* var. *chinensis*                        China
8–10 m

As for *C. kousa* but bracts slightly larger, 6–8 cm, and more colourful than *C. kousa*. 1907.

# CORYNOCARPUS

The name comes from *koryne* meaning a club and *carpos* meaning fruit, referring to the shape of the fruit. There are about five species occurring in New Zealand, New Caledonia, New Guinea and Australia.

*C. laevigatus*—**New Zealand Laurel**   E.H.      N.Z.
Seed
5–10 m      650 mm                                   Fast
Shade, parks, roadsides

Attractive, rather leafy tree with a rounded bushy crown. Semi-hardy. Withstands moderate coastal exposure. Leaves oval, large, 10–20 cm long, 5–10 cm wide, glossy, deep green; flowers small, greenish-white, in upright clusters, 10–20 cm long; fruit, plum-like, 2–4 cm long, orange, fleshy, believed to be toxic.

# COTINUS

The name is from the Greek name *Cotinus* for wild olive. The number of species is not clear as they are classified under *Rhus* by some authorities. There are probably only four or five all from North America.

*C. obovatus* (C. americana, Rhus cotinoides)—      USA
      **Smoke Bush**   D.H.
Seed, cuttings
5–10 m      700 mm                                   Fast
Autumn foliage, ornamental

Roundish shrub or small tree. Hardy. Wood yields the dyeing material 'young fustic'. Leaves rounded but tapered, 5–8 cm long, stalks long and slender, brilliant red, orange and yellow colours in autumn; flowers inconspicuous; fruit a small olive-like, dry berry, 3–5 mm diam. 1882.

*Crataegus monogyna*

*Crataegus monogyna—hedge*

# CRATAEGUS D.H.

Seed    650 mm    Well-drained soils    Moderate
Parks, hedges, hedgerows, roadsides, ornamental
*Crataegus* is from *kratos* meaning strength which refers to the strength of the wood. It is a variable genus. More than 1000 species have been described but many are of doubtful validity; probably only about 200 valid species exist. They are mainly from North America but also from Europe and Asia Minor. The genus comprises spreading, round, attractive, flowering trees with colourful autumn foliage and berries and a short trunk. They can withstand slight coastal exposure. Younger trees can be trimmed and older trees can be cut back severely. They have thorns. Fruit is a berry, produced in clusters. Foliage may be attacked by pear slug; easily controlled by spraying with maldison. Similar to *Pyracantha* but this genus is evergreen, *Crataegus* is deciduous.

*C. crus-gallii*—**Cockspur Thorn**        North America
4–5 m

Very colourful autumn foliage. Has long, thin thorns, 2–5 cm long. Leaves rounded, not lobed to any extent, finely-toothed, 4–6 cm; fruit deep red, 1 cm diam. 1656.

*C. monogyna* (C. oxycantha)—**Hawthorn**        Europe
5–6 m

Hardy, pink and red-flowering, easily grown (stratify seeds for 12 months before sowing). Thorns 2–4 cm long; leaves triangular, 5–7 cm across, 3–5 lobes, strongly dissected; orange-red berries, 6–8 mm diam., 1 seed per berry. (Some plants have 2 seeds per berry, sometimes referred to, incorrectly, as *C. oxycantha*.) Several horticultural varieties are available. L.C.
In 1965 this species was declared a noxious weed for Victoria (except the metropolitan area and in existing hedges) and so should not be planted in that State.

*C. phaenopyrum* (C. cordata)—**Washington Thorn**
5–8 m        Eastern USA
Handsome, round-headed tree, probably most attractive of the genus. Withstands slight coastal exposure. Leaves triangular, dissected to varying degrees, 5–7 cm long and across, 3–5 lobes; flowers pinkish-white; fruit orange, persistent. 1738.

*C. pubescens* (C. mexicana, C. stipulacea)—        Mexico
**Mexican Hawthorn**

5–8 m

Attractive, erect, small tree. Almost evergreen. Largely thornless. Leaves large, oval, 7–10 cm long, coarsely toothed, orange-red in autumn; flowers pinkish-white, 1–2 cm across, in heads of 10–15, 5–8 cm across; fruit yellow, globular, large, 2–3 cm diam., produced in large numbers, remain on the tree for most of the winter, edible. 1824.

*C. tanacetifolia*—**Tansy-leaf Hawthorn**        S. Europe,
4–8 m        Asia Minor
Attractive, almost thornless, small tree. Hardy, moderately frost-tolerant. Leaves 4–5 cm long, deeply cut into 5–7 lobes, hairy; flowers white, in fragrant, round clusters; fruit a large berry, 1–2 cm diam., yellow. 1789.

*Crataegus monogyna—pink flower form*

*Cryptomeria japonica* var. *elegans*

# CROTALARIA

*Krotalon* means a castanet; when the fruit is shaken the seeds rattle. Some 500 species found in tropical regions of most countries—more than a further one hundred are still undescribed.

*C. laburnifolia*—**Queensland Bird Flower**   E.H.   Qld,
Cuttings                                               W.A.
1–3 m        750 mm                                    Fast
Parks, ornamental

Pale green shrub, grown for its greenish-yellow, bird-like pea-flowers. Needs to be pruned each year otherwise becomes straggly and thin. Will shoot from a cut stump, with a tendency to sprout from root suckers in open, light soils. Moderately frost-resistant. Only common *Crotalaria* which can be easily transplanted. Leaves simple, 5–10 cm; flowers in terminal clusters from late spring to winter.

# CRYPTOMERIA

*Kryptos* means hidden and *meris* means part; all parts of the flower are hidden. Only one species from China and Japan but numerous varieties have been described.

*C. japonica* var. *elegans*—**Japanese Cedar**   E.C.   China,
Cuttings                                               Japan
5–10 m      750 mm    Deep, rich soils      Moderate
                                                   to fast
Parks, ornamental

Dense, compact, symmetrical, small tree with branches persistent to the ground. Greenish-bronze coloured foliage turning reddish-brown in autumn. Best in cool mountain areas. Widely planted. Frost-resistant. Tolerates smog. Retains the juvenile foliage; leaves (needles) 1–3 cm long, spreading, awl-shaped; cones globular, 1–2 cm diam., rarely seen on the var. *elegans*. The parent type *Cryptomeria japonica* is a valuable timber tree in Japan. The best stands occur in the high rainfall, summer fog belt and reach 40–50 m. 1861.

# CUPRESSUS—Cypress   E.C.

Seed, (cuttings)          Requires well-drained soils
Windbreaks, parks, hedges

*Cupressus* is taken from *kus* which means to produce and *parisos* meaning equal which refers to the symmetrical growth of *C. sempervirens*, the first species described. There are about fifteen species from America, Europe and Asia. The trees have dense, compact crowns and prefer cool, moist conditions but will grow on most soils. They are frost-resistant and wind-firm. Can be severely trimmed and form good hedges but they do not transplant easily. They can be raised from cuttings but are difficult to strike—misting and heated beds are required to give a reasonable strike but it may still take 2–3 years to produce planting stock. Types listed as varieties need to be raised from cuttings to ensure plants that are true to type. Timber is aromatic and durable. Foliage consists of soft branchlets; the actual leaves are small, 1–2 mm long, triangular and closely pressed to the stem. Cones are woody and rarely over 4 cm long.

*Cupressus funebris*

Left    *Cupressus glabra*

Below    *Cupressus glabra—blue form*

*C. benthamii*—**Mexican Cypress**          Mexico
12–15 m    650 mm                                Moderate
Windbreaks, parks, roadsides

Attractive tree with a fairly compact crown. Lower branches slightly drooping, giving good shelter to ground level. Withstands wet conditions better than most cypresses. Worthy of more extensive planting. Very similar to *C. lusitanica* but more elegant in appearance with a denser and narrower crown, and the tips of the leaf sprays more or less flattened. Some authorities consider this species to be only a geographical form of *C. lusitanica* but it is sufficiently different under local conditions to justify separate mention.

*C. funebris*—**Chinese**             Central China,
**Weeping Cypress**                         Nepal
10–20 m    650 mm    Prefers cool, moist    Moderate
                              soil
Roadsides, ornamental

Shapely, graceful, conical, greyish-green, branches not spreading and persisting to ground level. Will grow on lighter soils. Branchlets pendulous. Foliage flat or plate-like; cones globular, 5–10 mm, 8 scales. 1848.

*C. glabra* (C. arizonica)—**Arizona Cypress**    Arizona,
                                                           N. Mexico
8–10 m    450 mm                              Moderate to
                                                        slow
Roadsides

Distinctive, bluish-green foliage. Crown very variable, ranging from spreading to compact depending on seed source, often not quite as dense as other cypresses. Most drought-resistant of the cypresses. Withstands severe coastal exposure. Cones globular, 1–2 cm, 6–8 scales. 1882.

*C. lusitanica*—**Portuguese Cypress** or          Guatemala
**Mexican Cypress**
12–15 m    500 mm                              Moderate to
                                                              fast
Farm forests

Open grown trees have a spreading crown with pendulous branchlets; crown compact if planted at close spacings. Withstands wet conditions better than most cypresses but not satisfactory on alkaline soils. Good tree for adverse conditions. Leaf sprays spreading in all directions (not flat); cones globular, 1 cm diam., bluish, scales 6–8, each with a distinct point. 15th century.

*C. macrocarpa*—**Monterey Cypress**          California
10–15 m    650 mm    Will grow on          Fast
                              limestone soils

Semi-upright tree with a spreading, green crown; branches usually ascending at a sharp angle. Strong root system. Best cypress for trimmed hedges. Withstands severe coastal exposure—excellent for seaside planting. Leaves closely pressed, swollen tips; cones large oval, 2–4 cm long, 8–14 scales with numerous resin tubercles. Numerous varieties are available. During the mid-1950s the disease cypress canker killed many trees but its low level of infection over the last 10 years suggests that the remaining trees are less susceptible. 1838.

*C. macrocarpa* var. *lambertiana* (C. lambertiana var.
                                                        horizontalis)
                              —**Lambert Cypress**
                                                              Cultivar
10–15 m    600 mm                                        Fast

Distinguished by its almost horizontal branches, spreading crown. With open grown trees, spread often greater than height. Withstands severe coastal exposure—one of the best trees for seaside planting. Numerous horticultural varieties are available. 1875.

*Cupressus lusitanica*

**'Golden Cypress'**

Trees sold as Golden Cypress are usually golden coloured forms of the above two, viz: *C. macrocarpa* var. *aurea* and *C. macrocarpa* var. *lambertiana aurea*.

**C. sempervirens—Mediterranean Cypress or Italian Cypress**

| | | |
|---|---|---|
| 10–20 m | 500 mm | Mediterranean Fast |

Avenues, ornamental

Upright, narrow, columnar form, branches ascending very steeply. Very long-lived (centuries), old trees may exceed 30 m. Drought-resistant, withstands hot, dry conditions if irrigation water is available. Strong deep root system; cones 2–4 cm long, oval, 8–14 scales, no resin tubercles. L.C.

**C. sempervirens var. stricta—Roman Cypress, Pencil Pine or Candle Pine**  Cultivar

Similar to *C. sempervirens* but with very narrow columnar form; branches of two types—either almost vertical, close to the trunk, or short and horizontal.

**C. torulosa—Himalayan Cypress**

| | | |
|---|---|---|
| 8–12 m | 650 mm | Will grow on limestone |

W. Himalayas  Moderate

Ornamental, avenues

Usually very dense, compact, symmetrical, broadly-pyramidal tree, often with a spire-like top but form can be variable depending on seed source. Lighter green colour, one of the most attractive of the cypresses. Hardy, can be grown in warm, inland areas where irrigation water is available. Good for hedges as its compact growth requires less clipping. Branchlets on older trees become pendulous. Cones globular, 1 cm, scales 8–10. Often purplish when young. 1824.

*Cupressus macrocarpa—hedge*

*Cupressus macrocarpa* var. *aurea—hedge*

*Cupressus macrocarpa*—background; *Cupressus torulosa—*foreground.

# DACRYDIUM

The name is from *dakrudian* meaning a small tear which refers to the tear-like resin drops which are often exuded. There are about twenty species occurring in Australia, New Zealand, south-western Pacific, Malaysia and Philippines.

*D. cupressinum*—**Rimu**   E.C.                     N.Z.
Seed

| | | | |
|---|---|---|---|
| 8–15 m | 900 mm | Prefers moist, well-drained soils | Slow |

Parks, ornamental

Attractive, upright, tapering tree with slender, pendulous branchlets. Weeping appearance. Pyramidal when young. Good specimen tree. Grows to over 35 m in its native habitat producing high quality, durable, dark red timber. Growth too slow for farm forests. Prefers cool, moist, mountain conditions but will not stand snow. Suitable for a wide range of soils, including limestone, but not very satisfactory on wet sites. Root system shallow and restricted. Tolerant of shade. Not suitable for exposed positions, requires protection from strong winds. Plants up to 2 m can be transplanted. Hard to grow from seed or cuttings so plants are sometimes difficult to obtain. Good indoor plant provided soil is kept moist. Foliage bright green, reddish-brown in winter, cypress-like; leaves small, 3 mm long, scale-like, overlapping; fruit oval, 1 cm long, deep blue, on a red fleshy base, acorn-like; seed is short-lived, fertile crops only occuring every few years (male and female trees).

*Cupressus sempervirens*

# DAIS

*Dais*, meaning a torch, alludes to the flowers. There are six species from South Africa and Madagascar.

*D. cotinifolia*—**Pompon Tree**   D.H.           S. Africa
Seed

| | | |
|---|---|---|
| 3–5 m | 650 mm | Fast |

Parks, roadsides, ornamental

Narrow, upright tree when young; spreading, rounded crown when older. Attractive small tree but not widely grown. Deciduous in cold areas, almost evergreen in the warmer districts. Prefers a sunny position, tolerates dry conditions once established. Can be pruned for shape. Leaves simple, opposite, rich green, 4–10 cm long, oval, translucent; flowers small, pink, in dense rounded terminal clusters, 2–5 cm diam., often flowers in the second year from seed; fruit small, dry capsule; seed very small, black. 1776.

*Cupressus torulosa*

## DODONAEA—Hop Bush  E.H.

Seed                     Wide range of soils,         Fast
                         including heavy soils
Low shelter, hedges, ornamental

*Dodonaea* was named after R. Dodaens, a sixteenth-century medico and author on plants. There are about sixty species mainly from Australia but a few from North and South America and Africa. The dense, compact shrubs have foliage to ground level. They withstand frosts and moderate coastal exposure. Require seasonal trimming to prevent straggly growth and can be trimmed to form a hedge. Foliage is generally more or less sticky. Fruit are similar to hops. 'Hops' are of some fodder value to starving stock.

*D. angustissima* (D. attenuata)—**Slender**   W.A., S.A.,
                                **Hop Bush**   Vic., N.S.W.
                                               Qld, N.T.

2–3 m        250 mm
Fodder

More suitable for warmer, inland areas. Leaves narrow, 3–7 cm long, 2–5 mm wide; fruit red and yellow 'hops'.

*D. cuneata*—**Wedge Leaf Hop Bush**   S.A., Vic., Tas.,
                                        N.S.W., Qld

2–3 m        250 mm

Leaves wedge-shaped, 2–5 cm long, 5–8 mm wide. Flowers red and yellow; fruit bronze-coloured 'hops', in large clusters.

*D. microzyga*—**Scarlet Hop Bush**   W.A., S.A.

1–2 m        250 mm

Dense, rigid bush, with interlacing branches. More suited to hot, dry, inland areas than southern districts. Leaves pinnate, small, 3–7 very small leaflets, 7–9 mm long; fruit reddish-brown 'hops'.

*D. viscosa*—**Giant Hop Bush**   Widely spread (Australia,
                                  S. Africa, N.Z., North America,
                                  tropics)

2–6 m        450 mm
Fodder

Very variable shrub, under good conditions a small tree. Quick-growing. Leaves pale green, willow-like, 5–10 cm long, 5–8 mm wide; flowers inconspicuous, pale green; fruit, pale green 'hops'.

*D. viscosa* var. *purpurea*
2–3 m

Similar to *D. viscosa* but leaves slightly larger, 8–12 cm long, 1 cm wide, purplish-brown. Very attractive background, hedge or lawn specimen. Requires trimming periodically to maintain a compact form. 1930.

*Dodonaea viscosa* var. *purpurea*

## DOMBEYA

A genus named after J. Dombey, an eighteenth-century French botanist. There are about one hundred species from tropical and southern Africa.

*D. nataliensis*  D.H.                           Natal
Seed
4–6 m        600 mm                              Fast
Parks, ornamental

Spreading, compact, small tree. Moderately hardy and vigorous. Leaves large, poplar-like, 3–5 lobes, broadly oval, 10–15 cm long; flowers white, 2–5 cm across, in groups of 4–8, buttercup-like, autumn-flowering, the dead flowers remaining on the tree during winter.

## DRYANDRA   E.H.                               W.A.

Seed                                             Fast
Parks, roadsides, ornamental

*Dryandra* was named after J. Dryander, a Swedish botanist of the eighteenth century. There are about sixty species occurring in Western Australia. These upright, usually many-stemmed, rigid, large shrubs will grow on poor soils and are suitable for shallow soils over a clay hardpan. They require good drainage and will not tolerate wet sites. Growth is satisfactory on moderately but not strongly alkaline soils. They are hardy, easily grown, and will stand a lot of neglect once established. Can be trimmed for shape. Foliage is deeply serrated or strongly lobed. Flowers are yellow, terminal, winter-flowering and good for dry arrangements.

*D. formosa*
2–3 m        500 mm

Attractive, spreading, large shrub, moderately dense. Hardy, easily grown. Wind-firm. Withstands moderate coastal exposure. Leaves narrow, 10–20 cm long, 1–3 cm wide, deeply serrated into almost triangular lobes, grey underneath; flowers reddish-yellow, large, globular, 7–8 cm diam.

*D. nobilis*—**Great Dryandra** or **Tall Dryandra**

2–3 m      450 mm      Will grow on gravelly soils

Variable in shape, depending on seed source —usually broadly columnar and dense but can be spreading and rather straggly. Leaves 10–20 cm long, 2–3 cm wide, deeply serrated into triangular lobes, prickly; flowers large, 8–10 cm across, largest of the dryandras.

*D. sessilis*

3–5 m      400 mm      Prefers sandier soils

Upright, slightly open shrub, spreading to about one-third its height. Leaves wedge-shaped, 4–7 cm long, 1–2 cm wide, light green, serrated, prickly; flowers 4–5 cm across, mainly late winter flowering but some present most of the year.

# ELAEAGNUS

*Elaia* is the Greek word for an olive. There are about forty species from Europe and Asia.

*E. angustifolia*—**Russian Olive**   E.H.      S. Europe,
Seed, cuttings                                  Himalayas

5–7 m      500 mm      Requires well-      Fast
                        drained soil

Windbreaks, roadsides, parks

Attractive, small, spiny tree, usually with 2 or 3 crooked stems. Will grow on most soils, including calcareous and alkaline, provided they are well drained. Frost-resistant. Withstands light snow. Excellent for cold, dry plains but also suitable for the warmer districts. Coppices freely. Leaves 7 cm long, dull green, silver underneath; flowers small, 5 mm, bell-shaped, silver-yellow; fruit a berry, 5–10 mm diam., silver. Variegated forms are available. L.C.

# ELAEOCARPUS

*Elaia* is the Greek word for an olive and *carpus* is Greek for fruit. About 200 species occurring in Australia, New Zealand, Indonesia and Malaysia.

*E. dentatus*—**Hinau**   E.H.      N.Z.
Seed, cuttings

6–12 m      600 mm      Moderate
Parks, ornamental

Attractive, small tree, particularly in flower when it is covered with white flowers. Moderately frost-sensitive, otherwise hardy. Leaves oblong, 6–8 cm long, leathery, silky underneath; flowers white, prolific, like lily-of-the-valley, terminal clusters; fruit olive-like, with a hard, rough nut surrounded by purplish pulp.

*E. reticulatus*—**Blue Oliveberry**   E.H.      Vic., Tas.,
Seed                                              N.S.W., Qld

5–6 m      750 mm      Prefers goods soils      Moderate
Low shelter, parks, roadsides

Small, compact tree. Not suitable for alkaline soils. Not frost-resistant, otherwise hardy. Leaves 7–10 cm long, 2–3 cm wide, net-like venation; flowers cream to white, 2 cm diam. in loose clusters; fruit a blue olive-like berry, 1 cm diam.

# EREMOPHILA—Emu Bush   E.H.

Seed, cuttings

300 mm      Will grow on      Moderate
            limestone soils

Shade, low windbreaks, parks, roadsides, ornamental

The origin of *Eremophila* is doubtful (derivation suggests *eremos* meaning solitary and *philos* meaning love, but the allusion is obscure). There are about sixty species occurring in Australia. These attractive, flowering, small trees or large shrubs have a dense crown. They are suitable for hot, dry districts and are hardy, frost- and drought-resistant. Leaves are varied. Flowers are tubular, 2–4 cm long, with 5 lobes and usually pinkish to red. Fruit is globular, plum-like, fleshy and splits into 4 parts on drying.

*E. alternifolia*      W.A., S.A., N.S.W.

2–3 m

Very attractive, slender, dense, large shrub or small tree. Very showy. Leaves 1–3 cm long; flowers pink to red; fruit 1 cm diam.

*E. longifolia*—**Berrigan** or      All States except Tas.
**Native Plum Tree**

5–6 m
Fodder, gully erosion

Small tree with a slender trunk and drooping branches. Rounded crown. Suckers freely from the roots. Bark rough, pale brown, fissured; leaves long and narrow, 5–20 cm long, up to 6 mm wide, thick; flowers dull red to pink, 2–3 cm long, in groups of 1–3, some present most of the year; fruit 5–10 mm diam., dark purple when ripe.

*E. maculata*—**Spotted Emu Bush**      W.A., S.A., Vic.,
                                          N.S.W., Qld

3–4 m

Very attractive, large shrub or small tree. Leaves, flowers and fruit poisonous to stock. Leaves small, 1–2 cm long, pointed; flowers white to pink, spotted inside with purple to red spots, on a long slender stalk, free-flowering; fruit 1 cm diam. Several provenances are known.

# ERYTHRINA—Coral Tree D.H.

5–10 m                   Well-drained soils               Fast
Parks, roadsides, ornamental

*Erythrina* is derived from *erythros* meaning red which refers to the flower colour. There are about one hundred species from tropical countries and South Africa. The trees, planted for their colourful scarlet flowers, provide a magnificent display. They are hardy and are generally better in the warmer districts, but can be grown in areas with a climate similar to Melbourne. Moderately frost-resistant, only affected by severe frosts. Heavy pruning is necessary each year not only to stimulate flowering but also to reduce the leafy growth which obscures the flowers. The smooth leaves are in threes with oval leaflets and stem prickles. Flowers are pea-shaped, vivid red or scarlet, 5–7 cm long and in conspicuous, dense clusters. Fruit is a flat legume about 10–15 cm long.

*E. crista-gallii*—**Common Coral Tree**          Brazil
Seed              600 mm

 Hardiest of the coral trees. Flowers in summer producing long shoots of flowers. 1771.

*E. variegata* (E. indica)—**Indian Coral Tree**      India
Cuttings          650 mm

Not quite as hardy and vigorous as the previous species but produces a better display. Young stems prickly. Flowers in dense clusters, appearing in winter when the tree is leafless. 1814.

# EUCALYPTUS E.H.

Seed

*Eucalyptus* is derived from *eu* meaning well and *kalypto* to cover and refers to the cap (operculum) covering the stamens in bud. There are over 500 species and 150 varieties occurring mainly in Australia with a few in New Guinea, Indonesia, and a doubtful one from the Philippines. Many species hybridize, consequently there are many hybrids in addition to the approx. 650 known species and varieties.

Eucalypts are woody plants ranging from small shrubs 2 m tall to trees 80–100 m tall. They occur under most climatic conditions and on most types of soil. Root systems are generally strong and deep. Most species are reasonably frost-resistant, although some are frost-tender when young. They produce a large quantity of seed of good viability and longevity, and, except for a few species, sucker freely and can be pollarded. However they cannot be trimmed to form hedges. The crowns are usually not heavy; with the taller species, the lower branches are shed giving a clean trunk. They are easily raised from seed. Leaves are usually simple and entire (i.e. without serrated, etc. margins) and unless otherwise stated are lanceolate (i.e. shaped like a lance). The flowers do not have petals, the colour is due to the stamens. (If colour is not mentioned in the descriptions it can be assumed to be creamy to white.) Fruits are woody capsules with a cap or operculum when in bud.

*Eucalyptus accedens*

Dry-climate species, that is those shown as growing in rainfall areas of less than 500 mm, are frequently not satisfactory in southern districts. Bluish or mealy types are also more liable to attack by defoliating insects. In very wet summers some of the more susceptible species, particularly *E. obliqua*, *E. sieberi*, *E. regnans*, and *E. baxteri* may be killed by the root fungus *Phytophthora cinnamomii*. Trees of any age may be affected.

As a group, eucalypts display a great degree of similarity and identification requires specialist knowledge. Information on bark, leaves (juvenile, intermediate, and mature), buds, flowers and fruit and, in some cases, seed, is necessary for correct identification but the notes given in the descriptions which follow should enable the non-expert to identify most of the species he is likely to meet in practice. A recent publication *Eucalyptus Buds and Fruits* by Chippendale would be of value as it contains sketches of the buds and fruits of all eucalypts. Detailed botanical descriptions can be obtained from the companion volume *Key to the Eucalypts* by Blakely.

In popular terms eucalypts are often referred to by the nature of the bark: (a) gums—smooth usually light coloured and peeling off in ribbons or flakes; (b) stringybarks—fibrous brown, long and stringy more or less furrowed; (c) ironbarks—dark, rough, hard, deeply corrugated; (d) peppermints—like stringybarks but not so fibrous and not furrowed; (e) boxes—also like stringybarks and peppermints but harder and not so fibrous; and (f) bloodwoods—brown, fibrous, with a tile-like structure. Species from drier areas may be described as of mallee habit meaning that there is no single clearly defined trunk but rather several large branches arising from ground level. Mallees coppice vigorously from the root-stock or 'mallee root'.

*E. acaciaeformis*—**Wattle Leaf Peppermint**     N.S.W.
12–12 m     750 mm     Prefers heavier soils     Moderate
Windbreaks, shade, parks, roadsides

Attractive, densely-crowned, rather spreading, dark green tree. Not satisfactory on sandy soils. Hardy. Withstands light snow; more suitable for the cooler districts. Bark fibrous and hard; leaves narrow, 7–10 cm long, 2 cm wide; buds 3–7, small, 5 mm long and wide, cap conical, short; fruit hemispherical, 5 mm diam.

*E. accedens*—**Powder Bark**     W.A.
10–12 m     350 mm     Will grow on     Moderate
                       stony soils
Roadsides, parks

Small tree with a pinkish trunk, producing dense, heavy timber. Suitable for the warmer districts. Hardy. Frost- and drought-resistant. Leaves narrow, 6–12 cm long, 1–3 cm wide, thick; buds 5–10, 10–12 mm long, each cluster on a stalk 1–2 cm long, cap short, hemispherical; fruit slightly bell-shaped, 5–10 mm long, narrower at the apex, valves 4.

*E. aggregata*—**Black Gum**     Vic., Tas., N.S.W.
15–20 m     700 mm     Will grow in wet     Moderate
                       areas
Windbreaks, shade, roadsides

Upright tree with a dense, spreading crown. Branches often retained to within 1–3 m of the ground. Suitable for wet sites (non-saline), particularly at higher elevations. Frost-resistant. Withstands cold and light snow. Bark rough, dark, flaky; leaves 7–12 cm long, 2–3 cm wide, fairly prominently veined; buds 4–8, elliptical, 3–4 mm long, cap conical; fruit nearly globular, 5 mm diam.

*E. alpina*—**Grampians Gum**     Vic.
5–7 m     500 mm     Will grow on wet soils     Fast
Low windbreaks, parks, roadsides

Small, densely-crowned, spreading shrub or small tree; ranging from roundish to semi-upright. Grows well on granitic sands. Suitable for cold areas. Wind-firm. Withstands light snowfalls. Bark fibrous (stringy-bark); leaves thick, oval, 5–12 cm long, 3–8 cm wide; buds 3–5, often warty and irregular in shape, 1 cm long, cap round; fruit globular, sessile, often warty, 1–2 cm long and wide, valves protruding.

*E. amplifolia*—**Cabbage Gum**     N.S.W., Qld
8–15 m     650 mm     Will grow on     Moderate
                      heavy soils
Windbreaks, roadsides

Medium, moderately-dense tree. Bark smooth; leaves 10–20 cm long, 2–4 cm wide; buds 7–20, horn-shaped, 10–15 mm long, cap 3 times as long as the calyx; fruit globular, 5–6 mm diam., valves prominently exserted.

*Eucalyptus astringens*

*E. annulata*—**Open Fruit Mallee**     W.A.
5–7 m     450 mm     Prefers heavier but     Moderate
                     well-drained soils
Low windbreaks, parks, roadsides, ornamental, honey

Typical mallee. Compact, spreading, well-shaped small tree. Drought- and frost-resistant. Bark smooth, grey; leaves narrow, dark shining green, 7–12 cm long, 1–2 cm wide; buds 6–12, sessile, 15 mm long but distinctively constricted in the middle, cap long; flowers creamy-yellow; fruit globular, sessile, 7–10 mm diam. with the valves protruding prominently.

*E. approximans*—**Barren Mountain Mallee**     N.S.W.
2–3 m     650 mm     Will grow on     Moderate
                     granitic soils
Low shelter, parks     New introduction

Bushy, small tree often of mallee habit. Hardy, frost-resistant. Relatively new introduction. Bark smooth, dark grey; leaves narrow, 5–10 cm long, 5–7 mm wide, shining; buds 4–8, 6–7 mm long, cap round; fruit oval, 5–7 mm long.

*E. archeri*—**Archer Alpine Gum**     Tas.
3–5 m     750 mm     Moderate
Windbreaks, low shelter

Variable, small tree often with a twisted trunk. Withstands snow and exposure in alpine areas. Tolerates some wetness. Bark white, smooth; leaves 5–10 cm long, 1–3 cm wide, thick, leathery; buds in threes, sessile, 5–7 mm long and wide, cap short; fruit oval, 5–7 mm long, yellowish.

*E. astringens*—**Brown Mallet**  W.A.
12–18 m  450 mm  Requires well-drained  Fast
soil
Tall windbreaks, farm forests, high shade

Medium-sized, usually straight small crowned tree with good form. Good forms are suitable for uses similar to sugar gum but a multi-stemmed strain with a bushy crown also occurs. Very vigorous in the warmer districts. Withstands dry conditions and heavy soils but requires good drainage; growth on drainage flats can be poor. Tolerates saline conditions. Trunk slightly fluted; bark smooth, brown, shining, rich in tannin; leaves 7–12 cm long, 1–2 cm wide; buds 3–7, cylindrical, 2–3 cm long, 1 cm wide; fruit hemispherical, thick, 6–10 cm long and wide.

*E. baxteri* (E. capitellata)—**Brown Stringybark**  S.A., Vic., N.S.W.
20–35 m  600 mm  Fast
Farm forests, windbreaks

Tall, upright tree with a straight trunk and dense crown. Produces a general-purpose timber. Best growth on moist good soils but will form useful stands on sandy or poor soils. Not suitable for wet or saline sites. Hardy. Frost-resistant. Seeds and coppices freely. Survives severe coastal exposure but under these conditions may be a small, mis-shapen, dense bush, often only 1 m tall. Bark thick, reddish-brown, fibrous to the smaller branches; leaves 7–15 cm long, 2–5 cm wide, shining, slightly thick; buds 5–10, 6–8 mm diam., sessile, cap hemispherical, short, thick; fruit nearly globular, 5–8 mm diam. sessile, prominent brownish-red rim. 1889.

*E. bosistoana*—**Gippsland Grey Box**  Vic., N.S.W.
20–25 m  650 mm  Prefers good soils  Moderate
Farm forests, windbreaks, shade, parks, honey

Attractive tree with a tall, straight trunk and rather open crown. Produces valuable and durable timber. Prefers well-drained heavier soils but can tolerate limited periods of water-logging and dry periods. Grows well on coastal sands over clay. Bark rough, smooth on the branches; leaves thin, pale green, 8–18 cm long, 2–3 cm wide; buds 3–7, oval, 1 cm long, cap conical; fruit hemispherical, 6–8 mm long and wide.

*E. botryoides*—**Southern Mahogany**  Vic., N.S.W.
20–25 m  550 mm  Will grow on wet soils  Very fast
Shade, avenues, parks, stream banks

Widely-planted, hardy, spreading, dense, large-crowned tree. Produces useful timber. Versatile, will grow on poorly-drained areas, moderately saline soils and sandy soils. Wind-firm. Withstands smog and moderate coastal exposure. Frost-resistant. Can be pollarded. Will grow in 400 mm rainfall districts if irrigation water is available. Prone to borer damage in the more adverse sites. Bark rough and fibrous; leaves dark green, thick, sessile, 7–12 cm long, 3–5 cm wide, with parallel veins from the midrib; buds 6–10, angular, 1 cm long, cap round; fruit, barrel-shaped, 1–2 cm long, 10–12 mm wide, sessile. 1872.

*Eucalyptus botryoides*

*E. brockwayii*—**Dundas Mahogany**  W.A.
10–20 m  300 mm  Avoid deep sands  Fast
Windbreaks, shade, farm forests, parks, roadsides

Spreading, densely-crowned tree with branches ascending at a sharp angle. Relatively new introduction, producing good farm timber. Usually single-stemmed but multi-stemmed trees are not uncommon. Best on the lighter, loamy soils but will grow on heavier and alkaline soils. Vigorous. Drought-resistant. Moderately frost- and salt-tolerant. Bark smooth, reddish-brown; leaves up to 10 cm long, 1–2 cm wide; buds 8–12, sessile, 7–10 mm long; fruit small, in groups of 8–12, globular, 5–7 mm diam., very constricted at the opening.

*E. burdettiana*—**Burdett Gum**  W.A.
2–3 m  500 mm  Suitable for clay soils  Fast
Windbreaks, parks, ornamental

Dense, small tree or shrub, often with several stems. Will grow on stony soils. Tolerates some salinity but not poorly-drained sites. Frost-resistant. Withstands moderate coastal exposure. Bark smooth, greenish-cream; branchlets yellow; leaves 4–7 cm long, 1–2 cm wide, dark green, thick, veins very faint, diverging at 25°–30°; buds 3–5, sessile, cylindrical, 5 cm long; flowers yellow-green; fruit almost globular, 2–3 cm diam., sessile.

*E. burracoppinensis*—**Burracoppin Mallee**  W.A.
3–5 m  350 mm  Prefers lighter soils  Fast
Windbreaks, parks

Attractive, dense-crowned mallee, retaining branches to ground level. Drought-resistant, moderately frost-resistant. Bark smooth; leaves light green to bluish, 7–12 cm long, 1–2 cm wide, thick; buds in threes, pear-shaped, 2–3 cm long, slightly ribbed, cap hemispherical, thick; flowers whitish-yellow, large; fruit pear-shaped, 2–3 cm long, flared out towards the rim, disc 5 mm wide.

*E. caesia*—**Gungurru**  W.A.
4–5 m  400 mm  Will grow in sandy  Fast
and granitic soils
Low shelter, streets, parks, ornamental

*Eucalyptus caesia—flowers  R. Elliot*

Shapely, attractive mallee, one of the best of the small eucalypts. Easily grown but frost-tender when young. Not suitable for the cooler districts. Outer branches pendulous on older trees, branchlets mealy-white. Not very wind-firm, staking is recommended when young. Bark smooth, attractive, peeling off in reddish flakes, exposing fresh red or green bark underneath; leaves silver-grey, 5–10 cm long, 2–3 cm wide; buds in threes, oval, 2 cm long, cap strongly pointed; flowers very striking, reddish to pink in large clusters; fruit almost urn-shaped, 2–3 cm long and wide, mealy-white.

*E. calophylla*—**Marri**                     W.A.
6–10 m     600 mm     Not suitable for     Fast
                      heavy soils
Parks, streets, roadsides, ornamental, honey

Dense, round-headed, shapely tree producing showy masses of flowers in summer each year. One of the most prolific flowering of the eucalypts. Flowers white, often tinged with pink. Frost-tender when young, best in the 600–800 mm rainfall areas, but can be grown in drier parts with irrigation. Withstands moderate coastal exposure. Bark rough, flaky; leaves 7–15 cm long, 2–4 cm wide with pronounced divergent venation; buds 3–7, 10–15 mm long, cap round; fruit large, urn-shaped, 3–4 cm long, 2–3 cm wide.

*E. calophylla* var. *rosea*—**Red Flower Marri**   Cultivar
Similar to *E. calophylla* but flowers are red. Most of the trees in cultivation are derived from a red-flowered mutant originally in the Melbourne Botanic Gardens; they are not the same as *E. ficifolia* x var. *guilfoylei* as stated in some references. In appearance, very similar to *E. ficifolia* (W.A. red flower gum) and can only be distinguished by the seeds which with *E. calophylla* are black and wingless, whereas those of *E. ficifolia* are brown with a rudimentary menbraneous wing. Although still frost-sensitive this species is more frost-resistant than *E. ficifolia*.

*E. calycogona*—**Gooseberry Mallee**   W.A., S.A., Vic.
5–6 m     300 mm     Prefers heavier soils
Low windbreaks, parks, roadsides

Attractive, graceful, slender tree or shapely mallee with smooth, silvery-brown, deciduous bark and white flowers. Frost- and drought-resistant. Good tree for dry districts. Leaves pale green, thick, 5–10 cm long, 2–3 cm wide; buds 3–8, conical, short; fruit angular, square in cross-section, 1 cm long.

*E. camaldulensis* (E. rostrata)—**River**   All States
                                **Red Gum**
20–30 m   Indefinite  Best on deep, moist,   Fast
                      silty soils
Windbreaks, shade, parks, farm forests, honey

Well known and widely spread, the only eucalypt to occur naturally in all the mainland States. Medium-sized, spreading tree, narrow when young but spreading with age; often as wide as it is high if space permits; heavy branching on old open grown trees. In farm forests or other close formation it remains tall and narrow; extensively planted overseas as farm and industrial forests. Fast-growing when young. Good shade tree, grass grows to its base. Requires moist soil such as stream banks or areas subject to regular flooding; if these conditions are

*Eucalyptus calophylla*

*Eucalyptus camaldulensis*

*Eucalyptus camaldulensis—avenue*

*Eucalyptus camaldulensis—windbreak*

met it will grow in hot, dry areas with rainfall as low as 200 mm per annum; otherwise planting needs to be in above 500 mm rainfall areas. Withstands inundation for several months but will not survive continual wetness. One or more floodings each year is advantageous. Can be planted but not very satisfactory on calcareous soils. Drought- and frost-resistant. Does not withstand coastal exposure but will tolerate moderately saline conditions. Bark smooth, ashy-grey; leaves 12–20 cm long, 3–4 cm wide, branchlets on older trees pendulous; buds 5–10, conical, 8–10 mm long, cap conical, 1–2 times longer than the calyx; fruit globular, 5–7 mm, valves protruding prominently. *E. blakelyi* and *E. tereticornis* (E. unbellata) are closely related species of red gum which differ mainly in botanical characteristics and by being not so large or spreading in old age. 1872.

**E. campaspe—Silver Top Gimlet**  W.A.
6–8 m    250 mm    Grows in heavy soils    Fast
Windbreaks, light shade, streets, parks, ornamental

Slender tree, with a smooth, bronze-coloured fluted trunk and rounded crown. Smaller branchlets and twigs mealy-white. Fairly adaptable as regards soil, will grow on calcareous soils but growth only satisfactory in the warmer districts. Drought-resistant. Moderately salt-tolerant. Bark, rich in tannin; foliage bluish, leaves 5–10 cm long, 1–2 cm wide; buds 3–6, bluish, globular, 1 cm, sessile, cap round; fruit hemispherical, sessile, wider (1 cm) than long (6–8 mm).

**E. chapmaniana—Bogong Gum**  Vic.
12–20 m    1000 mm    Fast
Windbreaks, shade, parks

Spreading, short-boled, much-branched tree, growing to 30 m in its native habitat. Relatively new introduction and not yet widely planted. Frost-resistant, withstands prolonged snow, suitable for cold and alpine areas. Bark thick on trunk, smooth and white on the branches; leaves long and narrow, 15–25 cm long, 2–4 cm wide, slightly bluish, veins faint but diverge from the midrib; buds in threes, about 1 cm long, conical, sessile, cap hemispherical, short; fruits bluish, sessile, slightly bell shaped, 10–12 mm long, 8–10 mm wide, valves exsert.

**E. cinerea—Argyle Apple** or    Vic., N.S.W.
             **Silver Leaf Stringybark**
5–10 m    600 mm    Will grow on wet soils    Fast
Shade, low shelter, parks, streets, ornamental, honey

Compact, conical, small to medium tree with attractive silvery-blue foliage. Juvenile foliage is retained to maturity. Will tolerate adverse sites, particularly poorly-drained areas. Moderately drought- and frost-hardy. Widely planted as a street tree and ornamental in N.S.W. and A.C.T. Bark rough, fibrous, reddish-brown (stringybark); leaves variable, 7–12 cm long, 2–4 cm wide, bluish, almost sessile; buds in threes, sessile, 1 cm long, cap short; flowers cream; fruit hemispherical, sessile, 5–10 mm.
Should not be confused with the species formerly known as E. cinerea var. multiflora which is the common type in

*Eucalyptus chapmanniana*

*Eucalyptus cinerea*

Victoria. E. cinerea var. multiflora, now *E. cephalocarpa*, is much more variable in form and does not retain the silvery-blue juvenile foliage. The var. multiflora (*E. cephalocarpa*) is distinguished by the buds which are in large clusters of 4–10 compared with only threes in *E. cinerea*, and the leaves being long and green.

*E. citriodora*—**Lemon Scent Gum**                    Qld
15–20 m    500 mm    Well-drained loams    Fast
Parks, roadsides, avenues, farm forests, ornamental

Slender, graceful, upright tree with a very attractive, clean, long white trunk. Lightly-branched crown, thin, light green colour. Will grow in lower rainfall areas if irrigation water is available. One of the most popular of ornamental trees and one of the fastest-growing eucalypts in the drier districts. Very frost-tender when young, consequently only a small proportion of the trees planted survive; covering whenever there is a risk of frost until the tips are above the frost level (i.e. usually until the tree is about 2 m high) is necessary. Withstands slight coastal exposure. Bark chalky white to ground level; leaves slender, 10–18 cm long, 1–2 cm wide, rough and hairy when young, smooth and light green when mature, emitting a strong citrus odour when crushed; buds 3–5, oval, 1 cm long, cap round; flowers white, in dense sprays; fruit slightly urn-shaped, 8–10 mm long and wide.

*Eucalyptus citriodora*

*Eucalyptus cladocalyx—avenue*

*Eucalyptus cladocalyx—plantation*

*Eucalyptus cladocalyx* var. *nana*

established, slightly frost-tender when young. Drought-resistant. Wind-firm. Can be freely pollarded, copices vigorously from cut stumps. Older crowns become sparse and straggly but can be restored by pollarding. Best eucalypt for farm forests, good for high windbreaks but requires under-planting for low shelter. Bark greyish-white, smooth; leaves deep green, glossy, 8–15 cm long, 2–4 cm wide; buds 5–15, 1 cm long, cap short and fitting over the calyx like an acorn cup; fruit oval, 1 cm long, 6–8 mm wide, several ribs, constricted at the opening. 1872.

**E. cladocalyx var. nana—Dwarf Sugar Gum**
4–5 m      400 mm                                    Fast
Low shelter, background shrub, roadsides

Similar characteristics to *E. cladocalyx* but grows to a bushy, compact, small tree. Very good tree for under-planting with tall sugar gum.

**E. coccifera—Mt Wellington Peppermint**      Tas.
3–20 m      800 mm                            Moderate
Windbreaks

Variable tree, depending on its situation—twisted trunk and as little as 3–5 m tall in exposed sites, 15–20 m and straight in sheltered localities. Withstands severe cold and snow, one of the most cold-resistant eucalypts. Very suitable for alpine areas. Bark white, smooth; leaves bluish, small, 5–7 cm long, 1–2 cm wide, thick; buds usually in threes, bluish, wrinkled, 8–10 mm long, cap short; fruit sessile, more or less conical, 1 cm long, 7–10 mm wide, slightly ribbed.

**E. cornuta—Yate**                              W.A.
3–6 m      500 mm      Will grow on alkaline,      Fast
                       gravelly, and saline
                       soils
Low windbreaks, roadsides, parks, ornamental

Small, bushy, spreading tree suitable for the same conditions as sugar gums. Taller in the warmer inland areas. Frost- and drought-resistant. Withstands moderate coastal exposure. Will grow in lower rainfall areas but then tends to be short-lived. Bark rough, smooth on the branches; leaves narrow, 10–12 cm long, 1–2 cm wide; buds 5–15, sessile, 4–5 cm long, cap very distinctive finger-shaped, 3–4 cm long, 3–4 times longer than the calyx; flowers greenish-yellow; fruit oval, sessile, 8–10 mm wide, valves protruding conspicuously.

**E. corrugata—Rough Fruit Mallee**              W.A.
6–12 m      300 mm      Prefers sandy loams      Fast
Parks, roadsides, ornamental

Attractive, small tree (not a mallee) of good form and appearance. Will grow on stony and lateritic soils. Drought- and frost-resistant. Bark smooth, grey-brown; leaves 10–15 cm long, 1–2 cm wide, thick; buds 3–6, 10–15 mm long, cap hemispherical, corrugated; fruit hemispherical, 1 cm long, ribbed.

**E. cladocalyx—Sugar Gum**                      S.A.
20–25 m      400 mm      Not satisfactory on      Fast
                         very heavy soils
Farm forests, tall windbreaks, honey

Well known and widely planted. Upright, with an open crown; lower branches are shed rapidly giving a long clean trunk. Will grow on a wide range of soils but prefers lighter types. Will grow on limestone. Easily

*E. cosmophylla*—**Cup Gum**                                        S.A.
4–6 m       600 mm     Will grow on poorly-                Fast
                       drained sites
Windbreaks, low shelter, roadsides, honey

Upright, compact and neat tree when space permits; can be straggly if crowded. One of the best eucalypts for poorly-drained soils. Frost-resistant. Bark smooth, shed in flakes; leaves thick, oval, 12–20 cm long, 4–5 cm wide; buds 3–6, sessile, attractive, yellow with red spots, 10–15 mm long, cap round, beaked; flowers white; fruit hemispherical, cup-shaped, sessile, 10–15 mm long and wide, 2 ribs.

*E. crebra* (E. racemosa)—**Narrow Leaf Ironbark**
                                        N.S.W., Qld
12–25 m     500 mm     Prefers loam over          Moderate
                       clay
Farm forests, windbreaks, shade, roadsides, parks, honey

Symmetrical tree with a thin crown and slender, drooping branchlets. Open grown trees tend to be spreading. Good tree for warm, inland areas. Tolerates a wide range of soils and climates; grows well on heavy loams or sands over clay. Frost-hardy. Bark hard, dark reddish-brown, corrugated; leaves slaty-grey, narrow, 7–12 cm long, 10–15 mm wide; buds 4–9, small, 5–7 mm long, cap conical; flowers creamy; fruit almost globular, very small 4–6 mm diam.

*E. crenulata*—**Buxton Gum** or **Silver Gum**        Vic.
5–10 m      500 mm     Will grow on wet         Very fast
                       soils
Low shelter, roadsides, parks, ornamental

Usually a slender, upright, silver-grey tree. Some specimens tend to be more spreading, presumably a reflection of the seed type. Occurs in a very restricted swampy locality near Buxton in Victoria on private property, most of which is being cleared. Introduced only over the last 15 years or so, hence its characteristics under cultivation are not fully known. Results so far show it to be a vigorous, hardy tree suitable for a wide range of soils. Appears to withstand moderate coastal exposure. Bark smooth, grey, with silver streaks; leaves sessile, roundish, 3–5 cm, margin finely crenulated (only eucalypt likely to be seen in south-eastern Australia with this feature), young leaves at the ends of the branchlets bluish; buds 3–9, 5–7 mm long, oval, cap round; flowers white; fruit slightly bell-shaped, 4–5 mm long and wide. 1957.

*E. crucis*—**Silver Mallee**                          W.A.
4–5 m       250 mm     Prefers lighter          Moderate
                       soils
Low shelter, ornamental

Typically flat-topped, shrub-like mallee with slender, mealy-white branches. Suitable for granitic soils. Planted mainly for its bluish foliage, but green-foliaged strains occur. Drought- and frost-resistant. Withstands moderate coastal exposure. Can develop into a straggly bush; light pruning for shape is desirable. Bark smooth; leaves round, in pairs, covered in a mealy-white bloom, 4–5 cm diam; buds 3–7, oval, 1 cm long, cap round, longer than the calyx; flowers yellow; fruit barrel-shaped, 1 cm long and wide. This description is for the Westonia strain; the more normal type has lanceolate green leaves.

*Eucalyptus crenulata*

*Eucalyptus crucis*

*E. dawsonii*—**Slaty Box**                    N.S.W.

10–15 m      550 mm      Will grow on dry,      Fast
                          stony soils

Farm forests, windbreaks, roadsides

Upright tree with a clean trunk, slight tendency to be spreading. Suitable for farm forests where the site tends to be dry. Moderately frost- and drought-resistant. Bark smooth, slaty colour; leaves 7–12 cm long, 2–3 cm wide, bluish; buds 3–6, 5–6 mm long, cap hemispherical, short; fruit almost conical, 5–6 mm long, sessile.

*E. decipiens*—**Limestone Marlock**          W.A.

6–8 m       750 mm      Will grow on          Moderate
                          limestone

Shade, windbreaks

Spreading, densely-crowned tree, one of the few eucalypts which will grow well on limestone soils. Withstands moderate coastal exposure. Bark rough, fibrous, furrowed, persistent to the branches; leaves 4–10 cm long, 3–4 cm wide, thick; buds 6–15, 8–10 mm long, cylindrical, cap conical; fruit globular, 5–6 mm diam., sessile, valves exsert.

*E. delegatensis* (E. gigantea)—**Woollybutt**   Vic., Tas.,
                                                 N.S.W.

25–40 m     1000 mm     Good soils            Fast

Windbreaks, farm forests, honey

Tall, upright, excellent timber tree. Grows naturally in sub-alpine areas where some snow occurs each winter. Usually not planted privately in Australia but used fairly commonly overseas in cooler climates as a windbreak tree. When open grown it retains branches well down the trunk, and appears resistant to many insects. Worth planting in suitable areas. Very frost-resistant. Bark fibrous (stringybark); leaves 8–15 cm long, 3–5 cm wide; buds 7–15, almost globular, 7–9 mm long, cap round, short; fruit globular to pear-shaped, 1 cm long.

*E. desmondensis*—**Desmond Mallee**          W.A.

4–5 m       400 mm                            Moderate

Low shelter, ornamental

Sparsely-foliaged mallee with slender, silver-white stems and pendulous foliage. Stems rarely upright. Moderately drought- and frost-resistant. Bark smooth, white; leaves bluish, 7–12 cm long, 1–2 cm wide; buds 7–15, sessile, very bluish, 12–15 mm long, cap conical; flowers creamy-yellow; fruit almost globular, bluish, 1 cm diam.

*E. dielsii*—**Diel Mallee**                  W.A.

3–5 m       300 mm                            Fast

Low windbreaks, roadsides

Hardy, bushy mallee, retaining the erect branches close to the ground. Fairly recent introduction. Bark smooth, greenish-brown; leaves thick, 5–10 cm long, 5–10 mm wide, bright green, shining; buds 3–6, 12–16 mm long, cap bluntly conical; fruit urn-shaped, small, 6–8 mm long and wide.

*Eucalyptus eremophila*

*E. diptera*—**Two Wing Gimlet**              W.A.

6–8 m       250 mm      Not suitable for light soils   Fast

Low shelter, shade, parks, ornamental

Small tree, with short, fluted, smooth, bronze-coloured trunk. Tolerates saline soils and slight coastal exposure. Bark yellow-brown, thin; leaves thick, narrow, shining, 5–8 cm long, 1 cm wide; buds 3–5, 7–10 mm; fruit round, 1 cm diam., valves protruding. Bud, cap and fruit are unique in being 'flattened' to give two wings.

*E. diversifolia*—**Coast Gum** or **Soap Mallee**   S.A., Vic.

4–5 m       400 mm      Grows on sandy,       Moderate
                          calcareous or
                          shallow soils

Low shelter, parks, roadsides, honey

Bushy mallee, sometimes a small tree, useful for coastal sands over limestone. Withstands moderate coastal exposure. Bark greyish-white, ribbony; leaves thick, 7–12 cm long, 1–2 cm wide; buds 3–8, 8–10 mm, slender, cap conical, longer than the calyx; fruit globular, 1 cm, sessile.

*E. doratoxylon*—**Spearwood Mallee**         W.A.

3–4 m       400 mm      Prefers sandy to      Fast
                          sandy clay loam

Parks, streets, honey, ornamental

Attractive, showy, small tree or mallee with long, straight and slender stems (hence spearwood, used by the aborigines). Flowers well. Requires well-drained soils, tolerates some salinity. Moderately drought- and frost-resistant. Bark smooth, greenish-white; leaves dark green, shining, small and narrow, 5–8 cm long, 1 cm wide, opposite—one of the few eucalypts with opposite mature leaves; buds 3–8, cylindrical, 7–10 mm long, pendulous, cap long, beaked; fruit globular, 5–7 mm diam., contracted at the tip.

*E. dumosa*—**Dumosa Mallee**            S.A., Vic., N.S.W.

4–6 m       300 mm                            Moderate

Low shelter, honey

Large shrub or small tree. Useful for sandy, dry areas, adaptable to most soils including saline. Hardy. Wind-firm, Drought-resistant Bark rough, smooth on the branches; leaves narrow, 7–10 cm long, 1–2 cm wide; buds 3–7, 1 cm long, cap conical, short; fruit cup-shaped, faint ribs, 8–10 mm long, 5–6 mm wide.

*E. dundasii*—**Dundas Blackbutt**      W.A.

| | | | |
|---|---|---|---|
| 10–12 m | 300 mm | Prefers medium, well-drained soils | Moderate, then fast |

Farm forests, tall windbreaks, avenues

Attractive, single-stemmed, erect tree with a lightly-branched, spreading crown. Sometimes slow to develop a main shoot but growth is fast once dominance has been determined. Satisfactory on calcareous soils. Not suitable for poorly-drained sites. Bark rough, smooth, reddish-brown on branches; leaves narrow, 7–10 cm long, 1 cm wide, shiny green; buds 4–7, 12–15 mm long, cap conical, peaked; fruit barrel-shaped to cylindrical, 1 cm long, 5–7 mm wide.

*E. ebbanoensis*—**Sandplain Mallee**      W.A.

| | | | |
|---|---|---|---|
| 3–5 m | 250 mm | Prefers the lighter soils | Fast |

Windbreaks, roadsides                    New introduction

Bushy mallee. Drought-resistant. Moderately frost-resistant. Bark smooth; twigs red; leaves 7–10 cm long, 1–2 cm wide, rather thick; buds in threes, pear-shaped, 1 cm long, cap hemispherical, short, blunt; flowers cream; fruit hemispherical, 8–10 mm diam., valves exsert.

*E. elata* (E. lindleyana, E. numerosa, E. andreana)— **River White Gum**      Vic., N.S.W.

| | | | |
|---|---|---|---|
| 20–30 m | 750 mm | Prefers deep, moist, well-drained soils | Fast |

Farm forests, windbreaks, parks, avenues, ornamental

Attractive, upright, slender tree with a dense, pendulous crown. Isolated trees tend to be slightly spreading. Frost-hardy. Suitable substitute in appearance for Lemon Scent Gum (*E. citridora*) in frosty areas. Bark smooth, greyish-white, deciduous; leaves narrow, 10–20 cm long, 1–2 cm wide; buds 7–40, 5 mm long, club-shaped, cap round, very short; fruit hemispherical, 5–6 mm diam., clusters on a short stalk.

*E. eremophila*—**Tall Sand Mallee**      W.A.

| | | | |
|---|---|---|---|
| 5–7 m | 250 mm | Will grow on calcareous and very sandy soil | Fast |

Low windbreaks, parks, streets, roadsides, ornamental, honey.

Slender, upright, handsome mallee. May be single-stemmed but usually multi-stemmed. Showy. Versatile, suitable for a variety of soils, including heavy and saline soils but prefers sandy soils. In warm, low rainfall areas will grow on almost pure sand. Crown persists close

*Eucalyptus erythrocorys*

*Eucalyptus erythrocorys—buds and flowers*

to ground level for some years but lower branches are eventually shed. Drought- and frost-resistant. Withstands moderate coastal exposure. Bark smooth, polished, brown, slightly scaly; leaves narrow, fleshy, 7–12 cm long, 1–2 cm wide; buds 3–7, reddish, approximately 5–6 cm long, narrow, cap twice as long as calyx; flowers profuse, creamy-yellow, occasionally a pink tinge; fruit slightly bell-shaped, 10–12 mm long, 8–10 mm wide.

*E. erythrocorys*—**Red Cap Gum**      W.A.

| | | | |
|---|---|---|---|
| 5–7 m | 500 mm | Will grow in limestone, alkaline, sandy and heavy soils | Fast |

Low shelter, parks, streets, ornamental

Attractive, slender tree, often multi-stemmed. Becomes straggly with age; can be improved with a light pruning. One of the most ornamental of the eucalypts. Frost-tender when young. Large showy buds, cap

*Eucalyptus ficifolia*

*Eucalyptus ficifolia—flowers*

*Eucalyptus ficifolia—flowers*

vivid red, calyx emerald green; often planted solely for the ornamental value of the buds. Not suitable for exposed positions as it is liable to wind damage, otherwise withstands slight coastal exposure. Coppices readily. Bark more or less smooth, flaky; leaves leathery, 15–25 cm long, 1–4 cm wide, deep green, slender; buds in threes, on a strap-like stalk 2–3 cm long, sessile, large, 4–5 cm long, 2–3 cm wide, cap red, rough, much wider than the calyx; flowers greenish-yellow; fruit bell-shaped to hemispherical, large, up to 5 cm long and wide, ribbed, woody.

### E. erythronema—**Lindsay Gum**

| 5–7 m | 300 mm | Prefers light soils but will grow on heavier types | W.A. Fast |

Parks, roadsides, avenues, ornamental

Small, hardy, usually erect but sometimes a crooked tree of mallee habit. Suitable for hot, dry areas. Slightly frost-tender. Produces masses of red flowers during winter— very colourful, usually red but some trees white. Bark smooth, powdery, white to pinkish, rich in tannin; leaves glossy green, 5–8 cm long, 1–2 cm wide; buds 3–8, 2 cm long, on stalks 1–2 cm long, oval, cap conical, 2–2½ times as long as the calyx; fruit almost hemispherical, ribbed, 8–10 mm long, and wide. The varieties 'Urrbrae Gem' and 'Augusta Wonder' have deep red flowers.

### E. fasciculosa—**Pink Gum**

| 6–10 m | 450 mm | | S.A., Vic. Moderate |

Shade, parks, honey, ornamental

Small tree with a rounded, slightly open crown and irregular trunk—rarely straight. Attractive shape. Hardy. Will grow on most soils, including dry, sandy ridges. Pink timber. Bark smooth, pink turning white; leaves 7–10 cm long, 2–3 cm wide; buds 3–7, bell-shaped, approximately 1 cm long, numerous, occurring in large clusters, cap short, conical; flowers prolific, flowering over a long period; fruit pear-shaped, 6–8 mm long.

### E. ficifolia—**Western Australia Red Flower Gum**

| 5–8 m | 500 mm | Prefers moist, well-drained sandy soils | W.A. Moderate |

Shade, parks, avenues, ornamental, honey

Well-known, widely-planted tree. Variable in shape, can be upright but more often with a twisted and distorted trunk. Compact, round crown; very showy in middle to late summer when the whole tree can be covered by a mass of flowers; individual trees vary from deep red to almost white. Best in the warmer areas with 500–750 mm rainfall but can be grown where irrigation water is available. Very frost-tender when young. Withstands moderate coastal exposure. Wind-firm. Subject to a virus disease of the bark which can cause death by, in effect, ring-barking. Very similar to *E. calophylla* but usually a bit smaller; can only be definitely separated from this species by the seeds which are brown and winged

*Eucalyptus flocktoniae—coppice regrowth*

*Eucalyptus forrestiana—fruit*

compared with the black wingless seed of *E. calophylla*. Bark rough, persistent; leaves hard, divergent veins, 8–15 cm long, 3–4 cm wide; buds 3–7, 2–3 cm long on stalks 2 cm long, cap round, short; fruit urn-shaped, large, 3–4 cm long and wide.

*E. flocktoniae*—**Merrit**　　　　　　　　W.A., S.A.

| | | | |
|---|---|---|---|
| 10–12 m | 250 mm | Will grow on calcareous and saline soils | Fast |

Farm forests, windbreaks, roadsides, parks

Dense, pyramidal crown. Upright form good for the above purposes but can be of mallee habit if the seed source is incorrect. Promising tree for farm forests. Very drought-resistant, best in drier areas. Prefers sandy loams. Bark smooth, silvery-white, deciduous, suitable for tanning; leaves dark green, 8–15 cm long, 2–3 cm wide; buds 3–7, 15 mm long, cap sharply peaked, shorter and much wider than the calyx; fruit urn-shaped, 7–10 cm long.

*E. foecunda* (E. leptophylla)—**Slender Leaf**　W.A., S.A.,
　　　　　　　　　　　　　　　**Mallee**　　Vic., N.S.W.

| | | | |
|---|---|---|---|
| 3–5 m | 250 mm | Will grow on dry and heavy soils | Moderate |

Low windbreaks, shelter, ornamental, honey

Bright green, fine-foliaged mallee with slender stems. Prefers sandy loams but will grow on heavier types. Drought-resistant. Bark white, smooth, deciduous; leaves bright green, very narrow, 5–8 cm long, less than 6 mm wide; buds 3–6, 5–7 mm long, cap conical; fruit hemispherical, 5–7 mm diam.

*E. forrestiana*—**Fuchsia Mallee**　　　　　　W.A.

| | | | |
|---|---|---|---|
| 3–5 m | 400 mm | Will grow on heavy and sandy soils | Fast |

Low shelter, roadsides, streets, parks, ornamental

Shapely, bushy tree with fairly dense branches. Proving very satisfactory in the warmer areas. Slightly frost-tender when young, but later moderately frost- and drought-resistant. Withstands some coastal exposure. Attractive deep red or scarlet pendulous capsules, colourful long before flowering. Best displays on the heavier sandy loams. Bark grey, smooth; leaves narrow, shining, 5–10 cm long, 2–3 cm wide; buds single, 3–4 cm long, 1 cm wide, on long stalks up to 4 cm long, cap peaked, 1 cm long; flowers bright yellow, conspicuous; fruit bright red, large, 3–4 cm long, 2–3 cm wide, retaining its colour when dry, 4 small wings giving it a 'squarish' appearance.

*E. gardneri*—**Blue Mallet**　　　　　　　　W.A.

| | | | |
|---|---|---|---|
| 8–12 m | 350 mm | Will grow on gravelly and lateritic soils | Fast |

Farm forests, windbreaks, avenues, honey

Attractive, slender, upright tree with a long clean bole and bluish foliage. Easily grown, one of the most vigorous trees in the drier areas. Takes some time to form a distinct leader and so may be bushy for several years. Drought- and frost-resistant. Strong timber. High tannin content in the bark. Bark thick, flaking off in thin flakes, silver grey; leaves thick, bluish, 8–15 cm long, 2–3 cm wide; buds 6–10, 2–3 cm long, cap very long, 3–4 times as long as the calyx; flowers yellow; fruit oval to pear shaped, medium-to-large, 1 cm long, 5–6 mm wide.

*Eucalyptus gardneri—flowers　W. Middleton*

*Eucalyptus globulus*

**E. gillii—Curly Mallee**                        S.A., N.S.W.
5–7 m        250 mm        Will grow on         Moderate
                           limestone
Roadsides, parks, ornamental

Silvery-bluish mallee, branches and stems usually curly. Suitable for dry areas. Drought-resistant. Bark smooth; leaves bluish, almost sessile, heart-shaped, 4–5 cm long; buds 3–9, oval, 10–12 mm long, cap peaked, longer than the calyx; flowers pale yellow; fruit almost globular, 5–7 mm diam., valves protruding.

**E. glaucescens—Tingiringi Gum**               Vic., N.S.W.
4–10 m       650 mm                                    Fast
Low shelter, parks, ornamental

Bushy, compact, attractive shrub or small tree with silvery-bluish foliage; sometimes of mallee habit. Hardy, frost-resistant. Withstands cold climates and snow, suitable for sub-alpine sites. Bark smooth, white; leaves 10–15 cm long, 1–2 cm wide, thick; buds in threes, sessile, 6–7 mm long, cap very short; fruit top-shaped, bluish, sessile, 5–7 mm long.
On Mt Erica in Victoria a tall upright form occurs. Height exceeds 30 m, with a straight shaft-like trunk. Good farm forest tree in the high rainfall zones.

**E. globulus—Tasmanian Blue Gum**              Vic., Tas.
25–40 m      750 mm        Deep, well-drained   Very fast
                           loams
Farm forests, tall windbreaks, shade, parks, avenues, honey

Floral emblem of Tasmania. Tall, straight tree with a clean trunk producing strong construction timber. For a eucalypt it is densely-crowned with heavy branching. Will grow in drier areas, down to 500 mm but height then only 20–25 m. Not recommended for lower rainfall districts as it is subject to severe attacks by longicorn beetles after 15–20 years. Frost-resistant. Wind-firm. Young plants usually not eaten by stock or vermin. Bark smooth, deciduous in ribbons; young foliage distinctly bluish, on squarish stems; mature leaves dark green, very long, up to 40 cm and 6 cm wide; buds single, very bluish, rough and warty, 15–20 mm diam., cap short, round; flowers creamy; fruit large, 2 cm diam., very blue, rough and warty. 1872.

**E. globulus var. compacta**
3–5 m                                                  Fast
Similar to *E. globulus* but a dense compact form. Retains foliage to ground level. Horticultural variety originating from California.

**E. gomphocephala—Tuart**                          W.A.
10–20 m      450 mm        Prefers well-drained       Fast
                           soils
Windbreaks, shade, roadsides, parks, farm forests, honey

Large, spreading, often a much-branched tree, retaining branches to ground level when open grown. In farm forests is upright producing a reasonably straight, clean trunk. (Up to 45 m tall in its native habitat.) Grows well on coastal sands and limestone areas. Wind-firm, drought- and frost-resistant, suitable for drier inland areas. On heavy soils or those with poor vertical drainage may be short-lived. Withstands severe coastal exposure. Bark light grey, fibrous, persistent; leaves 12–18 cm long, 2–3 cm wide; buds in threes, 2–3 cm long, 1 cm wide, cap round, wider than the calyx, the whole bud shaped like a

*Eucalyptus globulus— windbreak*

young mushroom, distinctive; fruit sessile, bell-shaped, 1–2 cm long and wide.

*Eucalyptus incrassata—flowers* National Parks Service

### E. gracilis—**Yorrell** — W.A., S.A., Vic., N.S.W.

5–10 m  250 mm  Prefers sandy soils  Moderate
over limestone
Windbreaks, shade, roadsides, honey

Small, bushy-crowned, slender tree often with mallee form on poorer sites. Frost- and drought-resistant. Withstands saline conditions. Produces good quality honey. Bark hard, dark coloured; leaves 5–10 cm long, 1–2 cm wide, shining; buds 3–8, 5–6 mm long, cap round, small; fruit oval, small, 5–6 mm long.

### E. griffithsii—**Griffith Grey Gum** — W.A.

6–8 m  250 mm  Prefers loamy soils  Moderate
Low shelter, roadsides, ornamental

Small mallee-like tree, often crooked. Tolerant of saline areas. Drought-resistant. Bark rough, smooth on branches; leaves 10–12 cm long, 2–3 cm wide; buds in threes, ribbed, 1 cm long, cap round, giving the bud a flat topped appearance, very short, rough; flowers white; fruit bell-shaped, ribbed, 1 cm long and wide.

### E. grossa—**Phillip River Gum** — W.A.

3–4 m  250 mm  Prefers moist,  Moderate
heavy soil
Low shelter, ground cover, ornamental, honey

Straggly, mallee-like, rather open shrub with thick and often contorted branches. Very suitable for low shelter and ground cover. Slow-growing when young. Often spreads wider than its height; can be trimmed to give a dense shrub. Easily grown. Slightly frost-tender when young. Drought-resistant. Ornamental flowers. Bark roughish; leaves thick, broad, bright shining green, 5–10 cm long, 3–4 cm wide, strongly veined; buds 3–8 on a thick stalk, bronze-red, 2–3 cm long, cap conical; flowers large, yellow, conspicuous, on red stems; fruit cylindrical, 1–2 cm long.

### E. gummifera (E. corymbosa)—**Bloodwood** — Vic., N.S.W., Qld

15–30 m  650 mm  Fast
Farm forests, shade, tall windbreaks, parks, roadsides, honey

Spreading, well-shaped, densely-crowned tree. Good shade tree when open grown; tall and narrow when planted in close formation. Timber deep red colour. Hardy, grows on a wide range of soils. Flowers profusely for 2–3 months each year, worth planting as a highway or park tree for this alone. Copious red kino flows from the bark, hence the common name. Bark rough, tile-like flakes (tesselated), persistent; leaves 10–18 cm long, 2–4 cm wide, fine parallel venation; buds 4–8, 2 cm long, cap round, short; flowers white, dense; fruit distinctly urn-shaped, 9–12 mm long and wide; seed brown with wings. 1889.

### E. gunnii—**Cider Gum** — Vic., Tas.

15–20 m  750 mm  Will grow on  Moderate
heavy soils  to slow
Windbreaks, shade, farm forests

Attractive, tall straight tree with a clean appearance. Excellent for planting at high altitudes under sub-alpine conditions. Suitable for wet sites. One of the most frost-resistant of the eucalypts, widely planted in Britain and in cold climates overseas. Will also grow in hot, inland districts if a good supply of water is available. Can be pruned. Bark smooth, white, deciduous; leaves 5–7 cm long, 2–3 cm wide; buds in threes, bluish, 5–8 mm long, cap round, short; fruit hemispherical, 8–10 mm long and wide. (Some authorities consider this species to be a form of *E. glaucescens*.) 1850.

### E. incrassata—**Yellow Mallee** — W.A., S.A.

4–5 m  250 mm  Prefers sandy soils  Moderate
Low windbreaks, roadsides, parks, ornamental, sand-drift areas, honey

Dense, low mallee, with clean stems. Hardy. Drought- and frost-resistant. Tolerates saline soils and slight coastal exposure. Bark grey, smooth; leaves 6–10 cm long, 1–2 cm wide; buds 3–7, cylindrical, 10–12 mm long, cap beaked; flowers white, profuse; fruit barrel-shaped, slightly ribbed, up to 1 cm long.

### E. johnstonii—**Johnston Gum** — Tas.

15–40 m  1000 mm  Moderate
Farm forests, windbreaks  New introduction

Tall, narrow, densely-crowned tree with a shaft-like trunk. Up to 60 m in its native habitat. Withstands cold climates and snow, good sub-alpine tree. Bark smooth, bluish; leaves 5–10 cm long, 1–3 cm wide, thick; buds in threes, 12–15 mm long, 3–4 ribs, cap short, beaked; fruit hemispherical, 1 cm long, 2–3 ribs, valves protruding.

*Eucalyptus kitsoniana*

**E. kingsmillii—Kingsmill Mallee**  W.A.

2–4 m 300 mm Prefers sandy soils Moderate

Low shelter, parks, ornamental New introduction

Shapely, attractive, bushy, shrub-like mallee suitable for the warmer districts. Frost- and drought-resistant. Bark smooth; leaves 10–12 cm long, 2–4 cm wide; buds in threes, each group on a stalk 3–4 cm long, reddish-brown, large, 3–4 cm long, 8–10 ribs, cap 2–3 times longer than calyx, conical, ribbed, beaked; flowers creamy-white; fruit hemispherical, 15–20 mm diam., with 8–10 prominent ribs.

**E. kitsoniana—Bog Gum or Gippsland Mallee**  Vic.

2–8 m 600 mm Will grow on wet sites Fast

Windbreaks, roadsides, ornamental

Small, upright, compact tree on good sites; otherwise may be shrubby or of mallee habit. One of the few eucalypts suitable for wet or peaty soils. Will grow in sub-alpine areas. Coppices well. Rarely exceeds 2–3 m in wet areas, but reaches 7–8 m in well-drained soils. Frost-hardy. Bark smooth, grey; leaves leathery, 10–15 cm long, 2–4 cm wide; buds 5–8, sessile, yellowish, 1 cm long, cap round; fruit sessile, conical to hemispherical, 1 cm long and wide.

**E. kondininensis—Stocking Gum**  W.A.

7–10 m 400 mm Will grow on Fast
saline areas

Windbreaks, roadsides, avenues, parks, ornamental

Slender, attractive tree. Withstands saline and marshy areas better than most trees. Bark smooth, yellowish with purplish blotches; leaves narrow, 6–8 cm long, 1–2 cm wide, yellow-green; buds 4–7, 1 cm long, almost conical, cap conical; fruit cup-shaped, 5–6 mm diam.

**E. kruseana—Kruse Mallee**  W.A.

2–3 m 300 mm Will grow in sandy Moderate
and granitic soils

Parks, roadsides, ornamental

Attractive, bushy, rather straggly, bluish shrub with an open pyramidal crown. Fairly recent introduction but promising on a wide range of sites in the warmer districts. Grows naturally in arid sandy desert country, as low as 200 mm annual rainfall. Drought-resistant. Young plants may be damaged by frost but otherwise frost-hardy. Greenish-yellow flowers. Not yet as well known as it deserves to be. Bark smooth, red-brown, deciduous; leaves distinctive, sessile, round, bluish, small, 2 cm diam., crowded on the stem; buds 3–6 bluish-white, oval, 1 cm long, cap conical; flowers bunched; fruit oval, 5–7 mm long.

**E. laeliae—Butter Gum**  W.A.

12–15 m 500 mm Not suitable for Moderate
laterite soils

Shade, parks, ornamental New introduction

Attractive, upright tree with a moderately dense, frequently drooping crown when open grown. Trunk a bright butter yellow when freshly exposed, turning to bright white, persisting to the smaller branches. Very similar to *E. accedens* in appearance and botanical description except the fruit regularly has only 3 valves compared with 4 in *E. accedens*.

**E. lansdowneana—Crimson Mallee Box**  S.A.

5–7 m 400 mm Will grow on Moderate
limestone soils

Low windbreaks, roadsides, parks, ornamental

Usually a dense, compact, showy, small tree of mallee habit with very slender branches and pendulous foliage. Occasional straggly specimens are seen. Flowers in clusters on the branches, variable in colour, usually creamy but bright pink to crimson or almost reddish-purple types occur. Good for mass plantings such as low shelter, hedges, etc. Can be lightly trimmed. Bark smooth, deciduous; leaves narrow, 8–15 cm long, 2–4 cm wide, dark green, pendulous; buds 3–8, sessile, 6–8 mm long, cap round, short; fruit sessile, barrel-shaped, up to 1 cm long with 1–2 ribs.

**E. largiflorens (E. bicolor)—Black Box**  S.A., Vic.,
N.S.W., Qld

12–15 m 400 mm Prefers heavy Moderate
soils to slow

Farm forests, windbreaks, honey

Shapely, upright tree, spreading when open grown with an irregular light crown. Mature trees attractive. Very durable timber. Best on moist and heavy soils, such as clay depressions along river banks in warmer districts. Requires access to ground water for good growth. Wind-firm. Tolerates saline conditions and periodic flooding. Drought-resistant. Frost-hardy. Bark rough, fibrous, persistent; leaves 6–10 cm long, 2–3 cm wide;

buds 3–7, club-shaped, small, 4–5 mm long, cap round, short; fruit sessile, globular, 4–5 mm diam., one of the smallest of the eucalypts.

*E. lehmannii*—**Bushy Yate**                          W.A.
5–6 m        450 mm        Prefers lighter soils        Fast
Low windbreaks, parks, ornamental, honey

Compact, bushy, roundish, small tree with branches to ground level. Isolated trees subject to wind-throw; need to be planted in groups or close together for low windbreaks. Withstands severe coastal exposure. Drought- and frost-resistant. Flowers conspicuous. Bark rough; leaves 7–10 cm long, 2–3 cm wide; buds 7–15, fused together into a disc 6–8 cm diam. on a flat stalk about 1 cm wide, cap very distinctive, finger-like, green or pinkish, 3–4 cm long, 2–4 times longer than the calyx; flowers yellow-green, conspicuous amongst the foliage; fruit 5–8 cm wide disc, consisting of 7–15 capsules fused together, valves protruding prominently to form several points.

*E. leptopoda*—**Tammin Mallee**                          W.A.
4–5 m        300 mm        Prefers lighter soils        Fast
Windbreaks, parks, streets, roadsides

Attractive, slender mallee with graceful, narrow foliage. More suitable for the warmer districts. Hardy. Moderately frost- and drought-resistant. Bark smooth, white to red-brown; leaves dark green, 8–12 cm long, 5–10 mm wide; buds 5–12, oval, 1 cm long, cap conical, very long; fruit nearly globular, 5–7 mm diam.

*E. le souefii*—**Rock Gum**                          W.A.
8–12 m        350 mm        Prefers medium to        Moderate
                                     light soils
Windbreaks, parks, roadsides, farm forests, ornamental

Straight, upright, slightly bluish tree with a sparse, and rather open crown. Vigorous, easy to grow. Will grow on heavy soils. Drought-resistant. Slightly salt-tolerant. Bark rough at the butt, smooth on the branches; leaves narrow, 7–10 cm long, 1–2 cm wide; buds 3–7, 1 cm long, several prominent ribs, cap conical, ribbed; fruit hemispherical, 12–15 mm long, prominently ribbed. A similar tree with smooth bark (sometimes referred to as E. 'pterocarpa') is a hardier and more vigorous form of *E. le souefii*.

*E. leucoxylon*—**Yellow Gum**                   Vic., S.A., N.S.W.
10–25 m        400 mm        Prefers heavier soils        Fast
Farm forests, windbreaks, shade, parks, avenues, roadsides, ornamental, honey, birds

Usually a tall upright tree but can be bushy depending on seed source. Crown dense on young trees, more open when older. Grows well on heavy and alkaline soils. Easily grown. Tolerates limestone. Durable timber. Hardy. Frost-resistant. Wind-firm. Withstands moderate coastal exposure. Can be pollarded and coppices freely from low stumps. Useful honey tree but yield is not high. Bark smooth, white, deciduous (an 'ironbark' variety

*Eucalyptus leucoxylon—young tree*

*Eucalyptus leucoxylon—mature tree*

believed to be a hybrid with *E. sideroxylon* occurs naturally in central Victoria); leaves 8–15 cm long, 2–4 cm wide; buds in threes, 1 cm long, cap conical; flowers white; fruit oval, 10–15 mm long, on a long stalk (1–2 cm). 1872.

195

*Eucalyptus macrandra*

*E. leucoxylon* var. *rosea*—**Red Flower Yellow Gum**
As for *E. leucoxylon*

Similar to *E. leucoxylon* but flowers crimson to red. Not always true from seed. Being frost-hardy is preferred to the W.A. red-flower Gum (*E. ficifolia*) even though it does not flower quite so profusely.

*E. leucoxylon* var. *macrocarpa*
6–10 m

As for *E. leucoxylon* var. *rosea*, but flowers and fruit are larger (2 cm long, 2–3 cm diam.).

*E. linearis*—**White Peppermint**                     Tas.
8–12 m     650 mm     Prefers light soils          Fast
Windbreaks, roadsides, ornamental

Slender, graceful, spreading tree with a fine rather light crown. Will grow on poor soils. Hardy. Frost-resistant. Withstands light snow; good sub-alpine tree. Bark smooth, white, deciduous; leaves distinctive, very narrow, 5–12 cm long, 5 mm wide; buds 5–12, very small, less than 5 mm long, cap round, short; fruit sessile, globular to pear-shaped, under 5 mm diam.

196

*E. longicornis*—**Red Morrell**                       W.A.
15–25 m     300 mm     Prefers heavier soils         Fast
Farm forests, windbreaks, shade, parks

Tall, upright tree with a straight trunk. Good farm forest tree for dry climates. Suitable for clay flats, also sandy loams over clay or sands with a calcareous subsoil. Drought- and frost-resistant. Tolerates saline and alkaline sites. Bark rough, smooth on the branches, reddish-grey; leaves narrow, 7–12 cm long, 5–10 mm wide, shining; buds 5–10, nearly cylindrical, 1 cm long, cap conical, twice as long as the calyx; fruit almost globular, 5–6 mm diam.

*E. loxophleba*—**York Gum**                           W.A.
8–12 m     350 mm     Prefers lighter soils          Fast
Windbreaks, shade, parks, honey

Usually multi-stemmed, densely-crowned, small tree. Will grow on granitic and shallow soils. Drought- and frost-resistant. Bark rough, persistent; leaves pale green, 8–12 cm long, 10–15 mm wide, veins making a very acute angle with the midrib; buds 5–12, club-shaped, 6–8 mm long, cap hemispherical, short; flowers white, prolific; fruit pear-shaped, 8–10 mm long.

*E. macrandra*—**Flowering Marlock**                   W.A.
5–7 m     500 mm                                      Fast
Low windbreaks, parks, roadsides, ornamental, honey, birds

Compact, low-growing mallee which flowers profusely. Retains foliage to ground level. Good for mass plantings. Frost-resistent. Tolerates moderate coastal exposure, poor drainage, and saline soils. Bark smooth, light-coloured; leaves 7–10 cm long, 1 cm wide; buds 8–15, sessile, horn-shaped, distinctive, long, thin, 3–4 cm long, 5–6 mm wide, cap 3–4 times longer than the calyx; flowers yellow; fruit bell-shaped, 7–10 mm long, 5–7 mm wide.

Below   *Eucalyptus macrocarpa*—*flowers*   F. Collet

*Eucalyptus macrorhyncha*

*Eucalyptus macrorhyncha—mature tree*

*E. macrorhyncha*—**Red Stringybark**   S.A., Vic., N.S.W.
20–25 m   550 mm   Not in heavy soils   Moderate
Farm forests, windbreaks, shade, parks, honey

Upright, densely-crowned tree producing good milling timber. Isolated specimens very attractive, branchlets pendulous. Grows on a wide range of soils, particularly granitic and dry sites. Withstands high temperatures and occasional snow. Good paddock tree; provides shade and grass grows to the butt. Bark reddish, fibrous, persistent (stringybark); leaves 8–12 cm long, 1–2 cm wide; buds 6–12, 1 cm long, cap conical; fruit globular, 8–12 mm diam., broad rim, valves protruding. 1889.

*E. maculata*—**Spotted Gum**   Vic., N.S.W., Qld
15–20 m   450 mm   Will grow on   Fast
heavy soils
Farm forests, windbreaks, shade, parks, roadsides, ornamental, honey

Impressive, densely-crowned tree, with attractive mottled bark and branches persisting much lower than usual for tall eucalypts. Versatile tree, suitable for most purposes and areas. One of the best eucalypts for windbreaks, but can vary in shape from spreading to upright depending on seed source. Grows on a wide range of soils, including sandy and poorly-drained. Wind-firm. Slightly frost-tender when young. Tolerates smog. Timber used for tool handles. Bark smooth, mottled, light-coloured with greyish-purple patches; leaves 10–20 cm long, 2–4 cm wide, prominent divergent (almost transverse) veins; buds 3–5, oval, 1 cm long, cap conical, short; fruit urn-shaped, thick, 10–15 mm long.

*Below* *Eucalyptus maculata—shade tree and spotted bark of a mature tree*

*E. macrocarpa*—**Rose of the West** or   W.A.
**Mottlecah**
2–3 m   450 mm   Unsatisfactory in   Fast
heavy soils
Ornamental

Very variable, bluish, mallee shrub, ranging from a dense, compact shrub to a spindly, single-stemmed plant. Pruning will assist to keep it bushy. Dense forms suitable for low windbreaks, otherwise purely a novelty because of its large flowers, the largest among the eucalypts. Bark smooth, bluish; leaves sessile, very bluish, variable in shape usually oblong, up to 15 cm long and 8 cm wide; buds single, 5 cm long, bluish, cap blunt; flowers usually crimson, 10–15 cm diam.; fruit flat, large, 7–10 cm across, largest among the eucalypts.

*Eucalyptus mannifera* subsp. *maculosa*

**E. microcarpa** (E. hemiphlois)—**Grey Box** Vic., N.S.W., Qld

15–20 m     450 mm     Prefers better and     Slow
heavier soils

Farm forests, windbreaks, shade, honey

Somewhat spreading when open grown; upright in close stands. Timber very durable. Most satisfactory on heavier well-drained soils in the warmer districts, or on light soils with a clay subsoil. Frost-hardy. Coppices vigorously from low stumps. Good honey tree, flowers

*Eucalyptus melliodora*

**E. mannifera** subsp. **maculosa**—**Brittle Gum** Vic., N.S.W.

6–25 m     650 mm     Will grow on stony     Fast
and dry sites

Parks, roadsides, ornamental

Attractive but variable tree, trunk may be straight or very twisted depending largely on the seed source. Rather too variable to be recommended for tall windbreaks; planted mainly because of its attractive bark. Crown rather open, branchlets semi-pendulous. Suitable for adverse sites, particularly stony soils at higher elevations. Requires good drainage. Frost-resistant. Withstands light snow. Bark smooth, white, with bluish-grey to yellow patches; leaves 8–15 cm long, 1–3 cm wide, narrow; buds 3–7, nearly globular, 5–6 mm diam., cap short, rounded conical; fruit oval to hemispherical, 5–6 mm long and wide, valves protruding slightly.

**E. melliodora**—**Yellow Box**     Vic., N.S.W., Qld

15–25 m     500 mm     Prefers the better     Moderate
soils

Farm forests, windbreaks, shade, roadsides, parks, honey

Tall, graceful, spreading, densely-crowned, slightly pendulous, bluish tree, producing durable timber. Long-lived. Young trees narrow, symmetrical. Growth moderate to slow. Not very satisfactory on sandy, strongly calcareous, wet, or poor soils or in cold districts. Wind-firm. Coppices readily. One of the best honey trees, good yield and excellent quality, flowers profusely. Bark rough, fibrous, inner bark yellow; leaves 7–10 cm long, 10–15 mm wide, bluish; buds 3–6, small, 6–8 cm long, cap short; fruit globular, small, less than 6 mm diam.

*Eucalyptus melliodora—old tree*

freely each year. Bark fibrous, grey; leaves 7–10 cm long, 10–15 mm wide; buds 3–7, sessile, 6–8 mm long, cap conical; fruit globular, small, less than 5 mm diam.

*E. miniata*—**Melaleuca Gum** or W.A., Qld, N.T.
**Darwin Woollybutt**
5–8 m (?) 500 mm Prefers lighter soils Fast
Parks, roadsides, ornamental New introduction

Small bluish tree, variable in height, reaching 20–25 m in the northern coastal parts of Australia. Appears to be suitable for the warmer parts of south-eastern Australia, but little is known of its growth responses in the south. Worth trying. One of the most spectacular and showy of the flowering eucalypts, covered in masses of bright orange to red flowers during the dry season—better than the W.A. Red Flower Gum. Bark brown and fibrous on the trunk, white and smooth on the branches; leaves 7–15 cm long, 3–5 cm wide, bluish; buds 3–7, each group on a flattened stalk 2 cm long, oval, sessile, 2 cm long, 1 cm wide, cap conical, short; flowers prolific, conspicuous, 1–2 cm across, along and at the ends of branches; fruit barrel-shaped, 3–5 cm long, 2–3 cm wide, sessile, prominently ribbed.

*E. mitchelliana*—**Weeping Sallee** Vic.
5–10 m 750 mm Moderate
Windbreaks, low shelter, parks, ornamental

Attractive, spreading, densely-crowned small tree with weeping, willow-like foliage. Withstands severe frosts and prolonged snow, good tree for alpine regions. Bark smooth, whitish; leaves 8–15 cm long, 1–2 cm wide, diverging veins; buds 4–12, 5–6 mm long, cap conical; fruit sessile, globular, 5–6 mm diam.

*E. mitrata*—**Mitre Gum** W.A.
2–3 m 300 mm Prefers lighter soils Fast
Parks, ornamental, birds

Shapely, compact, widely-branched small tree or shrub suitable for the warmer districts. Showy cream flowers. Withstands frosts, dry conditions, and moderate coastal exposure. Bark smooth, grey; leaves variable, usually 7–12 cm long, 1–2 cm wide, thick, long, and tapering; buds 2–4, 2 cm diam., cap green, flat with a peak in the centre, strongly furrowed; flowers cream, prolific; fruit 3–4 cm long and wide, rich brown, deeply furrowed, valves an unusual rounded and distinctive shape ('eggs-in-the-nest').

*E. moorei*—**Little Sallee** N.S.W.
6–10 m 650 mm Tolerates slightly Moderate
wet soils
Low shelter, parks, honey

Multi-stemmed, almost mallee-like small tree, often spreading to 5–6 m. Suitable for sub-alpine, or cold, damp areas. Can be grown to form thickets. Similar to *E. stellulata*, except the latter is single-stemmed and tree-like. Bark smooth, olive green; leaves narrow, 5–6 cm long, 6–8 mm wide; buds 7–12, sessile, 7–10 mm long, cap conical; flowers white; fruit small, globular, 3–4 mm diam., in clusters of 7–12.

*Eucalyptus microcarpa*

*Eucalyptus mitchelliana*

*E. muellerana*—**Yellow Stringybark** S.A., Vic.,
N.S.W., Qld
20–30 m 600 mm Prefers heavier soils Moderate
Farm forests, windbreaks, shade, honey

Tall, erect, massive tree with a dense, spreading crown. Good quality, moderately durable timber, suitable for use in salt water. Will grow in heavy clay and poorly-drained soils; not satisfactory in deep, light, sandy soils. Good shade tree. Bark rough, brown, fibrous; leaves thick, shining, 10–16 cm long, 2–4 cm wide; buds 7–12, 6–8 mm, cap short; fruit globular, 6–8 mm diam. 1889.

*E. neglecta*—**Omeo Gum** Vic.
5–8 m 650 mm Will grow on wet soils Fast
Low shelter, roadsides, parks

Compact, bushy, densely-crowned small tree retaining its lower branches. Often multi-stemmed. Good for peaty or wet soils. Hardy. Wind-firm. Withstands light snow, suitable for sub-alpine localities. Bark smooth, greenish; leaves 7–12 cm long, 3–5 cm wide, roundish, dark glossy green; buds 5–8, sessile, 5 mm diam., cap hemispherical; fruit hemispherical, 4–6 mm diam., sessile.

*Eucalyptus muellerana—old tree*

*E. nicholii*—**Willow Leaf Peppermint**     N.S.W., Qld
8–10 m     650 mm     Will grow on poor soils     Fast
Shade, roadsides, streets, parks, ornamental

Attractive, decorative, slender-leaved, willow-like tree, with a rather compact crown. Young foliage purplish-bronze. Relatively new introduction, suitable for most sites. Will grow on drier, poorer, or gravelly soils. Frost-hardy. Bark blue-grey, smooth when young, fibrous, persistent later; leaves thin, very narrow, 8–15 cm long, less than 1 cm wide; buds 5–8, 5–7 mm long, cap conical; fruit globular, small, 3–5 mm diam.

*E. nitens*—**Shining Gum**     Vic., N.S.W.
25–40 m     1200 mm     Requires good,     Fast
                              moist soil
Farm forests

Tall, slender, upright tree with a long clean trunk. Suitable for the higher altitudes and cooler districts. Requires good, deep, moist soils. Withstands snow. Does not coppice. One of the ash group of eucalypts. Very selective as regards site, recommended that further advice be obtained before undertaking any large plantings. Bark smooth, ribbony; leaves 10–25 cm long, 1–2 cm wide, shining; buds 4–8, 8–10 mm long, 4–5 mm wide, sessile, cap conical, short; fruit conical, 5–7 mm long and wide, sessile.

*Eucalyptus nichollii*

*Eucalyptus obliqua*

*E. nutans*—**Red Flower Moort**                    W.A.

4–5 m        450 mm        Will grow on sandy        Moderate
                          loams and stony soils

Low shelter, roadsides, parks, ornamental

Bushy, spreading small tree. Moderately frost-hardy. Bark smooth, creamy; leaves 5–8 cm long, 2–3 cm wide, shining; buds 3–7 on a strap-like stalk, each bud sessile, 1–2 cm long, with 2–4 ribs, cap short, narrower than the calyx; flowers small, but prolific, rich red or crimson; fruit sessile, globular, 10–15 mm diam., ribbed.

*Eucalyptus occidentalis*

*E. obliqua*—**Messmate**        S.A., Vic., Tas., N.S.W.

20–25 m        650 mm        Well-drained soils        Moderate

Farm forests, tall windbreaks, shade, honey

Tall, straight tree, shedding the lower branches; clean trunk, crown reasonably dense. Hardy. Frost-resistant. Exceeds 50 m in high rainfall areas. Produces good quality, general-purpose, light-coloured timber. Only moderately durable. Requires well-drained soil; unsatisfactory if drainage is not good. Withstands light snow. May be attacked by the root fungus, *Phytophthora cinnamomii* in abnormally wet years. Bark fibrous, persistent to the smaller branches; leaves 10–18 cm long, 2–4 cm wide, glossy green, margins where they join the leaf stalk not opposite each other; buds 7–16, 6–8 mm long, cap short; fruit oval, 5–8 mm long. 1872.

*E. occidentalis*—**Swamp Yate**                    W.A.

12–15 m        400 mm        Prefers heavier soils        Fast

Farm forests, tall windbreaks, shade

Tree with erect branches giving a flat-topped rather narrow crown. Usually straight, but multi-stemmed trees occur—seed source should be carefully selected. Suitable for most soils. Will grow on wet or poorly-drained sites. Withstands slight coastal exposure. Moderately salt-tolerant. Frost-hardy. One of the faster-growing eucalypts in the drier districts. Bark thick, rough, persistent, branches smooth; leaves 12–15 cm long, 2–4 cm wide, thick; buds 3–7, 15–20 mm long, cap twice as long as calyx; flowers pale yellow; fruit bell-shaped, 1–2 cm long, slightly ribbed, valves protruding.

*Eucalyptus occidentalis—plantation*

*E. oleosa*—**Acorn Mallee** or        W.A., S.A.,
            **Oil Mallee**            Vic., N.S.W.

6–8 m        300 mm        Prefers sandy loams        Moderate

Shade, roadsides, eucalyptus oil, honey

Mallee, but more upright and taller than the typical mallee. Satisfactory on clay and limestone. Hardy. Drought- and frost-resistant. Tolerates saline conditions. Coppices freely. High oil content in the leaves. Bark rough; leaves narrow, 8–12 cm long, 1–2 cm wide, glossy green; buds 6–12, 8–10 mm long, cap conical; fruit globular, 6–8 mm diam. Very variable, has been divided into 6 species by some authorities.

**E. orbifolia—Round Leaf Mallee**  W.A., S.A.
4–5 m  350 mm  Medium to heavier  Moderate
soils
Parks, ornamental

*Eucalyptus paniculata*

Attractive, small, thin-stemmed mallee with reddish stems and mealy-white branchlets. Not as widely planted as it deserves to be. Bark colourful, red, flaky, peeling off to expose red or pale green fresh bark; leaves thick, bluish, round, 3–7 cm diam.; buds 2–5, 1 cm long, pink, cap conical, striped; flowers pale yellow; fruit slightly bell-shaped, 10 mm long, 15 mm diam., broad flat rim. Similar in most respects to *E. caesia* but is hardier and more frost-resistant.

**E. oreades—Blue Mountain Ash**  N.S.W., Qld
15–40 m  1200 mm  Requires moist, well-  Fast
drained loams
Farm forests

Tree with a shaft-like trunk and small crown. One of the ash group, produces good quality seasoning timber. Selective as regards site, only suitable for well-drained, good soils in the higher rainfall mountain districts. Sheds lower branches comparatively early. Frost-resistant. Withstands snow. Coppice, if occurring, very weak. Bark rough at the base, otherwise smooth, white, ribbony; leaves 10–15 cm long, 2–4 cm wide, pale green, shining, thick; buds 3–8, rounded, 5–7 mm long, cap hemispherical; fruit rounded conical, 5–7 mm long.

**E. ovata—Swamp Gum**  S.A., Vic., Tas., N.S.W·
12–20 m  650 mm  Will grow on  Moderate
wet sites
Windbreaks, honey

Densely-crowned, usually upright tree. Suitable for a wide range of soils in cool districts, particularly stream banks and wet sites. Withstands flooding longer than most eucalypts—several months at a time. Hardy, frost-resistant. Bark smooth, ribbony; leaves variable, generally oval, 7–12 cm long, usually 2–4 cm wide but may be only 1 cm wide; buds 4–8, 1 cm long, 5–6 mm wide, cap conical; fruit conical 8–10 mm long, 5–6 mm wide.

**E. paniculata—Grey Ironbark**  N.S.W.
5–25 m  500 mm  Moderate
Farm forests, windbreaks, shade, honey

Large-crowned, upright tree. In low rainfall areas and drier soils may be only 4–5 m tall, but where rainfall exceeds 900–1000 mm and soils are fertile height may be 25–30 m with 20 m of straight clean trunk. Fast-growing on moist sites. Open grown trees have a dense, bushy crown. Will grow on stony, dry sites, and sands over clay. Not suitable for hot districts. Slightly frost-tender. Produces a very heavy, dense and tough pink timber. Bark grey, hard, deeply furrowed, extending to the branches; leaves 8–12 cm long, 1–3 cm wide, sometimes slightly wavy; buds 3–9, oval, 10 mm long, 5 mm wide, arranged in large clusters, cap conical, short; flowers creamy; fruit hemispherical, 5–6 mm diam.

**E. pauciflora—White Sallee**  S.A., Vic., Tas., N.S.W.
10–20 m  550 mm  Will grow on poorly-  Moderate
drained soils
Windbreaks, shade

Variable tree with a rounded crown and several main branches. Wind-firm and hardy, tolerating some cold. Withstands moderate coastal exposure. Foliage eaten by stock when fodder is scarce. Good pollen tree. Bark smooth; leaves 8–15 cm long, 3–4 cm wide, thick, shining, veins almost parallel; buds 5–10, 6–8 mm long, cap short, hemispherical; flowers white, flowering profusely and frequently; fruit pear-shaped, 8–12 mm long, 6–8 mm wide.

**E. pauciflora, 'alpine form'** (E. pauciflora var. alpina, E. niphophila)—**Snow Gum**  Vic., Tas., N.S.W.
5–8 m  800 mm  Fast
Windbreaks, parks, ornamental, honey

Small tree with a curved or twisted, rarely straight trunk and heavy branches. Suitable for shelter in alpine areas; withstands long periods of heavy snow; occurring naturally to the tree line. Very hardy and frost-resistant. Coppices vigorously. Fibrous root system, one of the very few eucalypts which can be transplanted as large specimens. Planted as an ornamental because of its smooth, colourful, pink summer bark. Leaves thick, waxy, 5–10 cm long, 2–4 cm wide, veins thick, prominent, venation almost parallel; buds 3–7, sessile, 5–8 mm long, cap short; fruit globular, 8–10 mm diam., prominent rim.

**E. pauciflora var. nana—Wolgon Snow Gum**  N.S.W.
2–4 m  650 mm
Windbreaks

Slender mallee-like form of *E. pauciflora*, suitable for alpine areas. Tolerates cold and snow, but then only 2 m tall. Otherwise similar to *E. pauciflora*.

*E. pellita*—**Large Fruit Red Mahogany**   N.S.W., Qld
15–20 m     550 mm     Prefers loamy, well-     Fast
drained soils
Shade, windbreaks, parks, farm forests

Large, spreading tree when open grown but crowns restricted in close stands. Durable dark red timber. Frost-resistant. Bark rough, fibrous; leaves 10–15 cm long, 2–3 cm wide; buds 3–8, 2 cm long, cap rounded, wider than the calyx; fruit almost globular, 1 cm diam., valves protruding.

*E. perriniana*—**Spinner Gum**        Vic., Tas., N.S.W.
5–10 m     650 mm                        Fast
Low shelter, ornamental

Small, straggly snow gum. Very hardy. Frost-resistant. Will grow in alpine areas but usually planted at lower elevations as a garden novelty because of its silver grey foliage and round juvenile leaves. Juvenile leaves are opposite, and sessile, the stem appearing to pass through them so that dead pairs of leaves remain free to spin on the stem. More shade-tolerant than most eucalypts. Withstands some flooding. Bark smooth, lightly blotched; leaves favoured on cultivated specimens are juvenile leaves, mature leaves are 10–15 cm long, 1–2 cm wide and develop at about 4–6 years of age; buds in threes, bluish, 7–10 mm long, cap short; fruit hemispherical, 6–8 mm diam.

*E. pileata*—**Ravensthorpe Mallee**    W.A., S.A., Vic.
4–5 m     300 mm     Will grow on       Moderate
calcareous soils
Low shelter, parks, roadsides

Densely branched, bluish mallee; conspicuous because of its blueness. Moderately tolerant of saline soils. Drought-resistant. Bark rough; leaves thick, 5–10 cm long, 1–4 cm wide; buds 3–5, 7–9 mm, cap short, corrugated; fruit almost cylindrical, 8–10 mm long, 6–8 mm diam., corrugated.

*Eucalyptus pauciflora—alpine form*

*Eucalyptus platypus*

*E. platycorys*—**Boorabbin Mallee**              W.A.
5–8 m     400 mm     Prefers sandy soils     Fast (?)
Low shelter, roadsides, parks, ornamental

New introduction

Attractive, bushy tree. Tolerates very saline soils. Coppices readily. Bark smooth, grey green; leaves shining green, narrow, 5–10 cm long, 5–10 mm wide; buds 1–3, sessile, 1 cm long, slightly ribbed, cap conical, wider than the calyx; fruit hemispherical, 1 cm long and wide, ribbed.

*E. platypus*—**Round Leaf Moort**              W.A.
5–8 m     400 mm                        Moderate
Low shelter, shade, parks, roadsides, streets, honey

Usually a compact, bushy, neat, round small tree retaining branches to ground level. Dense shining green crown. Tends to be a variable species under cultivation. Grows well on heavier and poorly-drained soils but also suitable for sandy soils. Moderately frost-resistant. Tolerates competition and will grow under taller trees. Withstands moderate coastal exposure. Can be trimmed. Bark smooth, grey; leaves usually round, 5–8 cm diam., but may be 4–8 cm long and only 1 cm wide; buds 4–9 on a strap-like stalk, 3–4 cm long, horn-shaped, 2–3 cm long, cap twice as long as the calyx; flowers cream, prolific; fruit sessile, oval, 1 cm long with 2–3 ribs.

*E. platypus* var. *heterophylla*—**Coastal Moort**   W.A.
4–5 m     400 mm     Prefers lighter     Fast
soils
Low shelter, parks, roadsides, streets

Similar to *E. platypus* but withstands severe coastal exposure. Tolerates saline and limestone soils.

*E. polyanthemos*—**Red Box**        Vic., N.S.W.
10–15 m     500 mm     Will grow on poor     Slow
soils
Farm forests, windbreaks, shade, parks, honey

Open grown trees have a dense, low-branched spreading crown and short trunk; trees planted close together have a restricted crown and a longer trunk. Produces a light red, durable timber. Bluish-green foliage.

*Eucalyptus regnans forest*

Hardy. Frost-resistant, moderately drought-resistant. Wind-firm. Bark slightly fibrous to scaly; leaves mostly round, 5–8 cm diam., bluish; buds 10–12, in clusters, 6–8 mm long, cap short; fruit almost globular, 5–7 mm diam. 1889.

*E. polybractea* (E. fruticetorum)—**Blue**     S.A., Vic.,
                         **Mallee**        N.S.W.

5–7 m     300 mm     Prefers medium to     Moderate
                         heavier soils
Low shelter, roadsides, parks, eucalyptus oil, honey

Slightly bluish, hardy mallee. Drought-resistant. Coppices freely. One of the most important for eucalyptus oil production. Produces good quality honey. Bark smooth, light-coloured; leaves bluish-grey, narrow, 5–10 cm long, 1 cm wide; buds 5–12, small, 6–8 mm long, cap round, short; fruit globular, small, up to 4–5 mm diam.

*E. porosa*—**Mallee Box**                 S.A.
3–5 m     250 mm     Will grow on     Moderate
                         limestone
Windbreaks, honey

Multi-stemmed mallee with a dense, rounded crown. Occasionally a single-stemmed tree, reaching 8–10 m in height. Grows well on shallow calcareous soils but not on heavy clays. Bark rough; leaves fresh light green, shining, 5–10 cm long, 1–2 cm wide; buds 3–7, club-shaped, 5–7 mm long, cap short, conical; fruit pear-shaped, 4–6 mm long.

*E. preissiana*—**Bell Fruit Mallee**          W.A.
3–4 m     500 mm     Grows in poor, stony     Fast
                         or clay soils
Low shelter, parks, ornamental

Variable small tree ranging from compact to somewhat straggly. Moderately frost-resistant. Mainly a garden specimen; large yellow flowers, showy, 3–5 cm across, usually flowering when young. Bark smooth, brown; leaves thick, 10–12 cm long, 5 cm wide; buds in threes on a strap-like stalk, sessile, 2–3 cm diam., cap round, depressed at the centre not peaked as is usual; fruit bell-shaped, sessile, large, 3–4 cm long, 2–3 cm wide.

*E. pulverulenta*—**Silver Leaf Mountain Gum**     N.S.W.
3–8 m     600 mm     Prefers loams            Fast
Low shelter, parks, ornamental

Small, somewhat straggly, but bushy small tree. Hardy, frost-resistant, suitable for most soils. Planted mainly for its attractive silvery-bluish leaves. Branches more or less horizontal. Bark white, deciduous; leaves opposite, bluish, elliptical, 4–7 cm long, 2–4 cm wide; buds in threes, 12–15 mm long, very bluish, cap conical, short; fruit sessile, bluish, hemispherical, 10–12 mm diam.

*E. punctata*—**Grey Gum**           N.S.W., Qld
12–20 m     750 mm                       Moderate
Farm forests, shade, parks

Upright, straight, densely-crowned tree, spreading when open grown. Somewhat variable. Suitable for sandy to heavy clay soils. Does not tolerate shade. Timber red, durable. Bark dark grey, rough, shed in flakes, new bark bright orange; leaves 10–15 cm long, 1–2 cm wide, veins parallel and divergent; buds 3–7, 1 cm long, cap round to conical; fruit hemispherical, 1 cm diam.

*E. pyriformis*—**Large Fruit Mallee**         W.A.

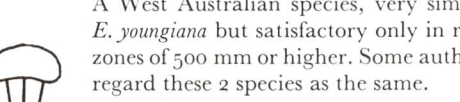

A West Australian species, very similar to *E. youngiana* but satisfactory only in rainfall zones of 500 mm or higher. Some authorities regard these 2 species as the same.

*E. radiata*—**Narrow Leaf Peppermint**    Vic., N.S.W.
15–30 m     700 mm     Not satisfactory     Fast
                         on sands
Farm forests, windbreaks, shade

Densely-crowned upright tree forming a long clean trunk in forest formation, but low-branching with pendulous foliage and spreading crown when open grown. Good paddock shade tree. Requires well-drained soil, best on loamy types. Hardy. Frost-resistant, Tolerates some snow. Bark fibrous, persistent; leaves 8–15 cm long, 1–2 cm wide; buds 8–16, small, 5–6 mm long, cap short; fruit hemispherical, 5–6 mm diam.

*E. regnans*—**Mountain Ash**  Vic., Tas.
30–60 m   1000 mm   Requires deep soil   Fast
Farm forests, roadsides, parks

Tall, straight, upright tree with a long, clean slender trunk. Crown small. Tallest hardwood, some trees exceeding 90 m. Produces good quality seasoning timber. Frost-sensitive when young. Very particular as regards site; good growth only obtained in cool mountain districts with high rainfall and deep, well-drained loams. Suitable sites in private ownership are limited, recommended that further advice be obtained before planting. One of the few eucalypts which does not coppice. May be attacked by the root fungus *Phytophthora cinnamomii* in abnormally wet years. Bark smooth, deciduous in long ribbons; leaves 10–15 cm long, 3–4 cm wide, veins distinct; buds 7–12, small, 8–10 mm long, cap short; fruit oval, 8–9 mm long.

*E. rhodantha*—**Rose Mallee**  W.A.
2–3 m   400 mm   Prefers sandy loams   Moderate
Low shelter, parks, ornamental

Usually a compact, bluish, bushy mallee shrub but can be variable and straggly in shape. Conspicuous large red flowers, flowering for long periods. Drought-resistant but slightly frost-sensitive. Similar to *E. macrocarpa* but hardier, not as tall and more reliable in shape. Bark smooth, bluish; leaves silvery-grey, sessile, thick, oval, 5–10 cm diam.; buds single, oval, large, 4–5 cm long, ribbed, cap conical; flowers red, large, occurring for most of the year; fruit large, 4–5 cm diam., ribbed.

*E. risdonii*—**Silver Peppermint**  Tas.
3–6 m   750 mm   Will grow on dry   Moderate
               sites
Low shelter, roadsides, parks, ornamental

Attractive, compact small tree with silvery-bluish foliage, branchlets pendulous. Withstands frost and light snow, good for sub-alpine areas. Bark smooth, deciduous; leaves bluish, sessile, opposite and joined, 4–5 cm long, 2–4 cm wide; buds 3–9, small, 5–6 mm long, cap short; fruit pear-shaped, 6–8 mm long.

*E. robusta*—**Swamp Mahogany**  N.S.W., Qld
12–20 m   650 mm   Will grow in wet   Fast
                areas
Windbreaks, roadsides, parks, honey

Impressive, spreading, densely-crowned tree. Will grow in swampy areas, alkaline and sub-saline soils in cooler districts Wind-firm. Withstands severe coastal exposure. Tolerates smog. Good tree for adverse sites where few other trees would survive. Bark rough, fibrous; leaves glossy, 10–12 cm long, 5–7 cm wide, divergent (parallel) veins, faint; buds 5–10, pear-shaped, 1 cm long, cap peaked; fruit cylindrical, 1–2 cm long, 1 cm wide, slightly constricted in the middle.

*E. rubida*—**Candlebark Gum**  S.A., Vic., Tas.,
               N.S.W., Qld
20–30 m   650 mm   Well-drained soils   Fast
Farm forests, windbreaks, parks, ornamental, honey

Upright, attractive tree with a long, clean trunk. Similar to *E. viminalis* but grows on drier sites. Pink streaks on bark in late summer attractive. Bark smooth, white, deciduous; leaves 12–20 cm long, 2–4 cm wide, glossy; buds in threes, 7–10 mm long, cap conical; fruit sessile, in threes, in one plane at right angles to each other, oval, 5–7 mm diam. Distinguished from *E. viminalis* by its round juvenile foliage whereas the latter has long and narrow foliage.

*E. saligna*—**Sydney Blue Gum**  N.S.W.
25–30 m   900 mm   Prefers deep, well-   Fast
               drained, heavier
               soils
Farm forests, windbreaks, parks, honey

Tall, straight tree with a clean trunk. Produces good quality, general-purpose timber. Moderately frost-resistant. Bark smooth, bluish; leaves 10–20 cm long, 2–4 cm wide; buds 3–9, sessile, 7–10 mm long, cap short; fruit small, 5–7 mm long. ('Blue' in the common name refers to the blue trunk; buds and fruit are usually not bluish as with other 'blue' gums.)

*Eucalyptus rubida*

*Eucalyptus salmonophloia*

**E. salmonophloia—Salmon Gum**                    W.A.
12–20 m    300 mm    Prefers loamy soils        Fast
Farm forests, tall windbreaks, shade, parks, streets, honey, eucalyptus oil

Tall, slender tree with attractive pink bark. Not densely-crowned but open grown trees develop spreading crowns. Prefers heavier soils and will tolerate poorly-drained, alkaline, and slightly saline sites. Extensive surface root system. Should not be planted for farm forests on unfavourable sites as its form can be poor if conditions are not good. Frost-resistant. Excellent specimen tree if open grown. Bark smooth, deciduous, salmon colour; leaves narrow, 6–12 cm long, 1–2 cm wide; buds 3–7, 6–8 mm long, cap conical; fruit globular, 5–6 mm diam., valves protruding.

**E. salubris—Gimlet**                    W.A.
10–12 m    250 mm    Prefers heavier soils      Moderate
Farm forests, streets, parks, ornamental

Graceful, slender, erect, willow-like tree. Crown not dense. Trunk distinctive; spirally fluted (particularly when young), with smooth shining light brown bark, attractive features which make it worth planting for ornamental purposes. Crown rather light so not very satisfactory for windbreaks or shade. Will grow on clay and calcareous soils. Drought- and frost-resistant. Leaves narrow, 5–10 cm long, 10–15 mm wide, shining, dark green; buds 3–8, 1 cm long, cap 1½ times as long as the calyx; fruit hemispherical, 4–5 mm diam.

**E. sargentii—Sargent Mallet** or                  W.A.
                 **Salt River Mallet**
5–7 m     300 mm    Prefers sandy loams       Moderate
Windbreaks, shade, roadsides

Compact tree with a globular umbrella-like crown. Very salt-tolerant and will grow on saline soils toxic to many other plants. Moderately drought- and frost-resistant. Bark rough on trunk, smooth and brown on limbs; leaves narrow, 5–10 cm long, 10–15 mm wide, thin; buds 3–7, 2 cm long, cap 3 times longer than the calyx; fruit hemispherical, 7–10 mm long, valves protruding.

**E. scoparia—Wallangarra Gum**              N.S.W., Qld
6–10 m     900 mm    Will grow on poor          Fast
                      sites
Windbreaks, parks, roadsides, ornamental

Attractive, slender tree with pendulous foliage. Moderately drought- and frost-resistant. Withstands light snow, suitable for high altitude planting. Long-flowering. Bark smooth, white; leaves 10–20 cm long, 10–15 mm wide, pale green; buds 3–7, oval, 5–7 mm long, cap round; fruit 5–6 mm long, oval, valves protruding.

**E. sepulcralis—Weeping Gum**                  W.A.
5–7 m     500 mm    Will grow in sandy        Moderate
                     and stony soils
Roadsides, parks, ornamental

Attractive, graceful tree with a very thin trunk, and sparse foliage. Long, bluish, pendulous, slender branchlets give it a weeping appearance. Frost-hardy. Withstands moderate coastal exposure. May be deformed if planted on an adverse site. Bark smooth, but striated, white; leaves vivid green, narrow, 7–10 cm long, 1 cm wide; buds 3–5, 1 cm long on long slender stems; flowers yellow; fruit large, 3–4 cm long, 2–3 cm wide, polished, oval, distinctive, constricted to a very small opening.

*Eucalyptus sargentii*

*Eucalyptus sideroxylon*

**E. sideroxylon—Red Ironbark**      Vic., N.S.W., Qld

15–20 m      500 mm      Will grow on gravelly      Slow
                    and lateritic soils

Farm forests, windbreaks, shade, parks, streets, ornamental, honey, birds

Upright, straight tree with an open crown,
shedding its lower branches rather earlier
than most eucalypts. Timber hard, very
heavy and durable. Grows well in heavy
soils. Frost-hardy. Moderately drought-
resistant. Coppices vigorously from low stumps. Bark
deeply furrowed, heavily impregnated with kino, often
5–6 cm thick; leaves 5–12 cm long, 10–15 mm wide,
dull; buds 3–7, 10–15 mm long, on long slender stalks,
cap short; flowers white; fruit globular, 7–10 mm diam.,
on a stem 1–2 cm long. 1889.

**E. sideroxylon var. *pallens*—Pink Flower**      S.A.,
                    **Red Ironbark**      Vic., N.S.W.

Similar to *E. sideroxylon* but flowers are pink
and has a thinner crown. Graceful tree,
planted as an ornamental.

**E. sieberi—Silvertop**      Tas., Vic.,
                           N.S.W.

15–30 m      650 mm      Will grow on poor      Fast
                    soils

Farm forests, windbreaks, shade, honey

Upright tree, shedding its lower branches.
Produces good quality timber. Best growth
on sandy soils overlying friable clay.
Moderately frost-resistant. May be attacked
by the root fungus *Phythophthora cinnamomii* in
wet years. Bark furrowed, hard; leaves narrow, 10–18 cm
long, 1–3 cm wide, shining green; buds 5–15, 6–8 mm
long, cap short; fruit pear-shaped, 1 cm long. Similar
uses as *E. obliqua*.

**E. socialis—Willow Leaf Mallee**      W.A., S.A., Vic.,
                           N.T.

3–4 m      300 mm      Moderate
Low shelter, honey

Usually a typical mallee, but in favourable
sites forms a small tree up to about 10 m
tall. Hardy, drought- and frost-resistant.
Bark rough at the base; leaves dull grey,
5–15 cm long, 1 cm wide; buds 3–8,
10–15 mm long, cap conical, long, finger-like; fruit nearly
hemispherical, 5–7 mm diam. Formerly *E. transcontinentalis*
(in part) and *E. oleosa* var. *glauca* (in part).

**E. spathulata—Swamp Mallet**      W.A.

6–8 m      400 mm      Will grow on      Fast
                    adverse soils
Low shelter, roadsides, parks, ornamental

Attractive, compact, small tree, with
ascending branches and very narrow leaves,
one of the narrowest among the eucalypts.
Will grow on heavy, saline calcareous or wet
soils. Moderately drought- and frost-
resistant. Good alternative for bushy sugar gum. Bark
smooth, attractive red-bronze colour, very high tannin
content; leaves 5–10 cm long, 3–8 mm wide; buds 3–7,
15 mm long, cap 2–3 times as long as the calyx; fruit
hemispherical to bell-shaped, 6–8 mm diam.

*Eucalyptus spathulata*

*Eucalyptus steedmanii*

*Eucalyptus st. johnii—dark foliage*

**E. steedmanii—Steedman Gum**                        W.A.
4–5 m        300 mm                                    Fast
Low shelter, roadsides, streets, parks, ornamental

Shapely, compact, rounded mallee with slender erect branches and dark green foliage to ground level for some years but eventually shedding the lower branches. Vigorous, hardy, but not suited to wet sites. Drought- and frost-resistant. Withstands moderate coastal exposure. Best is warmer districts. Conspicuous yellow flowers, 5–8 cm across for several months of the year, occasionally crimson. Buds and fruit distinctive with 4 wings running down the stems. Bark light brown, smooth, shining; leaves thick, 4–7 cm long, 6–10 mm wide; buds 3–5, 2–3 cm long, cap short, winged like the calyx; fruit slender, 2–3 cm long, 1 cm wide, valves protruding, 4–5 mm.

**E. stellulata—Black Sallee**                   Vic., N.S.W.
6–10 m        750 mm        Will grow on the      Moderate
                            wetter soils
Low shelter, shade, ornamental, honey

Spreading, much-branched tree. Frost-hardy. Wind-firm. Withstands snow. One of the most frost-resistant of the eucalypts. Good shelter tree for exposed high-altitude sites. Flowers prolifically each summer. Bark smooth, greyish, olive green with age; leaves leathery, 5–8 cm long, 2–3 cm wide; buds 7–16, sessile, 6–8 mm long, cap conical; fruit sessile, globular, 3–5 mm diam.

**E. st johnii** (E. bicostata)**—Blue Gum**        Tas., Vic.,
                                                      N.S.W.
20–35 m        900 mm        Deep, moist soils        Fast
Windbreaks, shade, farm forests, honey

Tall, straight tree with a dense crown. For best growth it requires a moist, well-drained position in the cooler districts. Will grow on heavy soils. Vigorous with a very deep, strong root system. Wind-firm. Coppices freely from cut stumps. Can be pollarded. Withstands severe frosts. Young growth, particularly coppice, is not seriously attacked by animals. Susceptible to serious borer attack if planted in unsuitable sites. Produces good heavy construction timber; good shade tree. Bark smooth; leaves dark green, long, 15–25 cm and 1–2 cm wide, young leaves very bluish; buds in threes, globular, 10–15 mm long, warty, sessile, bluish; fruit bluish, globular, 1 cm diam., sessile, warty, with 2 ribs.

**E. stoatei—Pear Gum**                                W.A.
6–8 m        400 mm        Prefers loamy soils        Fast
Windbreaks, roadsides, parks, streets, ornamental

Upright, dense tree with erect and persistent branches. More suitable for the warmer districts. Frost-hardy. Moderately drought-resistant. Can be lightly pruned. Showy buds, flowers and fruits, one of the more decorative eucalypts. Bark smooth, grey-green; leaves 7–10 cm long, 2–4 cm wide, thick; buds single, on long stalks, large, 3–4 cm long, slender, deeply ribbed, bright scarlet, cap short, similar to the calyx; flowers rich yellow, 4–5 cm diam., stamens remain bunched together; fruit scarlet, pear-shaped, large, 4–5 cm long, 3–4 cm diam., 10–12 ribs.

**E. striaticalyx—Cue York Gum**                 W.A., S.A.
6–10 m        350 mm        Will grow on            Fast
                            limestone soils
Windbreaks, roadsides, parks          New introduction

Attractive, small tree. Tolerates very saline soils, will grow where other vegetation fails. Bark dark grey; leaves light green, 7–10 cm long, 2–3 cm wide; buds 5–10, cylindrical, 1 cm long, cap hemispherical, peaked with 12–15 ribs; fruit cylindrical, 1 cm long.

**E. stricklandii—Goldfields Gum**                     W.A.
8–10 m        250 mm        Will grow on sandy        Fast
                            soils
Shade, roadsides, streets, parks, ornamental

Shapely, medium-sized tree. Best growth on loamy soils in warm districts but suitable for gravelly or hard soils. Will grow well under other trees. Hardy. Drought- and frost-resistant. Withstands moderate coastal exposure. Conspicuous large, yellow, scented flowers. Bark rough at butt, smooth and reddish-brown higher up the stem; leaves thick, shiny, slightly bluish, 12–20 cm long, 2–3 cm wide; buds 4–7, sessile, 2 cm long, slightly ribbed, cap dome-shaped; fruit cylindrical, 14–15 mm long, 10–12 mm wide, 2–3 ribs.

**E. tasmanica—Tall Silver Peppermint**               Tas.
10–15 m        650 mm                                  Fast
Windbreaks, farm forests, parks       New introduction

Attractive, upright, slightly bluish tree with a clean trunk and pendulous branchlets. Prefers well-drained soils. Frost-resistant. Withstands light snow. Bark smooth, deciduous; leaves smooth, shining, tapering at both ends, 5–15 cm long, 1–4 cm wide; buds 6–25, 5–9 mm long, cap short; fruit conical, 6–10 mm long.

*E. tetragona*—**White Leaf Marlock**　　　W.A.
3–4 m　　450 mm　　Prefers sandy soils　　Fast
Low shelter, parks, ornamental

Fairly compact shrub or small mallee with 'squarish' stems. Requires well-drained conditions; vigorous, frost-tender when young. Suitable for coastal districts, withstands moderate coastal exposure. Bark smooth, bluish; leaves very mealy, almost white, ornamental, 5–10 cm long, 2–5 cm wide; buds in threes, 1 cm long, 4 ribs, cap short; flowers white; fruit oval, 15–20 mm long, 4 ribs giving a 'squarish' cross section, very blue.

*E. tetraptera*—**Four Wing Mallee**　　　W.A.
2–4 m　　450 mm　　Will grow in poor,　　Fast
　　　　　　　　　　sandy soils
Parks, ornamental

Straggly mallee shrub with slender stems. Best in the warmer districts. Moderately drought- and frost-resistant. Can be shaped into a bushy shrub by careful pruning. Planted mainly for the ornamental values of its large massive angular, 4-winged, reddish fruits. Leaves large, very thick, up to 20 cm long, and 5 cm wide; buds single, large, 6–7 cm long, 3–4 cm diam., angular, on a thick strap-like stalk which curves downward, cap red, pyramidal; flowers pink; fruit large, 5–7 cm long, 5–6 cm diam.

*E. torquata*—**Coolgardie Gum**　　　W.A.
4–7 m　　300 mm　　　　　　　　　　Fast
Roadsides, streets, parks, ornamental, birds

Graceful, well-shaped, lightly-crowned, single-stemmed, spreading, small tree. Grows on a wide range of soils including sandy, gravelly, and heavy but not wet sites. An outstanding ornamental tree, widely planted. Drought-resistant. Frost-tender when young, later moderately resistant. Flowers pink to red, occasionally white, in pendulous clusters, flowering for long periods each year. Often flowers at 2 years. Bark dark, rough, persistent; leaves 10–15 cm long, 1–3 cm wide; buds 3–8, 2–3 cm long, corrugated, cap wider than calyx,

its base corrugated and point smooth; fruit cylindrical, 10–12 mm long, 8–10 mm wide, base swollen and corrugated, top smooth.
The hybrid 'Torwood' (*E. torquata* x *E. woodwardii*) has orange-pink flowers but being raised from seed the colour is not reliable.

*E. urnigera*—**Urn Fruit Gum**　　　Tas.
5–10 m　　650 mm　　　　　　　　　Moderate
Windbreaks, shade, parks, ornamental

Small tree with drooping branches. Withstands light snow and exposure at high altitudes but also suitable for lower areas. Frost-resistant. Flowers white, large and showy. Bark smooth, deciduous; leaves narrow, 8–15 cm long, 2–3 cm wide, glossy green; buds in threes, 1 cm long, cap short, broader than the calyx; fruit distinctly urn-shaped, 15–18 mm long, 8–10 mm wide.

*E. viminalis*—**Manna Gum**　　　S.A., Vic.,
　　　　　　　　　　　　　　　　Tas., N.S.W.
20–35 m　　700 mm　　Prefers good, deep,　　Fast
　　　　　　　　　　moist soils
Farm forests, windbreaks, parks, avenues, ornamental, honey.

Attractive, upright tree with a long, clean white trunk and slender branchlets. Good specimen tree. Best in sheltered gullies in the higher rainfall districts; under good conditions can reach 50 m tall. Produces useful timber. One of the most frost-resistant of the eucalypts, withstands 10°C of frost. Bark smooth, deciduous, ribbony; leaves slender, 10–20 cm long, 1–3 cm wide, pale green; buds in threes, sessile, 6–8 mm long; cap conical; flowers most of the year but not prolific; fruit sessile, globular, 5–7 mm diam., arranged in threes in the one plane at right angles to each other. Can be variable, the coastal form is spreading, much-branched, 5–20 m tall, with rough bark persistent to the larger branches. Usually not propagated by reputable nurseries. 1889.

*Eucalyptus torquata—flowers　W. Middleton*

*Eucalyptus viminalis—coastal form*

209

*Eucalyptus viminalis—mountain form*

*E. viridis*—**Green Mallee**                    S.A., Vic., N.S.W., Qld

5–7 m        400 mm        Prefers heavier        Moderate
                                          soils
Low windbreaks, parks, roadsides, ornamental, eucalyptus oil, honey

Dense, compact mallee with slender stems and branches. Will grow on a wide range of soils including gravelly and sandy sites but not poorly-drained areas. Frost-resistant. Coppices freely. Showy, small, white flowers. Bark dark, thin; leaves very narrow, 5–10 cm long, 5–7 mm wide, dark green; buds 4–8, 6–8 mm long, cap short; fruit globular, 4–6 mm diam.

*E. wandoo* (E. redunca var. elata)—**Wandoo**        W.A.
12–20 m    450 mm                            Moderate
Farm forests, windbreaks, shade, honey

Large spreading tree when open grown. Best growth on soils with a clay subsoil. Growth slow for the first few years then moderate to fast. High tannin content in the bark. Frost-resistant. Bark white, deciduous; leaves 6–10 cm long, 2–5 cm wide, thick, bluish; buds 4–10, horn-shaped, 2–3 cm long, cap 1½ times as long as calyx; fruit pear-shaped, 8–10 mm long.

*E. websterana*—**Webster Mallee**        W.A.
2–4 m        250 mm                            Moderate
Low shelter, parks, ornamental

Bushy, attractive mallee or small tree, slender red branches retained to ground level. Foliage sparse but well-distributed. Moderately frost-resistant. Drought-resistant, suitable for dry districts. Bark reddish-brown, smooth; leaves bluish, 2–4 cm long, 1–2 cm wide; buds 3–6, oval, 1 cm long, cap broadly conical; fruit hemispherical, 5–7 mm diam., broad rim.

*E. woodwardii*—**Lemon Flower Gum**        W.A.
8–12 m    300 mm    Requires deep, well-        Fast
                              drained soils
Shade, parks, streets, ornamental

Upright, narrow but open-crowned, slightly straggly tree with pendulous branches and bluish leaves. Suitable for sandy soils. Showy, large clusters of bright yellow flowers causing the branchlets to bend down. Good street and avenue tree. Flowers when very young. Best in warm, dry districts. Hardy, drought- and frost-resistant. Bark smooth, deciduous; leaves thick, very bluish, 7–12 cm long, 3–5 cm wide; buds 3–6, pear shaped, 15–18 mm long, cap peaked; fruit bell-shaped, 15 mm long, 10 mm wide, slightly ribbed.

*E. youngiana*—**Large Fruit Mallee**        S.A.
4–6 m        350 mm    Prefers sandy loams    Fast
Low shelter, parks, roadsides, ornamental

Attractive, small mallee-like tree with slender stems. Frost-sensitive when young, otherwise very hardy. Drought-resistant. Bark smooth, light brown; leaves slightly bluish-green, 5–12 cm long, 3–5 cm wide; buds in threes, very large, 5–6 cm long, 3–4 cm wide, strongly ribbed; flowers yellow or red, very large and conspicuous; fruit flattened, globular, 4–6 cm across, prominently ribbed.

# EUCRYPHIA

*Eucryphia* is derived from *eu* meaning well and *kryphios* meaning covered, which refers to the cap formed by the sepals fusing at their tips. Four species, two from Chile and two from Australia, are known.

*E. cordifolia*—**Brush Bush**   E.H.                Chile
Cuttings
6–10 m        750 mm                            Fast
Parks, ornamental

Attractive, small tree, suitable for cool districts. Tolerates some lime. Hardy, moderately frost-resistant. Leaves oval, 5–8 cm long, 3–4 cm wide, wavy margins; flowers conspicuous, solitary, 5–8 cm across, white with prominent red stamens, showy; fruit woody, pear-shaped capsule.

*E. lucida*—**Leatherwood**   E.H.                    Tas.
Seed

| 6–10 m | 900 mm | Moist, well-drained loams | Moderate |

Parks, roadsides, honey

Spreading large shrub or small tree, more suitable for the cooler mountain districts. Requires good soil. Moderately frost-resistant. Withstands light snow. Leaves 1–5 cm long, elliptical, rounded ends, white underneath; flowers white, 1–2 cm across, somewhat rose-like; fruit woody, pear-shaped capsule, 1–2 cm long.

# EUGENIA

Genus named after Prince Eugen of Savoy, a sponsor of botanical exploration. There are about 1000 species occuring in all tropical and sub-tropical countries.

*E. australis* (E. paniculata, E. myrtifolia)   N.S.W., Qld
         —**Brush Cherry**   E.H.
Seed

| 6–15 m | 750 mm | Requires well-drained soils | Fast |

Windbreaks, shade, parks, roadsides

Attractive eucalypt-like tree. Widely planted. Good for coastal areas, withstanding moderate exposure and sandy soils. Not very frost-resistant but otherwise hardy. Bark fibrous, tesselated; leaves glossy, oval, 7–10 cm long, coppery-brown when young; flowers white, myrtle-like; fruit dark red, oval berries, 1–2 cm long.

*E. smithii* (Acmena smithii)—**Lilly Pilly**   E.H.   Vic.,
Seed                                                N.S.W., Qld

| 4–8 m | 650 mm | Requires well-drained soils | Fast |

Low windbreaks, hedges, parks

Shapely, compact, upright small tree retaining branches to ground level. Widely planted. Hardy. Frost-tender when young. Withstands moderate coastal exposure. Can be trimmed to form a hedge 1 m tall or allowed to grow to tree size—15 m on good sites but 5–6 m is more usual. Reaches 30 m in its natural habitat. Leaves glossy green, pointed, 5–8 cm long, 2–4 cm wide; flowers cream; fruit soft, white or pale violet berry, nearly globular, 1 cm diam.

*E. ventenatii*—**Weeping Lilly Pilly**   E.H.   N.S.W., Qld
Seed

| 7–10 m | 650 mm | Will grow in sandy soils | Moderate |

Parks, hedges, ornamental

Attractive, stately, slightly spreading tree with a pendulous habit. Tolerates slight coastal exposure. Can be trimmed to form a hedge. Leaves oval, 7–12 cm long, pointed, wavy margin; flowers myrtle-like, white, small; fruit a berry, 1 cm diam., white then brown.

*Eugenia smithii*

*Eugenia smithii—flowers*

# FAGUS

This is the old Latin name. About ten species from Europe, Asia and North America are known.

*F. sylvatica*—**Common Beech**   D.H.        Europe,
Seed                                          Asia Minor

| 12–25 m | 750 mm | Prefers the better soils | Moderate to slow |

Windbreaks, shade, parks, avenues

Large, shapely, spreading tree with a dense, rounded crown when open grown. Good specimen tree—one of the best for parks. Attractive spring and autumn foliage. Will grow on most types of soils including limestone but not sands, poorly-drained, or saline sites. More suitable for the cooler, mountain districts. Very hardy and easily grown. Will grow in sub-alpine areas but not in

*Fagus sylvatica—purple foliage form*

exposed or coastal positions. Has a deep tap root, plants more than 3–4 years old are difficult to transplant. Tolerates shade. Largely disease-free, foliage is rarely attacked by insects. Despite its relatively slow growth it is much hardier and easier to grow than generally believed; worth more extensive planting. Bark smooth, light-grey; leaves oval, 3–8 cm long, slightly toothed, 5–10 pairs of parallel veins from the midrib, autumn foliage golden, may remain attached for the winter; flowers inconspicuous; fruit a bristly 4-lobed husk, 2–3 cm diam. with bristles 1–2 cm long, containing 2 triangular nuts. Numerous horticultural varieties are available. L.C.

## FICUS

This is the old Latin name. There are about 600 species from the warmer regions of all countries.

*F. macrophylla*—**Moreton Bay Fig**    E.H.    N.S.W.,
Seed                                             Qld
15–20 m    650 mm                                Fast
Shade, windbreaks, parks

Large, massive, heavily-branched, densely-crowned, spreading and symmetrical tree, formerly widely planted. Declined in popularity because it restricts growth of other plants and the fruits (figs) are messy on hard paths, etc. Excellent, fast-growing shade tree so is now being planted more frequently in large, open areas. Hardy, suitable for most soils, including poorly-drained and saline. Damaged by severe frost when young. Bark grey, smooth; leaves large, oval, 15–25 cm long, glossy, thick midrib, rusty-brown underneath; fruit a purplish fig, 1–2 cm diam. 1869.

## FLINDERSIA

Genus named after Capt. Matthew Flinders (1779–1814). There are about six species from Australia and the Moluccas.

*F. maculosa*—**Leopardwood**    E.H.    N.S.W., Qld
Seed
3–5 m    350 mm                        Slow, then
                                       moderate
Shade, parks, ornamental, subsistence fodder

One of the most attractive of the native trees. Hardy, drought- and frost-resistant. Suitable for most soils. Growth straggly and bushy with long thin stems for a few years, then it forms a strong leader. Distinctive bark, not apparent until the tree form has developed. Difficult to propagate but fairly easily established. Worth persisting with, particularly for ornamental purposes. Bark spotted, yellow and brown ('leopard'); leaves usually single, 3–7 cm long, opposite or nearly so; flowers white, 5–7 mm diam., in clusters; fruit a pod, 3–4 cm long, splitting into 5 parts, winged seeds.

## FRAXINUS—Ash    D.H.

Summer shade, windbreaks, parks, roadsides, autumn foliage

*Fraxinus* is the old Latin name. There are about sixty species from Europe, Asia and North America. They are attractive, shapely, upright and generally wide-crowned trees. Although hardy and suitable for a wide range of well-drained soils, most species are not very satisfactory on deep sands or calcareous soils. They are widely planted. Winter buds are felted. Foliage is pinnate, opposite and in pairs with a terminal leaflet. Flowers are small, white and in sprays. Fruit is a winged seed which is usually in bunches.

*F. americana*—**American Ash** or **White Ash**    E. North
Seed                                                America
15–20 m    650 mm                                   Fast
Farm forests

Tall, erect and attractive tree. Dark green foliage, purplish tones in autumn. Hardy, suitable for a wide range of conditions. Will grow in slightly alkaline soils. Shade-tolerant when young. Tolerates some flooding. Buds dark brown; leaves 20–30 cm long, leaflets 5–9, 10–15 cm long, silver underneath, slightly purplish in autumn; seed 4–5 cm long. 1724.

*F. angustifolia*—**Narrow Leaf Ash**    Mediterranean
Seed
12–15 m    550 mm                        Fast

Attractive, tall, fairly compact tree. Not widely planted, usually raised as root-stock for grafting. Buds dark brown; leaves 20–35 cm long, leaflets 7–13, 6–8 cm long, irregularly toothed; seed 2–3 cm long. 1880.

*Fraxinus ornus*

**F. excelsior—English Ash**          Central Europe
12–20 m    600 mm                          Fast
Streets, avenues, farm forests

Spreading tree with a rounded crown. Will grow in heavier soils than the other *Fraxinus* sp. but best on deep soils in cool, moist areas. Strong-growing. Frost-hardy. With-stands some smog. Does not tolerate shade. Slightly slower growth rate than desert or claret ash. Buds black; leaves 20–30 cm long, leaflets 11–15, 7–10 cm long; seed 1–3 cm long.    L.C.

**F. excelsior var. *aurea*—Golden Ash**
Grafts
6–10 m    500 mm                          Fast

Large-crowned, moderate-sized tree. (Usually grafted onto *F. oxycarpa* root-stock.) Foliage turns rich golden-yellow in autumn. Orange-yellow twigs. Stands warm conditions with an occasional watering. Otherwise similar to *F. excelsior*.

**F. lanceolata (F. pennsylvanica)—Green Ash**
Seed
                                        North America
10–12 m    650 mm    Prefers moist loam    Fast

Upright tree, with a rather broad, irregular crown and short trunk. Best growth along stream banks but will grow on a wide range of soils. Fibrous root system. Aggressive on good sites. Frost-resistant and withstands light snow. Will tolerate moderately dry sites and slightly alkaline soil. Large specimens can be transplanted. Buds rusty-brown; leaves 15–20 cm long, leaflets 7–9, 7–10 cm long, 2–3 cm wide, elliptical, serrated, same green colour on both surfaces; seed 2–4 cm long. 1824.

**F. ornus—Manna Ash** or **Flowering Ash**          Europe,
Seed                                        Asia Minor
8–12 m    600 mm                          Fast
Streets, avenues

Shapely, dense, compact crown. One of the flowering ashes with small, heavily-perfumed, whitish flowers in terminal sprays. Root system less vigorous than other *Fraxinus* sp. Buds grey-brown; leaves 12–20 cm long, leaflets 5–9, 8–10 cm long, only slightly toothed; seed 2–3 cm long. Before 1700.

**F. oxycarpa (F. oxyphylla) Desert Ash**    Mediterranean
Seed
7–12 m    450 mm                          Fast
Streets

Shapely tree, similar to *F. excelsior*. Hardy, tolerates hot and dry conditions. Leaf growth appears early (late winter). Buds brown; leaves 20–30 cm long, leaflets 7–11, 7–10 cm long, yellow in autumn; seeds 1–2 cm long. 1815.

**F. x 'raywoodii'—Claret Ash**          Hybrid
Grafted
8–15 m    500 mm                          Very fast
Streets, avenues

Very handsome, pyramidal, compact tree. Australian hybrid (one parent *F. oxycarpa*, other unknown). Usually grafted on *F. excelsior* root-stock. Foliage turns rich wine-colour in autumn, is retained until late in the season. Frost-hardy. Best in cool districts but will grow in warm areas. particularly if an occasional watering can be given in the first year or two. Buds brown; leaves usually in threes, leaflets variable, 5–11, narrow, 7–10 cm long.

**F. syriaca—Syrian Ash**          Syria, W. and
Seed                                        Central Asia
5–7 m    500 mm                          Fast
Low windbreaks

Attractive, compact small tree. Suitable for a wide range of sites. Very hardy, frost-resistant. Buds brown; leaves usually in whorls of 3 crowded on the branches, 15–20 cm long, leaflets 3–5, 4–6 cm long, tapering towards the tip, nearly sessile. 1880.

**F. velutina—Velvet Ash**          Arizona,
Seed                                        New Mexico
8–10 m    400 mm    Will grow on alkaline    Fast
                    and saline soils

Upright tree, with a slightly broad crown. Suitable for the warmer and drier districts. Tolerates alkaline and saline conditions better than most trees. Moderately frost- and drought-resistant. Intolerant of shade. Large specimens can be transplanted. Young growth covered in grey down. Buds brown; leaves 15–30 cm long, velvet, leaflets usually 5–7, 5–10 cm long, 3–4 cm wide, oval, faintly serrated; seed 2–3 cm long. Before 1890.

# GEIJERA

The origin of the name is not known. There are about three species from Australia.

*G. parviflora*—**Wilga**  E.H.    W.A., S.A., Vic.,
Seed                          N.S.W., Qld
3–5 m      350 mm           Moderate to slow
Windbreaks, shade, roadsides, honey, fodder

Shapely, roundish tree, with a dense, spreading crown and semi-pendulous foliage. Suitable for most soils in warm, inland districts. Usually a good fodder tree but some forms are not eaten by stock. Very hardy. Deep-rooted. Drought-resistant. Bark rough; leaves narrow, 8–15 cm long; flowers white, small, bell-shaped, in terminal clusters; fruit a hard, oval 'nut', 5–7 mm diam., containing 1 seed.

# GINKGO

This is the Chinese name. The only species occurs in China and it is doubtful if it survives naturally anywhere. It has been maintained in cultivation as a holy tree around Chinese temples.

*G. biloba*—**Maidenhair Tree**  D.C.    China
Seed, cuttings
10–15 m    750 mm    Prefers rich, well-    Slow
                      drained soil
Parks, streets, ornamental, autumn foliage

Very attractive, spreading, rather open-crowned tree. Deciduous conifer. Spring growth fresh light green; autumn foliage deep, rich golden. Believed to be the ancestral form of all conifers, but does not look like a conifer. Trees either male or female. Will grow on a wide range of soils. Easily transplanted when leafless. Tolerates smog. Leaves fan-shaped, divided into two lobes, 5–10 cm across, on stalks up to 8 cm long, poplar-like, growing on short spurs; fruit egg-shaped, 2–3 cm long, yellow, fleshy, enclosing a woody shell (i.e. the cone, which contains the seed), seeds nut-like, 1 cm long. (Male and female trees are required to give fertile seed.) 1730

# GLEDITSIA

Genus named after J. G. Gleditach, an eighteenth-century Director of Berlin Botanical Gardens. There are about twelve species from North America.

*G. triacanthos*—**Honey Locust**  D.H.    Central North
Seed                                      America
12–20 m    650 mm                         Fast
Windbreaks, summer shade, roadsides streets parks, hedges, fodder, gully control

Attractive, open-crowned, spreading tree. Prefers good, moist, well-drained soils but will grow on heavy and limestone soils. Suitable for the warmer, inland districts. Drought-resistant once established. Moder-

*Grevillea asplenifolia—flowers*

ately tolerant of saline conditions. Easily grown, hardy, withstands slight coastal exposure and light snow. Not tolerant of shade. Suckers freely. Can be trimmed to form a stock-proof hedge; has long (5–8 cm) stiff thorns which discourage stock. Foliage variable, usually bi-pinnate, 15–20 cm long, 4–7 pairs pinnae, leaflets oval, 2–5 cm long (pinnate leaves with 15–30 leaflets also occur), turns golden-yellow in autumn; flowers inconspicuous; fruit a reddish-brown pod, 20–45 cm long, 2–3 cm wide, lined with a sweetish pulp, of fodder value when green. 1700.

*G. triacanthos* var. *inermis*

As for *G. triacanthos* but is almost free of thorns. Not so useful for stock-proof hedges. Tolerates some flooding.

# GREVILLEA  E.H.    Australia

Seed, cuttings    Fast
Genus named after C. F. Greville (1749–1809), one of the founders of the Royal Horticultural Society. There are about 250 species from Australia, New Guinea and New Caledonia.

A large group of mainly Australian shrubs but there are a few taller growing species. They are characterized by the typical 'cat's-claw' or 'spidery' flowers. Hardy and frost-resistant and will grow in most soils except poorly-drained or deep sterile sands. Most species require acid soils; only a few will grow under alkaline conditions. They generally attract birds. Leaves are variable and the fruit is an oval, woody capsule.

*G. asplenifolia*—**Fern Leaf Grevillea**    N.S.W.
2–4 m    650 mm    Will grow on heavy
                   soil
Low windbreaks, roadsides, parks

Compact, low-growing shrub with spreading branches. More suitable for cool districts but will grow in drier areas. Frost-resistant. Leaves stiff, long, 15–25 cm, narrow, 1 cm wide, deeply serrated, almost to the midrib, sticky brown hairs underneath; flowers reddish-pink. This species is very similar to *G. longifolia*.

*G. banksii*—**Banks Grevillea**          Qld
2–3 m          550 mm
Low shelter, parks

Semi-erect, moderately dense large shrub or small tree. Adaptable to a wide range of sites, especially in the warmer, inland districts. Slightly frost-resistant. Leaves large and finely divided, fern-like, 10–20 cm long, leaflets narrow, light green; flowers, red, 7–10 cm long, in large, loose, terminal clusters. White-flowered forms are more common naturally, but are not normally propagated. 1868.

*G. barklyana*—**Gully Grevillea**          Vic., N.S.W.
3–10 m          650 mm          Prefers moist, well-drained sites
Parks, roadsides

Small dense tree or large shrub. Grows up to 8–10 m in good sites. Prefers some shade and cool soil. Vigorous and fairly adaptable as to site. Leaves of two types—narrow, entire, 10–20 cm long, or lobed, oak-like, 10–20 cm long with 3–7 triangular segments; flowers pink, in clusters 5–8 cm long, flowering early spring; fruits small, less than 1 cm long.

*G. confertifolia*—**Grampians Grevillea**          Vic.
Cuttings
2 m          500 mm
Low shelter, ground cover

Attractive, dense, spreading shrub. Leaves narrow, 2 cm long, pointed; flowers reddish-pink, in compact, large clusters. A prostrate form only about 20–30 cm high is also known. If the taller type is required, propagation by cuttings from suitable plants is necessary.

*G. glabrata*—**Smooth Grevillea**          W.A.
2–3 m          500 mm          Will grow on granitic soils
Low shelter, parks, roadsides

Attractive, spreading shrub with pendulous branches. Spread often exceeds its height. Hardy, frost-resistant. Easily grown. Can be pruned. Leaves 1 cm long, greyish, pointed, with 3 lobes (fan-shaped); flowers white, spikes 4–5 cm long, in sprays 12–20 cm long.

*G. hookeriana*—**Tooth Brush Grevillea**          W.A.
2–3 m          550 mm
Low shelter, parks, roadsides

Attractive, bushy but spreading shrub. Leaves 10–20 cm long with 5–12 segments, like a fishbone fern; flowers crimson, 5–7 cm long. 1886.

*G. ilicifolia*—**Holly Grevillea**          S.A., Vic.
2 m          400 mm          Prefers sandy soils
Low shelter, ground cover

Compact but variable bush. Leaves thick, 3–5 cm long, broad, wedge-shaped, prickly (holly-like); flowers greenish-red.
Various forms occur, the most extreme being a prostrate form with needle-like foliage.

*G. lavandulacea*—**Lavender Grevillea**          S.A., Vic., N.S.W.
Cuttings
2 m          400 mm
Low shelter, ground cover

Attractive, low-growing, dense, roundish shrub, often spreading 2–3 m across. Useful for covering stumps, low banks, etc. Leaves silver-grey, 1–2 cm long, narrow; flowers reddish-pink, produced in profusion for about two months in late winter. 1848.

*G. longifolia*          N.S.W.

As for *G. asplenifolia* but leaves are only serrated about one-third to the midrib or are entire.

*G. nematophylla*—**Water Bush**          W.A., S.A., N.S.W., N.T.
2–5 m          300 mm          Prefers light soils
Low shelter, parks, roadsides, stream banks

Large shrub or small tree, retaining branches to ground level. Will grow wherever water is available, even on sandhills if rain infiltrates. Leaves long, narrow, 8–20 cm long and less than 2 mm wide, sharp point; flowers creamy, in clusters 5–8 cm long; fruit 1 cm long, dark brown.

*G. robusta*—**Silky Oak**          N.S.W., Qld
15–25 m          500 mm          Prefers good, deep soil but will grow on heavy soils
Windbreaks, roadsides, parks, light shade, honey, birds

Upright tree with a relatively thin crown. Semi-deciduous in southern Australia. Best for warmer areas. Frost-sensitive when young. Drought-resistant once established. Withstands slight coastal exposure. One of the few grevilleas which will grow on alkaline soils. Leaves much divided, pinnate, 15–25 cm long, silvery underneath; flowers massed, bright orange, more colourful and prolific in dry years, one of the better flowering trees; fruit, 2–3 cm long.

*G. rosmarinifolia*—**Rosemary Grevillea**          Vic., N.S.W.
Cuttings
2–3 m          500 mm
Low shelter, hedges, birds

Compact, dense shrub with slender branches. Widely planted. Can be severely trimmed; an excellent hedge plant. Easily grown, frost-hardy, and suitable for most soil conditions. Foliage dark green, almost needle-like, 3–5 cm long; flowers red with a long flowering period each year.

*Grevillea robusta*

*G. striata*—**Beefwood**                                 S.A., N.S.W.,
                                                              Qld, N.T.
6–10 m       400 mm       Prefers heavy       Moderate
                           soils                to slow
Windbreaks, parks, subsistence fodder

Shapely, small to medium tree suitable for hot, dry inland areas Adaptable to most soils. Drought-resistant. Can be trimmed. Bark thick, dark, rough; leaves long, 15–40 cm, narrow, 3–6 mm, leathery, with about 10 striations on each leaf; flowers cream, 5–7 cm long, in slender clusters; fruit 15–20 mm long, oval, with a prominent beak.

*G. thelemanniana*—**Spider Net Grevillea**   Vic., N.S.W.
2 m          450 mm
Ground cover

Low-growing, hardy shrub with pendulous branchlets. Foliage needle-like, 2–4 cm long, soft; flowers red. A prostrate form is also available. 1838.

*G. vestita*—**Hairy Grevillea**                    W.A.
2–3 m    .    450 mm

Attractive, bushy shrub. Hardy, but sometimes a little hard to establish. Will grow on limestone and granitic soils. Leaves wedge-shaped, 2–4 cm long, 1–2 cm wide with 3 terminal lobes; flowers white, in clusters, abundant.

*G. victoriae*—**Royal Grevillea**            Vic., N.S.W.
2–3 m       750 mm

Attractive, bushy shrub; prefers good, moist soils in cooler districts. Leaves variable, usually 5–10 cm long, narrow; flowers deep red, in profuse drooping clusters.

# GRISELINA

Genus named after F. Griselini (1717–1783), an Italian botanist. About six species from New Zealand and Chile are known.

*G. littoralis*—**New Zealand Broadleaf**   E.H.        N.Z.
Cuttings
5–8 m       650 mm                                    Fast
Windbreaks

Bushy shrub or small tree. Hardy. Wind-firm. Withstands severe coastal exposure, good for protection from sand blast. Resistant to drying out once established. Good for coastal plantings. Leaves thick, fleshy, glossy green, broadly oval, 5–8 cm long; flowers inconspicuous, yellow-green; fruit greenish-black berry, 5–7 mm diam. Golden-foliaged forms are available. 1872.

# HAKEA  E.H.

Seed                                                   Fast
Low windbreaks, roadsides, streets, parks, honey, birds
Genus named after Baron Hake (1745–1818), a German patron of botany. There are over one hundred species. This is a largely Australian group of bushy shrubs. Although slightly frost-tender when young, they are hardy after the first year or so. Requiring reasonably well-drained sites, they do not grow well on poorly-drained, heavy soils. Large plants cannot be transplanted but will tolerate light pruning. Leaves are variable. Flowers are pin-like and form clusters. Fruit is woody, oval to globular, and separates into two halves.

*H. bucculenta*                                         W.A.
3–4 m       400 mm       Requires well-
                          drained soil

Erect shrub. One of the showiest of the hakeas. Worth more extensive planting. Leaves needle-like, 10–15 cm long, less than 2 mm wide; flowers scarlet, 8–10 cm long brushes, showy; fruit globular, 1–2 cm diam.

*H. elliptica*—**Oval Leaf Hakea**                    W.A.
4–5 m       500 mm

Dense, erect shrub. Withstands moderate coastal exposure. Very adaptable, grows on granitic soils. Leaves oval, 5–8 cm long, shining; flowers white, in dense clusters; fruit about 2 cm diam.

*H. eriantha*—**Tree Hakea**  Vic., N.S.W., Qld
3–6 m    750 mm    Moderate

Small tree. Prefers cool, moist sites but will withstand some dryness. Fairly wind-firm, suitable for exposed sites. Leaves 7–15 cm long, narrow, veins (except the mid-vein) inconspicuous; flowers white or faintly pink; fruit 2–3 cm long, 1 cm wide, smooth.

*H. francisiana*  S.A., W.A.
3–4 m    500 mm    Not suitable for    Moderate
wet soils

Upright, compact, small tree. Frost-resistant. Leaves 15–25 cm long, narrow, 5–7 prominent veins; flowers numerous in 15–20 cm racemes, pink to red.

*H. laurina*—**Pincushion Hakea**  W.A.
3–5 m    400 mm

Compact, rounded, small tree with dense foliage to ground level. Not so densely crowned in coastal areas. Requires well-drained soil. Isolated specimens not wind-firm; staking necessary. Prefers open position. Tolerates moderate coastal exposure and smog. Leaves narrow, 8–12 cm long, 1–2 cm wide, several veins (no midrib); flowers conspicuous, red, in globular heads, 4–5 cm diam.; fruit oval, 2–3 cm long. 1830.

*H. leucoptera*—**Silver Needlewood**  S.A., Vic.,
3–5 m    400 mm    N.S.W., Qld

Small tree or shrub, rather open crown. Will grow on poor, sandy and calcareous soils. Suitable for erosion control on sandy soils. Can be used as a 'fence' or hedge—the stiff 2–5 cm long needle-like leaves discourage stock. Leaves are useful as famine fodder for sheep. Flowers white, often sparse, but showy when in quantity; fruit 1–2 cm diam.

*H. muellerana* (H. flexilis)—**Flexile Hakea**  Vic., N.S.W.
3–5 m    550 mm    Moderate

Tall, upright, moderately dense shrub or small tree. Satisfactory on a wide range of soils. Suitable for stock-proof hedges. Hardy. Leaves needle-like, 3–7 cm long, spreading outwards horizontally; flowers yellow to white, small, along the stem; fruit oval, 1 cm long, beaked.

*H. multilineata*—**Pink Bottlebrush Hakea**  W.A.
3–4 m    300 mm

Erect, compact but variable shrub. One of the more attractive hakeas. Grows under hard conditions, but sometimes difficult to establish. Leaves 16–15 cm long, 1 cm wide, gum-like, several prominent veins (no midrib); flowers rosy-pink, in dense showy spikes 7–10 cm long at end of branches. A hybrid of this and *H. laurina* is available commercially at *H. 'paynei'*. (*H. multilineata* species has recently been sub-divided into five species one of which still retains the name *H. multilineata*.)

*H. nodosa*—**Yellow Hakea**  S.A.,
3–4 m    550 mm    Vic., Tas.

Compact shrub. Withstands wet sites better than most hakeas but preferably should not be planted in them. Leaves 2–3 cm long, narrow, soft needle-like; flowers yellow, thickly clustered on the branches, up to 1 cm long; fruit oval, often warty, 1–2 cm long.

*H. petiolaris*—**Sea Urchin Hakea**  W.A.
3–5 m    400 mm

Dense, spreading, shapely, small tree. Will grow on granitic soils. Leaves 5–10 cm long, 3–6 cm wide, oval, bluish-green; flowers purplish-pink, globular, 4–5 cm diam.; fruit 2–3 cm long, oval.

*H. preissii*—**Needle Tree** or **Christmas Hakea**  W.A.
2–3 m    350 mm    Prefers heavier soils    Moderate
Ornamental

Rounded, single-stemmed, large shrub or small tree, spreading as wide as it is tall. Crown slightly open; foliage retained to ground level. Often rather difficult to establish. Leaves 1–2 cm long, needle-like, rigid, dull green; flowers 2–3 cm across, yellow, in dense bottlebrush-like spikes, prolific, almost covering the plant; fruit 2 cm long, grey, with 2 small horns.

*H. pubescens*  N.S.W., Qld
2–3 m    500 mm

Bushy shrub. Suitable for low stock hedges. Leaves needle-like, sharp, 3–7 cm long; flowers pink.

*H. salicifolia* (H. saligna)—**Willow Hakea**  N.S.W.
2–5 m    600 mm    Will grow on wet sites

Shrub or small tree, height very variable according to site. Prefers good, moist soils. Very fast-growing. Can be heavily trimmed to form a dense hedge. Leaves narrow, 7–12 cm long, willow-like; flowers white, in dense clusters; fruit 1–3 cm long, with wart-like protuberances. 1791.

*H. sericea*—**Silky Hakea**  Vic., Tas., N.S.W.
2–4 m    650 mm

Tall shrub or small tree. Hardy. Withstands long, dry summer. Leaves needle-like, rigid, slender, pointed, 3–7 cm long; flowers variable, from white to red; fruit 2–3 cm long, rough, with a small beak. Naturalized in New Zealand, noxious weed in South Africa.

*H. suaveolens*—**Sweet Hakea**  W.A.
2–4 m    450 mm

Dense, compact, spreading shrub retaining branches to ground level. Good low windbreak or for use with taller species. Hardy, easily grown. Although wind-firm, in very exposed sites branches may occasionally be

*Hakea suaveolens*

*H. carolina* (H. tetraptera)—**Snowdrop Tree**   D.H.
South-eastern USA

Seed, root cuttings

5–10 m        900 mm        Moist well-                    Moderate
drained soil

Low shelter, parks, roadsides, ornamental

Attractive, bushy, rather spreading tree. Old trees may reach 15 m, with arching branches. Not widely planted nowadays but worth reviving. Planted extensively overseas. More suitable for the cool, mountain districts. Very hardy but not satisfactory in exposed windy positions. Can be moderately pruned. Withstands shade. Large plants can be transplanted. Remarkably disease free. Leaves oval, pointed, 7–10 cm long, bright green; flowers white, 1 cm long, snowdrop-like, in 12–15 cm hanging clusters, very showy; fruit pear-shaped, 2–4 cm long, with 4 wings. 1756.

broken off by the wind. Only moderately frost-resistant. Withstands severe coastal exposure. Will grow on dry or adverse sites. Disliked by stock. Can be lightly trimmed to form a dense hedge. Leaves 7–10 cm long, 4–6 cm wide, stiff, divided into numerous needle-like points; flowers showy, creamy-white, fragrant, in terminal clusters; fruit 2 cm long, rough, with 2 small horns. 1803.

*H. ulicina*—**Furze Hakea**        S.A., Vic., Tas.,
N.S.W.

Similar to *H. muellerana* but leaves less needle-like and 10–20 cm long. Some authorities regard *H. ulicina* as a variety of *H. muellerana*, the only difference being in the length of the leaves.

*H. victoriae*—**Royal Hakea**        W.A.

2–3 m        400 mm        Prefers warmer, well-
drained sites

Erect, showy shrub. Will grow on gravelly sites. Can be difficult to establish. Leaves distinctive, round, 8–10 cm diam., pressed closely to the stem, colour changes from green to brown during the year; flowers yellow.

*H. vittata*—**Hooked Needlewood** or        S.A., Vic.,
**Striped Hakea**        N.S.W., Qld

2–3 m        400 mm        Will grow on
hard sites

Small tree, somewhat open-crowned. Similar to *H. leucoptera*. Leaves stiff, needle-like, 4–7 cm long; flowers white, small; fruit 2 cm long, with a black stripe along the junction of the two halves.

# HALESIA

Genus named after S. Hales (1677–1761), the author of *Vegetable Staticks*. There are about four species from North America and one from China.

# HETERODENDRUM

*Heterodendrum* is from *heteras* meaning variable and *dendrum* meaning a tree which alludes to the variable nature of the individual plants. About five species from Australia are known.

*H. oleifolium*—**Cattlebush**   E.H.        W.A., S.A.,
Vic., N.S.W., Qld

Seed

4–6 m        300 mm        Moderate to slow

Windbreaks, shade, parks, roadsides, fodder, honey

Variable but usually rounded shrub or small, neat tree with an erect stem. Varies in height (2–10 m) depending on the site. Responds to trimming, can be formed into an attractive tree or open hedge. Drought- and frost-resistant. Subsistence fodder only, but young shoots believed to be toxic to stock. Leaves greyish-green, 5–15 cm long, thin, flat; flowers inconspicuous, greenish, in clusters; fruit fleshy, globular, 8–10 mm diam., black glossy seed.

# HOHERIA

Genus named after *Horhera* the Maori name. There are four species from New Zealand.

*H. populnea*—**New Zealand Lacebark**   E.H.        N.Z.

Seed

4–5 m        650 mm        Fast

Parks, roadsides, streets, ornamental

Compact, upright tree. Hardy. Can be pruned; old trees can be restored by heavy pruning. Large plants can be transplanted. Leaves broad, 7–10 cm long, 4–5 cm wide, serrated; flowers white, 2 cm across. Variegated-leaf forms are available.

*Homalanthus populifolius—foliage*

*Hymenosporum flavum*

## HOMALANTHUS

The name is derived from *homalas* meaning smooth and *anthos* meaning flower. There are about thirty species from Australia, Pacific islands and Malaysia.

*H. populifolius*—**Bleeding Heart Tree**   E.H.     N.S.W.,
Seed                                                           Qld
4–5 m        750 mm      Well-drained soils           Fast
Low windbreaks, shade, parks, roadsides, ornamental

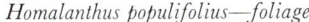 Attractive, bushy, small tree. Moderately frost-resistant. Leaves smooth, heart-shaped, 8–15 cm long, some turning bright red, on stalks up to 10 cm long; flowers small, yellow, 5–8 cm in terminal catkins; fruit purple, in clusters.

## HYMENOSPORUM

*Hymenosporum* is derived from *hymen* meaning a membrane and *sporos*, a seed. This alludes to the membrane on the seed. One species from Australia is known.

*H. flavum*—**Native Frangipani**   E.H.     N.S.W., Qld
Seed
5–8 m        750 mm       Deep, well-drained           Fast
                                    soils
Shade, parks, roadsides, streets, ornamental

Attractive, shapely, upright, rather open tree. Moderately hardy. Not very frost-resistant. Best in coastal areas but can be grown inland on most soils. Suckers from cut roots. Leaves deep green, oval, 10–15 cm long, 3–4 cm wide; flowers pale yellow, fragrant, tubular, 2–4 cm diam. in terminal clusters 10–20 cm across; fruit a capsule, flattened, 5 cm long, 2–3 cm across; seeds brown with papery wings.

## HOVENIA

Genus named after D. Hoven, a parliamentarian of Amsterdam. There is one species from Japan.

*H. dulcis*—**Japanese Raisin Tree**   D.H.     India
Seed
6–8 m        750 mm      Grows well in        Moderate
                                    sandy soils
Shade, parks, autumn foliage

Attractive, small, spreading tree, with erect branches. Extensively cultivated in Japan. Suitable for shade. Prefers cool climates but tolerates hot summers. Moderately hardy although slightly frost-sensitive. Leaves heart-shaped, 10–15 cm long, 7–12 cm wide; flowers greenish-white, small, in clusters 5–7 cm long; fruit in bunches, each fruit a small pip at the end of a thickened, fleshy stem with red pulp. Stems are edible, sickly sweet and relished by orientals. 1812.

## IDESIA

Genus named after E. Y. Ides, a thirteenth-century Dutch traveller in China. Only one species from China is known.

*I. polycarpa*—**Wonder Tree**   D.H.     China
Seed
6–10 m        750 mm      Prefers loams           Fast
Shade, parks, streets, avenues, ornamental

Very attractive, spreading tree with almost horizontal branches. Will not stand severe frosts or exposure to strong winds. Probably the best of the berry trees. Male and female trees—both required for good display of berries. Stands light trimming only. Leaves large, heart-shaped, 8–15 cm long, 10–12 cm diam., deep green; flowers yellow-green, inconspicuous; fruit crimson berries, produced prolifically in large grape-like clusters remaining until following spring, very conspicuous and showy after leaf fall. Berries not touched by birds. 1864.

# ILEX

*Ilex* is the old Latin name. There are about 400 species from most tropical and temperate regions.

*I. aquifolium*—**Holly**   E.H.          Europe, Asia, N. Africa
Seed, cuttings
6–15 m      650 mm                            Moderate
Parks, streets, hedges, ornamental

 Dense, upright, pyramidal, decorative tree. Long-lived, old trees may exceed 20 m. More suitable for the cooler, mountain areas but will grow in the warmer districts. Hardy. Very frost-resistant, withstands prolonged snow. Can be trimmed severely to form a dense hedge. Bark silver-grey; leaves glossy green, rigid, wavy margin with sharp points (absent on mature plants), oval, 4–8 cm long; flowers inconspicuous; fruit red berry, colourful. Male and female trees—female trees raised from cuttings should be planted for berry display. Numerous horticultural varieties available. L.C.

# JACARANDA

This is a Brazilian name. There are about fifty species from Central and South America.

*J. acutifolia* (J. mimosaefolia)—**Jacaranda** D.H.   Brazil,
                                                      Bolivia, Argentine
Seed, cuttings
3–15 m      650 mm      Well-drained loams      Fast
Shade, parks, roadsides, ornamental

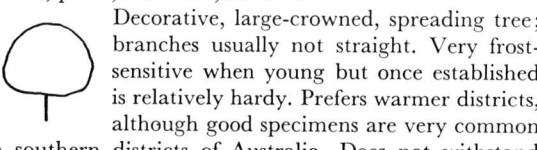 Decorative, large-crowned, spreading tree; branches usually not straight. Very frost-sensitive when young but once established is relatively hardy. Prefers warmer districts, although good specimens are very common in southern districts of Australia. Does not withstand strong winds. Can be lightly trimmed. Beautiful tree, good specimens are amongst the most attractive of all trees. Foliage dark green, fern-like, pinnate, 15–20 pinnae, 10–15 pairs leaflets; flowers blue, 4–5 cm long, produced in prolific clusters 15–20 cm long, 40–80 flowers per cluster.

# JUGLANS

*Juglans* is adapted from the old Greek name *Jovis glans*. There are about sixteen species from Europe, Asia, North and Central America.

*J. nigra*—**Black Walnut** D.H.   Central North America
Seed
20–25 m      1000 mm      Prefers deep, moist,      Fast
                          well-drained loams
Farm forests, windbreaks, summer shade, parks

 Large, spreading tree when open grown; tall and straight with a narrow crown when planted close together. Produces a valuable furniture timber. Will grow on good alluvial soils in lower rainfall zones, and on soils derived from limestone. Soil is much more critical than climate. Deep root system. Not tolerant of shade. Wind-firm. Withstands severe frosts when leafless but may be damaged by late spring frosts. Coppices lightly. Tolerates some flooding. Not many plants will grow underneath it due to its production of a toxic substance, juglone. Bark grey; leaves pinnate, large, 30–50 cm long, 15–25 leaflets 7–10 cm long, 2–3 cm wide, shining, serrated; fruit an edible nut, 2–4 cm diam. enclosed in a thick, yellow-green husk 4–5 cm diam., 1–3 in a cluster. 1656.

# JUNIPERUS

This is the old Latin name. About sixty species from the cooler regions of Europe, Asia and North America are known.

*J. communis*—**Common Juniper**   E.C.      Europe, Asia,
                                                      America
Seed
6–10 m      650 mm      Will grow on              Slow
                        calcareous soils
Windbreaks, roadsides, parks, hedges, ornamental

Dense, compact, small tree. Variable, but usually not spreading. Stands clipping, forms good hedges; the prickly foliage discouraging stock. Frost-resistant. Wind-firm. Few plants have such a wide natural distribution or grow on such a wide range of soils. Seeds often dormant in the soil for a year or so; soaking in hot water for a few minutes improves germination. Bark reddish-brown; leaves narrow, 6–12 mm long, close to the branchlets and tapering to a fine sharp point; fruit globose, fleshy and soft, bluish-black, 6–10 mm diam. Does not look like a conifer cone as the 'cone scales' are fleshy rather than woody as in other conifers. Numerous horticultural varieties available. L.C.

# KOELREUTERIA

Genus named after J. G. Koelreuter (1733–1806), a German professor of natural history. There are about eight species from eastern Asia.

*K. paniculata*—**Golden Rain Tree** or D.H.      China,
              **Pride of India**                 Korea, Japan
Seed
5–8 m      500 mm      Will grow in strongly   Moderate
                       alkaline soils
Low windbreaks, parks, roadsides, ornamental

Attractive, bushy tree with a round, rather spreading crown. Retains the lower branches. Tolerates warm dry conditions. One of the best trees for strongly alkaline soils. Large specimens can be transplanted during winter. Leaves pinnate, 20–40 cm long, leaflets 7–15, 4–7 cm long, oval, margin irregularly crenated, rich gold in autumn; flowers yellow-green, 1 cm across, 4 petals, in broad clusters, 15–25 cm long; fruit a 3-sided papery capsule, 4–5 cm long, yellow-green, seeds black, pea-shaped. 1763.

*Laburnum anagyroides*

*Lagunaria patersonii—young tree*

# KUNZEA

Genus named after G. Kunze (1793–1851), a German physician and botanist. There are about thirty species, all from Australia.

*K. ambigua*—**White Kunzea** E.H. Vic., N.S.W.
Seed
3–5 m    600 mm                     Fast
Parks, roadsides, ornamental

Tall, erect shrub with dense foliage. Hardy. Tolerates moderate coastal exposure. Leaves small, narrow, pointed, 8–12 cm long; flowers prolific, white to pink, 5–8 mm diam.; fruit globular, woody, 5–6 mm diam.

# LABURNUM

*Laburnum* is the ancient Latin name. There are three species from southern Europe and south-western Asia.

*L. anagyroides* (L. vulgare)
          —**Laburnum** or   D.H.  Southern Europe
          **Golden Chain**        South-western Asia
Seed, cuttings
5–7 m    650 mm   Will grow on limestone.   Fast
Parks, roadsides, ornamental

Upright small tree with slender, ascending branchlets drooping at the ends. Very showy when in flower. Versatile tree suitable for most soils except poorly drained and coastal sites. Not suitable for the warmer districts. Hardy, frost resistant. Tolerates some shade. Can be heavily trimmed. Easily transplanted even when large. Fairly disease free. Bark olive-green, polished; branchlets slightly silky, slender, drooping; leaves trifoliate, leaflets elliptical to round, 3–5 cm long, entire, slightly silky underneath, leaf stalk long and thin, 4–8 cm

long, little or no colour change in the autumn; flowers golden-yellow, pea shape, about 2 cm long, in conspicuous pendulous racemes, 20–30 cm long; fruit a legume, 5–7 cm long, with a narrow wing; seeds black. All parts of the plants, particularly the seeds, are poisonous. The arrangement of the leaflets in threes and the long slender leaf stalks are distinctive. Numerous horticultural varieties are available. (1560).
*L.* × *"Vossii"* (L. watereri) (*L. anagyroides* × *L. alpinum*) Grafted.
Similar to *L. anagyroides* but racemes longer, 25–40 cm long, and showier. Semi-weeping when mature. Rarely produces seeds. (1864).

# LAGUNARIA

*Lagunaria* is similar to *Lagunaea* which was named after Lagina (1494–1560), a Spanish botanist. There is a single species from Australia, Norfolk Island and Lord Howe Island.

*L. patersonii* (Fugosia patersonii)—  E.H.  N.S.W., Qld,
   **Pyramid Tree** or **Norfolk Island**        Norfolk Is.,
   **Hibiscus**                        Lord Howe Is.
Seed
6–8 m    400 mm   Will grow on wet sites   Fast
Windbreaks, shade, parks, roadsides

Attractive, dense-crowned, pyramidal tree. Will grow in most soils including limestone, slightly saline and wet areas. Wind-firm. Withstands severe coastal exposure. Frost-hardy. Satisfactory in both the cooler and warmer districts. Grows well on irrigated soils. Tolerates smog. Leaves grey-green, oval, 5–10 cm long; flowers attractive, 5–7 cm across, pink at first, later white, produced over long periods each year; fruit a pod, 1–2 cm long and wide, can cause irritation ('cow-itch') because of the fine silica spicules or 'splinters' in the seed covering.

*Lagunaria patersonii—mature tree*

*Lagunaria patersonii—flowers—F. Collet*

# LARIX

*Larix* is the old Latin name. There are about ten species from Europe, Asia and North America.

*L. decidua*—**Common Larch**  D.C.                    Europe
Seed
20–30 m    900 mm    Deep soil                        Fast
Farm forests, summer shade, parks, roadsides, ornamental, autumn foliage

Attractive, symmetrical, conical tree. One of the few deciduous conifers. Foliage fresh light green in spring, turning golden in autumn. Useful and decorative tree which should be planted more extensively in the cooler districts. Develops a deep tap root, hence requires deep soils. Hardy and frost-resistant. Timber durable. Leaves needle-like, 1–4 cm long; cones oval, 2–3 cm diam., scales thin, rounded ends, 6–10 mm wide. For farm forests the hybrid *L. x eurolepis* (Dunkeld larch) is preferable because of its faster growth. 1629.

*Larix decidua plantation—Alnus glutinosa in front*

*L. x eurolepis*—**Dunkeld Larch** or            Britain
                 **Hybrid Larch**  D.C.
Seed
15–25 m    800 mm                                  Fast

Hybrid between *L. decidua* (European larch) and *L. kaempferi* (Japanese larch). Displays hybrid vigour (faster growing than either parent type) and tolerates poorer conditions. Botanical characteristics variable between those of the parents otherwise similar to *L. decidua*. Growth characteristics come true to type from seed. 1904.

# LAURUS

*Laurus* is the old Latin name. Five species from Europe and Canary Islands are known.

*L. nobilis*—**Sweet Bay** or            Mediterranean
             **Laurel**  E.H.
Cuttings
8–10 m    650 mm    Well-drained soils          Fast
Windbreaks, parks, roadsides, streets, hedges

Compact, pyramidal to broadly columnar tree. Hardy. Moderately frost-resistant. Withstands moderate coastal exposure. Can be readily trimmed to form a hedge. Transplants easily. Leaves oval, 7–10 cm long, thick, glossy, aromatic, used as flavouring in cooking; flowers inconspicuous, greenish-yellow; fruit shining, black berry.  L.C.

# LEPTOSPERMUM  E.H.

Seed                                               Fast
Windbreaks, parks, low shelter, hedges, honey
*Leptospermum* is derived from *leptos* meaning slender and *sperma* meaning seed; referring to the slender seeds.

*Larix decidua foliage and cones*

*Larix decidua in spring*

There are about fifty species from Australia, New Zealand and Malaysia.

Vigorous large shrubs or small trees, suitable for a wide range of soils. Hardy, frost-resistant. Coppice freely. Can be trimmed to form a hedge. Bark grey, papery; leaves small; flowers white, prolific, with 5 large conspicuous petals; fruit globular, usually about 5–8 mm diam., woody; seed very small.

*L. laevigatum*—**Coast Tea-tree**    Vic., Tas., N.S.W., Qld

3–6 m    600 mm
Small tree with dense foliage, trunk often leaning or crooked in exposed sites but otherwise upright. Wind-firm. Withstands severe coastal exposure, occurring commonly around the coast; good for planting on coastal sands and for stabilizing drifting sand. Leaves oval, grey-green, leathery, 1–2 cm long; flowers 2 cm diam.

*L. laevigatum* var. *minus*    Vic.
2–3 m    300 mm    Grows on deep sands
Hedges
Bushy, multi-stemmed, compact shrub. Hardy, drought- and frost-resistant. Coppices vigorously, stands severe trimming, forms a good hedge. Leaves slightly shorter than *L. laevigatum*, otherwise similar. An inland form occurs in the Sunset Country of north-western Victoria.

*L. lanigerum* (L. pubescens)—**Woolly Tea-tree**    S.A., Vic., Tas., N.S.W.
3–5 m    550 mm    Will grow on wet soils
Erect, dense, tall shrub with hairy branchlets and greyish foliage. Tolerant of a wide range of soils from very acid to strongly alkaline. Suitable for heavy alluvial soils and where vertical drainage is poor. Withstands moderate coastal exposure. Leaves 1–2 cm long, oval, silky, with a sharp point; flowers 2 cm diam., calyx silky. 1784.

*Larix decidua in autumn*

*L. phylicoides* (L. ericoides, Kunzea peduncularis) —**Burgan** E.H.    Vic., N.S.W., Qld
2–3 m    550 mm    Poorly-drained soils
Multi-stemmed, bushy shrub. Useful for poorly-drained sites particularly shallow soils over impermeable clays. Coppices very freely. Leaves grey-green, 1–3 cm long, pointed; flowers 1–2 cm diam., prolific.

223

*Leptospermum scoparium—original form*

*Leucadendron argenteum*

*L. scoparium*—**Manuka**        Vic., Tas., N.S.W., Qld, N.Z.

3–6 m     550 mm

Multi-stemmed shrub. Often a bit straggly but can be trimmed to any desired shape. Grows on dry and poorly-drained sites. Leaves narrow, pointed, 1–3 cm long; flowers 1–2 cm diam., massed along the stem. 1792.

Parent type of the ornamental tea-trees commonly grown. First originated late last century, there are now more than 35 varieties available. Varieties usually have pink to red flowers.

# LEUCADENDRON

*Leucadendron* is derived from *leukas* meaning white, and *dendron*, a tree. This alludes to the silvery leaves of the Silver Tree. About seventy species from South Africa are known.

*L. argenteum*—**Silver Tree**   E.H.       S. Africa
Seed
8–10 m     650 mm     Requires good, well-drained soils     Moderate
Parks, roadsides, ornamental

Attractive, shapely, moderately densely-crowned tree. Planted mainly for its ornamental silvery foliage. Requires good, well-drained soil; will not stand lime, strongly alkaline or heavy soils. Frost-sensitive when young. Withstands moderate coastal exposure. Leaves 7–12 cm long, 2–3 cm wide, silvery-white with furry surfaces; flowers inconspicuous; fruit a small nut. 1693.

# LEUCOPOGON

*Leucopogon* is derived from *leukos* meaning white and *pogon* meaning a beard which refers to the white fringed flowers. There are about 150 species from Australia, Malaysia and New Caledonia.

*L. parviflorus*—**Coast Beard-heath**   E.H.       All States
Seed
3–5 m     600 mm              Fast
Low windbreaks, shade

Low-growing, much-branched shrub; usually mallee-like in habit but upright in sheltered positions. Hardy. Withstands severe coastal exposures, occurs commonly along the coast. Can be lightly trimmed to form an open hedge. Not widely planted but is worth more attention. Bark rough, furrowed, grey; leaves narrowly oval, 2–3 cm long, dark green, pointed; flowers white, inconspicuous; fruit globular, white, 6–8 mm diam., fleshy. 1822.

# LIQUIDAMBAR

The name is from *liquidas* meaning liquid and *ambar* meaning amber which alludes to the resin exuded by some species. There are about six species from Asia and North America.

*L. formosana*—**Chinese Liquidambar**   E.H.       China, Taiwan
Seed
7–15 m     650 mm              Fast
Summer shade, parks, autumn foliage

Attractive, upright, symmetrical tree. Hardy, frost-resistant, suitable for both cool and hot districts provided moisture is not limiting. Foliage colours well, turning to orange, red then deep purple in late autumn to winter with the leaves remaining on the tree until the following spring. Branchlets corkless; leaves maple-like, 5–10 cm diam., thin, three pointed compared with 5–6 in *L. styraciflua*, young leaves glossy purple; flowers inconspicuous; fruit globular, 2–3 cm diam., covered in soft, fleshy spines. Selected grafted specimens are occasionally available. 1884.

*L. styraciflua*—**Liquidambar** or  E. North America,
**Sweet Gum**  D.H.  Mexico
Seed
10–20 m  650 mm  Moderate
Summer shade, parks, avenues, autumn foliage

Broadly pyramidal, shapely tree. Widely planted, stands wet sites better than most deciduous trees. Best growth on deep soils. Hardy, frost-resistant. Coppices freely. Wind-firm. Tolerates smog. Can be grown in irrigated areas but best growth and displays of coloured foliage are obtained in the cool, mountain districts. Stands light pollarding to give a bushier crown. Corky wings on the branchlets. Leaves dark green, maple-like, alternate (maples opposite), 10–15 cm across, 5–6 lobes; flowers inconspicuous; fruit globular, 2–3 cm. covered in soft, fleshy, short spines on a stalk 3–8 cm long. Autumn colours variable, predominantly yellow in some trees, red and brown in others. Variety 'Festeri' is one of the best with rich red-brown autumn colouring. 1681.

*Liquidambar styraciflua in autumn*

*Liquidambar styraciflua — good form of autumn foliage — some forms are yellow only*

# LIRIODENDRON

*Liriodendron* is from *lirion* meaning lily and *dendron* a tree. This refers to the flowers. There are two species, one from North America and the other from China.

*L. tulipifera*—**Tulip Tree** or  Southern USA
**Yellow Poplar**  D.H.
Seed
15–35 m  750 mm  Good, well-drained soils  Fast
Summer shade, parks, roadsides, autumn foliage

Upright, broad-leaf tree retaining branches to ground level. Prefers deep fertile soils in cool, mountain districts. Frost-hardy. Coppices freely. Intolerant of shade. Easily killed by earth fills. Leaves large, 10–20 cm across, usually 4 lobes, lobed like maples but distinctive in that there is no central lobe at the apex; autumn foliage rich yellow; flowers large, attractive, greenish-yellow, tulip or cup-shaped, orange at the centre, 5–7 cm across, usually only appearing when the tree is about 10 years old; fruit, upright cluster of maple like seeds, 5–8 cm long. 1663.

# MACLURA

Genus named after W. Maclure, a nineteenth-century American geologist. About twelve species from North America, Asia and Africa are known.

*M. pomifera* (M. aurantica)—**Osage**  Southern USA
**Orange**  D.H.
Seed
6–15 m  400 mm  Will grow on limestone  Fast
Windbreaks, protective hedgerows

Small tree, erect but spreading and retaining branches to ground level when open grown. Crown rather irregular, with characteristic curved branches. More suitable for the warmer districts. Strong, deep root system. Coppices freely. Thorny, thorns 1–2 cm long, shaded branchlets thornless. Stands severe pruning, can be trimmed to form a dense stock-proof hedge. Not widely planted nowadays but still a useful species. Timber bright yellow, useful for bows. Leaves bright green, 7–12 cm long, on a slender stalk, oval, dark green above, turning clear yellow in autumn; flowers green, in round clumps, inconspicuous male and female flowers on separate plants; fruit green, globular, 10–15 cm diam., like a large orange, not edible. Milky sap exuded on bruising may be irritating to some people. 1818.

225

*Liriodendron tulipifera*

*Liriodendron tulipifera—flower*

rich soil for best growth. Strong, spreading root system. Hardy, frost-resistant except at high altitudes. Leaves large, oval, 15–25 cm long, 5–12 cm wide, glossy, stiff, rusty-brown underneath; flowers large, 15–20 cm diam., conspicuous, white, perfumed, summer-flowering; fruit cone-shaped, fleshy, 7–10 cm long, rusty-brown. 1734.

## MALUS—Crab Apples  D.H.  Cultivars

Seed, grafts, cuttings
2–5 m          650 mm                                        Fast
Parks, roadsides

*Malus* is the old Latin name. There are about thirty-five species from Europe, Asia and North America.

 These are small trees suitable for the cooler districts. They are mainly decorative. Numerous varieties are available. Leaves are typical apple leaves. Flowers are white, red, purple. Fruit are small apples about 1–2 cm in diameter and various colours.  L.C.

## MELALEUCA—Honeymyrtles and Paperbarks  E.H.

Seed, cuttings                    Prefer acid soils    Fast
Low windbreaks, parks, ornamental, birds

The name is from *melas* meaning black and *leukos* meaning white which alludes to the black trunk and white branches of some species. Over one hundred species occur, mainly in Australia but a few in Indonesia and Malaysia.

These versatile shrubs or small trees are suitable for adverse conditions, and will grow under wetter conditions than most species. They are generally frost-hardy and withstand some degree of coastal exposure. Can tolerate moderate trimming and annual pruning for shaping is

## MAGNOLIA

Genus named after P. Magnal (1658–1751), a French botanist. About eighty species from North and Central America, Asia and Indonesia are known.

*M. grandiflora*—**Southern Magnolia**  E.H.    Southern
Seed, layers                                      USA
10–25 m     650 mm                             Moderate
Parks

 Large, spreading, dome-shaped tree. Well known and widely planted. Pyramidal when young, spreading with age. Stands warmer conditions better than the deciduous magnolias. Good specimen tree but should have plenty of space. Requires moist, deep,

*Magnolia grandiflora*

desirable. May be formed into 'open' hedges. Often not wind-firm as single specimens in exposed sites and should be planted in rows or groups in such areas. Large specimens do not transplant readily. The honey is satisfactory although not particularly good. Bark is usually papery but sometimes hard and furrowed. Leaves are small, usually less than 5 cm long, and often almost needle-like. Flowers are colourful and arranged like a bottlebrush. Fruit are small, globular, woody and sessile, and remain on the stem for some years.

*Melaleuca armillaris—flowers*

*M. acuminata*—**Mallee Honeymyrtle**　　W.A., S.A., Vic., N.S.W.

2–3 m　　250 mm

Drought-resistant shrub with slender branches. Leaves very narrow, 6–10 mm long less than 5 mm wide; flowers white, grouped on the stem; fruit small, about 3 mm diam. Spring-flowering.

*M. armillaris*—**Giant Honeymyrtle** or　　Vic., Tas., **Bracelet Honeymyrtle**　　N.S.W., Qld

3–5 m　　450 mm

Compact, strong-growing, graceful, bushy shrub; slimmer crown when grown in shade. Suitable for a wide range of soils, including poor, sandy, slightly saline or heavy clay soils and wet areas (see *M. ericifolia*). Tolerates long wet or dry periods. Withstands severe coastal exposure and smog. One of the most useful of small trees. Bark whitish-brown, papery; leaves needle-like, 2–3 cm long; flowers creamy-white in 7–10 cm spikes; fruit 5 mm diam., closely packed. Spring to summer-flowering. 1788.

*M. crassifolia* (M. laxiflora)—**Mauve**　　W.A. **Honeymyrtle**

2 m　　400 mm　　Will grow on sandy soils

Low spreading shrub. Sometimes straggly but responds well to trimming. Flowers mauve. Winter to spring-flowering.

*M. decussata*—**Cross Leaf Honeymyrtle**　　S.A., Vic.

2–4 m　　400 mm　　Will grow in poorly-drained soils

Vigorous, straggly shrub; occasional trimming necessary to obtain a good shape. Tolerates smog. Leaves flat, 6–12 mm long, 2–3 mm wide, soft, opposite to each other, forming a cross when the stem is viewed 'end-on'; flowers blue-purple, in short spikes up to 3 cm long; fruit very small, less than 3 mm diam. Late spring to summer-flowering. A form with almost needle-like leaves occurs in the Grampians but is not often seen in cultivation. 1803.

*M. diosmifolia*　　W.A.

2–4 m　　500 mm

Spreading, rounded, rigid shrub often spreading more than its height. Suitable for poorly-drained sites. Withstands moderate coastal exposure. Foliage attractive, distinctive, leaves oval, crowded together, 1 cm long; flowers yellow, in brushes, 2–4 cm diam.; fruit 5–6 mm diam. 1794.

*M. elliptica*—**Granite Honeymyrtle**　　W.A.

2–3 m　　500 mm　　Will grow on granite soils

Attractive, bushy, easily-grown shrub. Leaves oval, grey-green, about 1 cm long, opposite; flowers 7–8 cm long, red, showy; fruit 5–6 mm diam., closely packed. Spring to early summer-flowering.

*M. ericifolia*—**Swamp Paperbark**　　Vic., Tas., N.S.W., Qld

3–5 m　　500 mm　　Suitable for poorly-drained or saline soils

Bushy shrub, similar to *M. armillaris* but more suitable for wet and saline sites. Withstands severe coastal exposure. Leaves needle-like, but blunt, 5–15 mm long; flowers 2–3 cm long, pale yellow. Spring to early summer-flowering. 1790.

*M. fulgens*—**Scarlet Honeymyrtle**　　W.A.

1–2 m　　450 mm　　Requires well-drained soils

Attractive, showy, compact, easily-grown shrub, with ascending branches. Leaves very narrow, opposite, grey-green, 2–3 cm long; flowers 7–8 cm long, scarlet, formed throughout the foliage; fruit 6–9 mm, spaced apart. Spring-flowering. 1803.

*M. gibbosa*—**Slender Honeymyrtle**    S.A., Vic., Tas.

2–3 m    500 mm

Dense, low, round, wiry shrub, often spreading 2 m across. Will grow on wet soils but prefers moist, well-drained sites. Fairly drought-resistant once established. Leaves oval, about 5–7 mm long; flowers usually in globular heads, 2 cm diam., mauve to reddish-purple; fruit 5–7 mm. Summer-flowering.

*M. glomerata*    W.A., S.A., N.S.W.

3–5 m    250 mm    Prefers loose, sandy soils
Roadsides

Compact, large shrub or small tree, very suitable for hot, dry districts. Drought-resistant, good desert plant. Leaves very narrow, 2–5 cm long; flowers cream, in globular clusters.

*M. halmaturorum*—**Salt Paperbark**    S.A., Vic.

5–7 m    400 mm    Will grow on swampy
and saline soils

Large shrub or small tree, trunk crooked, short, with grey-white papery bark. Very hardy and versatile. One of the best trees for planting on swampy or saline soils, will grow down to the tidal zone in bays, estuaries, etc. Wind-firm. Withstands severe coastal exposure. Leaves very small, narrow, 5–7 mm long; flowers creamy-white in rounded heads. Spring-flowering.

*M. huegelii*—**Chenille Honeymyrtle**    W.A.

2 m    450 mm    Will grow on
calcareous soils

Attractive, rather open, upright, small shrub. Suitable for saline soils; withstands moderate coastal exposure. Can be trimmed to form a hedge. Foliage soft, leaves oval, less than 6 mm long, closely pressed to the stem; flowers white in large, dense 8–10 cm long spikes; fruit up to 3 mm diam. Summer-flowering.

*M. hypericifolia*—**Red Flower Paperbark**    N.S.W.

2–3 m    450 mm    Will grow on heavy soils
Spreading, loosely-branched shrub with semi-pendulous branchlets. Responds well to trimming—with trimming it forms a dense compact bush. Easily grown. Withstands moderate coastal exposure. Leaves larger than most melaleucas, 3–5 cm long, 7–12 mm wide; flowers deep red in spikes, 7–10 cm long through the foliage, flowers abundantly and for long periods each year (winter or summer); fruit 8–10 mm diam., closely packed. 1792.

*M. incana*—**Grey Honeymyrtle**    W.A.

2–3 m    450 mm    Will grow on swampy soils
Greyish shrub, 'woolly' in appearance with semi-pendulous branchlets. Does not withstand severe frosts. Tolerates smog. Leaves grey-green, very narrow, about 1 cm long; flowers yellow in terminal spikes, 2–3 cm long, often 5–9 spikes in a cluster. Winter to spring-flowering. 1817.

*Melaleuca linariifolia*

*Melaleuca linariifolia—flowers*

*M. lanceolata* (M. pubescens)    W.A., S.A., Vic.,
—**Moonah** or    N.S.W., Qld
**Black Paperbark**

5–7 m    500 mm

Bushy, dense, small tree with a round crown. Very versatile; grows well on swampy, saline, limestone or sandy soils and in the hotter districts. Wind-firm. Withstands severe coastal exposure and smog. Can be trimmed. One of the hardiest of the melaleucas. Bark hard, dark; leaves greyish-green, 6–12 mm long; flowers creamy, in spikes 3–5 cm long; fruit up to 3 mm diam. Spring-flowering.

*M. lateritia*—**Robin Redbreast Bush**    W.A.

2–3 m    450 mm

Erect, bushy, spreading, much-branched shrub. Easily grown. More suitable for the warmer districts. Leaves needle-like, 1 cm long; flowers vivid orange-scarlet, in spikes 5–8 cm long and 3–4 cm wide; fruit 5–7 mm diam., closely packed. Summer-flowering.

*M. linariifolia*—**Flax Leaf Paperbark**     N.S.W., Qld
5–10 m     500 mm     Will grow in heavy soils
Shade, streets

Shapely, slightly spreading, densely-crowned tree. Tends to form several stems. Very good park specimen tree. May grow to 10–12 m in warmer districts. Hardy, easily grown. Suitable for wet and saline sites. Deep-rooted. Tolerates some flooding. Withstands moderate coastal exposure. Bark white, papery; leaves narrow, 3–4 cm long; flowers white, prolific, often covering the entire crown; fruit up to 3 mm diam., slightly bell-shaped. Summer-flowering. 1793.

*M. longicoma* (M. macronycha)     W.A.
     **—Long Hair Paperbark**
2–3 m     300 mm     Will grow on gravelly soils

Small, showy shrub. Suitable for granitic sands in warm areas. Leaves narrow, 3–5 cm long; flowers red, in 3–5 cm long spikes; fruit 8–10 mm. Summer-flowering.

*M. neglecta*—**Mallee Honeymyrtle**     S.A., Vic.
2 m     400 mm

Bushy shrub with corky bark, and slender but sometimes tortuous branches. May become straggly and open when older. Salt-tolerant. Leaves narrow, 5–7 mm long; flowers white, in short spikes through the foliage; fruit about 3 mm diam. Spring-flowering.

*M. nesophila*—**Showy Honeymyrtle**     W.A.
2–3 m     450 mm     Grows well on sandy,
                                    coastal soils

Attractive, multi-stemmed, hardy, easily-grown shrub or small tree. Can be trimmed to form a dense, compact hedge. Withstands moderate coastal exposure. Can be heavily trimmed. Worth planting more extensively. Leaves glossy, light green, oval, 2–3 cm long; flowers showy, mauve, in globular terminal clusters, 2 cm diam., prolific. Summer-flowering.

*M. pentagona*—**Oval Leaf Honeymyrtle**     W.A.
2 m     450 mm     Will grow on sandy and
                                    saline soils

Shrub with small sharply-pointed leaves 1–2 cm long; flowers, purplish-pink, globular heads, 1 cm diam. Spring-flowering.

*M. quinquenervia* (M. leucadendron) W.A., N.S.W., Qld,
     **—Broad Leaf**     N.T., Indonesia,
     **Tea-Tree**     Malaysia
10–15 m     500 mm     Will grow on calcareous,
                                    wet and saline soils
Shade

Small to medium-sized spreading tree. Wind-firm. Withstands wet and saline conditions better than most trees; one of the few which can be grown on brackish soils in exposed sites. Frost-sensitive.

*Melaleuca styphelioides—young tree*

Withstands severe coastal exposure and hot, dry conditions once established. Bark thick, spongy, conspicuous, peeling off in sheets; leaves oval, 8–15 cm long; flowers creamy-yellow, in spikes, 10–20 cm long. Winter-flowering. Only *Melaleuca* occurring extensively outside Australia. 1796.

*M. radula*—**Graceful Honeymyrtle**     W.A.
2 m     500 mm

Attractive shrub with slender but rather open pendulous branches. Leaves narrow, 2–5 cm long; flowers mauve, in globular heads, 2–3 cm diam. throughout the foliage; fruit small, up to 3 mm diam. Winter to spring-flowering. 1906.

*M. squarrosa*—**Scent Paperbark**     S.A., Vic.,
                                            Tas., N.S.W.
3–6 m     650 mm     Will grow in very
                                    wet areas

Attractive, erect small tree requiring trimming occasionally to obtain a good shape. One of the best plants for very wet and swampy sites; can stand inundation for several months. Tolerates saline conditions. Withstands severe coastal exposure. Can be trimmed. Leaves triangular, 6–12 mm, bright green, arranged in a distinctive cross pattern; flowers yellow, in oval, dense spikes 2–5 cm long; fruit about 3 mm diam. Summer-flowering. 1794.

*Melaleuca styphelioides—mature tree*

*Melia azedarach—flowers*

*M. styphelioides—***Prickly Paperbark**     N.S.W., Qld
6–10 m     500 mm
Streets, shade

Attractive, shapely, symmetrical, dense-crowned small tree. May reach 12 m in warmer districts. Versatile, hardy, easily grown, vigorous. Suitable for poorly-drained brackish or saline soils. Deep-rooted, grass grows to its base. Wind-firm. Withstands moderate coastal exposure. Tolerates smog. Widely planted as a street tree; will grow in most districts including the warmer areas. Bark white, papery; leaves bright green, prickly, oval, 6–8 mm long; flowers creamy-yellow, in dense spikes, 4–5 cm long. Summer-flowering. 1793.

*M. wilsonii—***Violet Honeymyrtle**     S.A., Vic.
1–3 m     400 mm     Will grow well on
                                            calcareous soils

Bushy, spreading attractive shrub but can become straggly if not pruned occasionally. May be only 1 m in cooler climates, but up to 3 m in sandy depressions. Can be trimmed to form a compact hedge. Leaves very narrow, almost needle-like, 6–10 mm long; flowers crimson; fruit about 3 mm diam. Spring-flowering. 1861.

# MELIA

*Melia* is the old Latin name for Ash, alluding to the ash-like foliage. About five species from southern Asia and Australia are known.

*M. azedarach—***White Cedar** or     N.S.W., Qld
              **Umbrella Cedar**  D.H.
Seed
6–10 m     450 mm                         Fast
Summer shade, roadsides, streets, parks, specimen tree

Fairly dense-crowned, spreading tree, shedding its lower branches. Drought-resistant. Slightly frost-sensitive when young. Widely planted, more suitable for the warmer, inland districts; will grow in

hot districts if irrigation water is available. Withstands moderate coastal exposure. Can be pollarded and pruned. Produces root suckers. Nut-like seeds can be a nuisance on hard footpaths. Leaves deep green, ash-like, pinnate, 30–45 cm long, 15–20 cm wide, leaflets 2–7 cm long, toothed; flowers 2 cm long, pale blue in loose, upright clusters; fruit globular, yellow, 6–12 mm diam., fleshy, containing a hard nut which often remains on the tree after the leaves have fallen. Geographical types occur through Indonesia to India and Iran but the local form is usually only raised locally. Asiatic forms cultivated since before 1600.

# MESPILUS

This is the old Latin name. Only one species from Europe and south-western Asia is known.

*M. germanica—***Medlar**  D.H.     S. Europe, Asia Minor
Seed
5–6 m     750 mm     Well-drained soils     Moderate
Low shelter, parks, roadsides, ornamental

Small, somewhat branching tree. Uncommon in Australia but widely planted overseas. Very hardy, withstands severe frost. Relatively large specimens can be transplanted. Leaves elliptical, 5–8 cm long, autumn foliage very colourful; flowers white to pink, hawthorn-like, large, 2–3 cm diam.; fruit pear-shaped, 2–5 cm long, brown, edible when well ripened. L.C.

# METASEQUOIA

The name is from *meta* which means like, and *sequoia*, sequoia. This alludes to its similarity to *Sequoia*. Only one species from China is known.

*M. glyptostroboides*—**Dawn Redwood**   D.C.     China
Seed, cuttings
12–15 m     750 mm     Requires well-
                       drained soil          Moderate
Parks, roadsides, probably windbreaks, shade, farm
forests

Neat, erect deciduous conifer suitable for the cooler districts. Rediscovered in 1945, previously only known as a fossil tree. Being a recent introduction its characteristics (height, etc.) under cultivation are not yet fully known; promising as a shade, windbreak and farm forest tree. Hardy, frost-resistant. Evolutionary ancestor of the redwoods (*Sequoia*). Foliage pinnate, light green, 6–8 cm long, leaflets opposite, 20–30 pairs, in one plane, 8–15 mm long, 1–2 mm wide, turning bright orange-yellow in autumn; cones globular, about 2 cm diam., tending to be 'squarish' in appearance. 1948.

# METROSIDEROS

*Metrosideros* is from *metra* meaning middle and *sideros* meaning iron, which alludes to the hard heartwood. About sixty species from New Zealand, Australia, Malaysia, South Africa and Chile are known.

*M. excelsa* (M. tomentosa)—
              **New Zealand Christmas Tree** or     N.Z.
              **Pohutukawa**   E.H.
Seed
5–7 m     550 mm     Not suitable for        Moderate
                     wet sites
Windbreaks, parks, streets, roadsides

Upright, compact, dense tree with ascending branches when young, spreading crown when mature. Withstands severe coastal exposure and is one of the best for coastal planting in southern Australia. Very hardy. Wind-firm. Tolerates smog. Somewhat slow for the first 2–3 years then moderate. Juvenile foliage slightly frost-sensitive; mature foliage more resistant but does not appear until plant is about 1 m tall. Fibrous rooted, transplants easily. Can be trimmed to form a hedge. Leaves smooth, deep green, grey underneath, oval, 3–5 cm long; flowers deep red, eucalypt-like, flowering mid-summer, prolific in cooler districts; fruit like a small eucalypt fruit. 1840.

# MORUS

*Morus* is the old Latin name. There are about twelve species from Europe, Asia and North America.

*M. alba*—**White Mulberry**   D.H.       Temperate Asia
Seed
5–8 m     650 mm                          Moderate
Parks, birds, ornamental

Spreading, rounded, bushy-crowned small tree. Formerly widely planted for silkworms. Suitable for shallow, but not alkaline soils. Hardy. Frost-resistant. Coppices. Birds are very fond of the fruit. Leaves light green, oval, 7–15 cm long, often lobed, serrated; flowers in small spikes, up to 3 cm long; fruit a berry, 1–2 cm long, usually purplish-violet, sweet. Numerous horticultural varieties are available.   L.C.

# MYOPORUM   E.H.

The name is from *myo* the Latin verb to shut, and *poros*, a pore. This refers to the translucent leaf pores. About thirty species from Australia, New Zealand, Malaysia, China and Japan are known.

*M. floribundum*—**Slender Myoporum**       Vic., N.S.W.
Seed
2–3 m     550 mm                            Moderate
Parks, roadsides, ornamental

Attractive, slender small tree with arching branches. Rather uncommon although it has been available for some years. Withstands moderate coastal exposure. Can be lightly trimmed. Leaves very narrow, 6–8 cm long, dark glossy green, drooping; flowers white, 1–2 cm across, prolific, clusters along the branches.

*M. insulare*—**Boobialla**       W.A., S.A., Vic.,
                                   Tas., N.S.W.
Seed, cuttings
3–5 m     500 mm     Will grow in heavy soil     Fast
Low windbreaks, hedges

Dense, compact, rounded shrub or small tree, retaining foliage to ground level. Hardy, easily grown; growing well on most soils except very wet types. Tolerates moderately saline sites. Wind-firm. Drought-resistant. May be damaged by heavy frosts. Withstands severe coastal exposure. Can be trimmed to form a compact hedge. Widely planted. Leaves 5–10 cm long, 1–2 cm wide, bright green; flowers white, inconspicuous; fruit a small purplish berry.

*M. platycarpum*—**Sugarwood**   S.A., Vic., N.S.W., Qld
Seed
5–6 m     300 mm     Prefers lighter soils      Moderate
Low windbreaks, shade, parks, fodder, honey

Compact and symmetrical tree. Retains its shape to old age provided it is protected from stock—otherwise it can be crooked and mis-shapen. Will survive on limestone soils. Frost- and drought-resistant. Can be trimmed to form a hedge. Useful subsistence fodder. A sugary solution exudes from trunk wounds. Bark rough and fissured; leaves thick, shining green, 3–8 cm long, 1–2 cm wide, a few minute teeth towards the tip, almost sessile; flowers prolific, small, whitish, under 1 cm diam., in clusters of 4–5; fruit oval, 5–6 mm long, slightly flattened.

231

*Nothofagus fusca*

# NOTHOFAGUS

The name is from *notho* meaning false, and *fagus* meaning beech. About thirty-five species from Australia, New Guinea, New Zealand and South America are known.

*N. cunninghamiana*—**Myrtle Beech**　E.H.　　Vic., Tas.
Seed
15–25 m　1000 mm　Prefers deep,　　Moderate
　　　　　　　　　　moist soils
Windbreaks, parks, roadsides

Large, upright, densely-crowned, attractive tree producing fine grained reddish-pink timber popular for furnishings. Young spring growth rich bronze in colour. Most suitable for the cooler mountain districts. May be reduced to a large shrub in very exposed positions. Frost-hardy, withstands light snow. Can be pollarded and trimmed to form a hedge. Specimens up to several feet tall can be transplanted provided the roots are balled. Does not withstand sustained dry periods. Leaves deep green, small, rounded-triangular, 10–15 mm long, serrated; flowers inconspicuous; fruit, a prickly ball, 1–2 cm diam.. containing 3-winged seeds.

*N. fusca*—**Red Beech**　E.H.　　N.Z.
Seed
5–8 m　650 mm　Prefers deep,　　Moderate
　　　　　　　　moist loams
Parks, roadsides, streets

Attractive, upright compact small tree. Hardy, easily grown. More suitable for the cooler districts. Leaves small, oval, 2–5 cm long, coarsely serrated. About 1880.

*N. obliqua*—**Chilean Beech**　D.H.　　Chile
Seed
15–20 m　900 mm　Prefers loamy soils　　Fast
Windbreaks, parks, roadsides, autumn foliage

Attractive, upright, tree with semi-pendulous, fine foliage. Hardy, withstands cold and frost. Not tolerant of shade. Grass grows to the base of the tree. Easily grown. Autumn foliage red and orange, decorative, holding well into the winter. Leaves 5–8 cm long, oval, slightly lobed; flowers inconspicuous; fruit, 4-winged husk, containing 3 small triangular seeds. 1849.

# NYSSA

*Nyssa* is a water nymph, as some species grow in swampy situations. There are about ten species from North America and southern Asia.

*N. sylvatica*—**Tupelo**　D.H.　　E. North America
Seed
15–25 m　900 mm　　　　　Moderate to fast
Windbreaks, shade, honey, ornamental, autumn foliage

Attractive, shapely upright tree, with slender branches, drooping in older trees. Prefers deep rich, well-drained soils but will grow in slightly wet sites. Hardy, frost-resistant, easily grown. Moderately shade-tolerant. Coppices. Autumn foliage brilliant red and yellow; one of the most colourful in cool mountain districts. Not widely planted but worth more attention. Bark rough, fissured, brown to black; leaves oval, 7–12 cm long, 3–7 cm wide, smooth, pale green underneath; flowers small, greenish-yellow, not conspicuous; fruit blue-black berry, 1 cm diam. Before 1750.

# OLEA

*Olea* is the old Latin name. There are about fifty species from Europe, Asia, Indonesia, Australia, New Zealand and South Africa.

*O. europaea*—**Olive**　E.H.　　Mediterranean
Seed, cuttings
5–7 m　400 mm　　　　　　Slow
Windbreaks, hedges, ornamental, fruit

Compact, low-branching, small tree. Long-lived. Hardy, will grow on sandy and calcareous soils. Deep-rooting, not suitable for shallow soils. Drought-resistant. Can be trimmed to form a hedge. More suitable for the hotter and drier areas; can be grown in very low rainfall areas if irrigation water is available. Leaves grey-green, oval, 5–8 cm long; flowers creamy-white; fruit globular, 2 cm wide—olive of commerce. Selected horticultural varieties should be planted if fruit production is important. L.C.

*Olea europaea—hedge*

# OXYDENDRUM

The name is from *oxys* meaning sour and *dendrum*, a tree. This alludes to its ability to grow on sour (acid) sites. One species from North America is the only one known.

*O. arboreum*—**Sorrel Tree**   D.H.          North America
Seed, cuttings
3–5 m       900 mm     Requires acid soil     Moderate
Low windbreaks, roadsides, parks, ornamental

Attractive, compact, small tree, more suitable for cool, mountain districts. Requires lime-free soil. Frost-resistant. Hardy, but should not be planted in exposed positions. Fibrous-rooted. Moderately large specimens can be transplanted. Colourful autumn foliage, orange, red, and scarlet tones predominating. Leaves oval, 10–15 cm long, 3–5 cm wide; flowers small, white, in clusters, 15–25 cm long. 1747.

# PARROTIA

A genus named after P. W. Parrot (1792–1841), a German naturalist. Five species from Iran and the Caucasus are known.

*P. persica*—**Persian Witch Hazel**   D.H.          Iran
Layers (seed, if available)
6–8 m       750 mm     Requires well-     Moderate
                      drained soil
Parks, roadsides, ornamental, autumn foliage

Wide, rather straggly, small tree, suitable for cool, mountain districts. Planted mainly for its vivid yellow-brown autumn foliage. Hardy, frost-resistant, easily grown. Moderately drought-resistant. Coppices readily. Large specimens can be transplanted. Light trimming is beneficial. Bark smooth, grey, flaky; leaves 7–12 cm long, diamond-shaped, bronzy-green, unevenly toothed; flowers consist of reddish-brown stamens, appear in quantity in late winter on the branchlets and give the tree a hazy appearance. 1840.

# PAULOWNIA

Genus named after Anne Paulowna (1795–1865), daughter of the Russian Tsar. There are about seventeen species from eastern Asia, mainly China.

*P. tomentosa* (P. imperialis)
          —**Royal Paulownia**   D.H.          China
Seed
9–12 m       750 mm                          Fast
Parks, shade, roadsides, ornamental

Upright, attractive tree, suitable for the cooler districts. Planted mainly for its clusters of deep mauve foxglove-like flowers. Young plants can be damaged by late frosts (stems pithy), otherwise hardy. Withstands moderate coastal exposure. Sprouts freely from stumps and roots. Specimens several years old can be transplanted. Leaves large, roundish, 20–30 cm long, 12–15 cm wide, dull green, grey underneath; flowers upright, terminal clusters, 15–30 cm 'tall', each flower about 4–5 cm long; fruit pod, 5–8 cm long, often remaining on the tree over winter. L.C.

# PERSOONIA

Genus named after C. H. Persoon (1755–1837), an English botanist. About sixty species from Australia and one from New Zealand are known.

*P. pinifolia*—**Pine-leaf Geebung**   E.H.          N.S.W.
Seed, cuttings
3–4 m       650 mm                          Fast
Parks, streets, roadsides

Attractive, bushy shrub, occasionally a small tree, with arching, graceful branches Requires good drainage and an open position. Can be trimmed. Leaves pine-like, 2–3 cm long; flowers yellow, conspicuous, in narrow, conical terminal clusters, 7–12 cm long; fruit berry-like, about 2 cm long.

# PHEBALIUM

The name is from *Phibole*, myrtle, referring to its similarity to myrtle. There are about thirty species from Australia and New Zealand.

*P. squameum* (P. billardieri)          Vic., Tas.,
          —**Satinwood** or          N.S.W., Qld
          **Satin Box**   E.H.
Seed
9–12 m       900 mm     Well-drained soils     Fast
Windbreaks, parks, roadsides, hedges

Very erect, narrow, box-like tree; useful for quick shelter in higher rainfall areas. Not widely planted but worth more attention. Frost-resistant. Can be trimmed to form a hedge. Not easily transplanted. Withstands moderate coastal exposure. Leaves 8–12 cm long, 1 cm wide, satiny underneath; flowers white, small, inconspicuous. 1822.

*Picea abies*

# PHOTINIA

The name is from *photeinos* meaning shining which refers to the leaves. There are about sixty species from Asia, Indonesia and North America.

*P. serrulata*—**Chinese Hawthorn**   E.H.        China
Seed, cuttings
5–6 m        550 mm                        Fast
Low windbreaks, parks, roadsides, hedges, birds

Compact, small tree, developing a spreading crown later. Hardy, easily grown. Wind-firm. Can be trimmed to form a hedge. Foliage has been suspected as causing the death of cows. Leaves 10–12 cm long, 4–5 cm wide, shining, deep green, serrated, leaves on young shoots much larger, rich red-bronze colour; flowers white, 6–8 mm across, in large conspicuous clusters, 10–15 cm across; fruit a red berry, liked by birds. 1804.

Where space is limiting and particularly for hedges the smaller (2–3 m) *P. robusta* and *P. glabra rubens* are more commonly planted. (The latter produces rich red leaves after pruning.)

# PICEA—Spruce   E.C.

Seed                Require well-      Slow when young,
                    drained loams     then moderate
Windbreaks, farm forests, parks
The name is from *pix* meaning pitch which refers to the resin content. About thirty species from Europe, Asia and North America are known.

These are attractive, pyramidal, compact trees, with a single tapering trunk and branches arranged in whorls. The lower branches are retained to ground level except in dense stands when they are suppressed and eventually shed. They are more suitable for the cooler districts and are not satisfactory on sandy soils, preferring deep clay loams. Wind-firm on deep soils even though the root system is not deep. Although hardy and frost-resistant, young spring shoots may be damaged by a severe late frost. They can withstand cold and snow and are shade-tolerant when young. Bark is scaly. Branchlets have small stubs left by fallen leaves. Foliage is needle-like, single and square or rhomboid in cross-section. Cones are pendulous and brown when ripe and the scales are thin. *Picea* is distinguished from other single-needle conifers by (a) leaf 'stubs' on the branchlets, (b) pendulous cones which remain on the tree, and (c) cones do not break up after seed fall. Cone sizes given below are for closed cones (not open as after seed fall).

*P. abies* (P. excelsa)—**Common** or        Central and
                        **Norway Spruce**     N. Europe
15–30 m        750 mm

Tall tree with a narrow conical crown. Will grow to over 40 m in favourable localities. Root system widespread. Best growth in partial shade. Withstands heavy snow. Leaves 1–3 cm long, glossy-green, soft, arranged spirally, only slightly pointed; cones 10–15 cm long, thin scales. Numerous horticultural varieties are available. Before 1548.

*P. engelmannii*—**Engelmann Spruce**  W. North America
15–25 m        1000 mm

Erect tree with a spire-like crown, growing to 50 m in its native habitat. Requires moist but well-drained soil. Withstands severe cold and snow, good alpine species. Tolerant of shade, can stand several years of suppression. Very susceptible to smog—good specimen and park tree in smog-free areas. Bark reddish, rich in tannin; young shoots very hairy; leaves 2–3 cm long, soft, horny tip, disagreeable odour when crushed; cones variable, 3–7 cm long, about 2 cm wide, scales thin, toothed margin. 1864.

*P. omorika*—**Serbian Spruce**        S.E. Europe
15–20 m        750 mm     Will grow on
                        calcareous soils

Spire-like tree, branches more restricted and crown narrower than most conifers; more suitable for restricted areas. Not widely planted. Hardy; unlike most conifers it will tolerate some atmospheric pollution. Stands warmer conditions than other spruces. Leaves 1–2 cm long, blunt; cones 4–5 cm long, narrow. About 1880.

*Picea pungens*

*Picea pungens—blue form*

*P. sitchensis*—**Sitka Spruce**　　　W. North America
25–30 m　1200 mm　Requires rich, moist,
well-drained soil

Tall, slightly bluish tree, growing to over
30 m in favourable localities (over 70 m in
its native habitat). Only suitable for the
cooler, higher rainfall mountain districts.
Produces valuable timber. Leaves stiff,
sharply pointed, 1–2 cm long, bluish; cones 5–10 cm
long, scales very thin and papery. 1831.

*Picea sitchensis*

*P. pungens*—**Colorado Spruce**　　　W. North America
15–20 m　700 mm

Attractive, symmetrical tree with dense
foliage, stiff branches and needles. More
tolerant of dry sites than most other spruces.
Leaves 1–3 cm long, spreading around the
shoot, prickly; cones cylindrical, 5–10 cm
long. In most batches of seed a few plants with prominently bluish foliage appear; these are sold as the
Colorado blue spruce. They usually grow to 6–10 m
only, and are popular as garden specimens. Several
horticultural varieties are available—var. 'Kosteriana'
is one of the best blue trees available. Blue strains should
be raised from cuttings but striking cuttings is difficult
and slow, often requiring two years in the cutting bed
before potting. 1862.

*Picea sitchensis—blue tips of young shoots*

*Pinus canariensis*

*P. canariensis*—**Canary Island Pine**    Canary Islands
20–25 m    450 mm    Will grow on slightly    ·Slow to
alkaline soils    moderate

Parks, roadsides

Branches arranged in regular whorls. Drought-resistant, growing in lower rainfall areas than most pines. Wind-firm. Coppices, but not vigorously—so usually survives light to moderate fires. Needles densely crowded, giving the tree a 'tufted' appearance. Bark distinctive, reddish-brown, furrowed; needles in threes, slender, 18–30 cm long; cones 15–25 cm long, 6–8 cm wide, nut-brown, scales thick, without a prickle. 1888.

*P. contorta*—**Lodgepole Pine**    W. North America
20–25 m    650 mm    Will grow on adverse sites    Fast

Hardy tree, suitable for dry, or gravelly sites but not calcareous soils. Worth planting more often. Growth fast and to 35 m in favourable areas. Withstands severe cold and snow. Not tolerant of shade. Bark reddish-brown, flaky; needles in twos, 3–7 cm long, dense, mid-green; cones 4–6 cm long, scales thin with a short prickle. Several varieties are available, the most suitable for timber production is the var. *contorta* from the coastal (but not beach areas) of California and Oregon. 1855.

# PINUS—Pine    E.C.

Seed
Farm forests, windbreaks, shade
*Pinus* is the old Latin name. About seventy species from Europe, Asia, and North America are known.

These single-stemmed, upright trees are pyramidal when young and flat-topped when old. Branches are arranged in whorls and retained to ground level in open grown trees but suppressed and eventually shed when forest grown. They require well-drained soil and are not satisfactory in swampy or wet areas. Not tolerant of shade but are frost-resistant and wind-firm. Do not respond satisfactorily to trimming. The timber is useful but not durable. Leaves are 'needles' which are long and release a turpentine odour when crushed. Cones are woody and the scales are usually thick. (Cone size given refers to the unopened cone.) The true pines (*Pinus* sp.) can be distinguished from other conifers by (a) the long leaves (needles) and (b) the leaves usually being arranged in bundles of 2, 3 or 5 with a silky, broad sheath, 5–20 mm long, which binds them together at the base. *P. cembroides* var. *monophylla* has single needles and the var. *parryana* bundles of 4 needles. Both forms are very rarely planted in southern Australia; it is unlikely that specimens would be seen outside botanical gardens. All other pines have their needles arranged in bundles of 2, 3 or 5.

*Pinus halepensis*

*P. halepensis*—**Aleppo Pine**         Mediterranean
10–20 m     400 mm     Will grow on         Moderate
                        limestone soils
Roadsides

Bushy tree, with a narrow, compact and dense crown when young, shedding lower branches with age. Most drought-resistant of the pines; will stand long rainless periods during the summer. Suitable for hot, dry districts and most soils except tight clays. Deep-rooted and wind-firm. Withstands moderate coastal exposure. Young trees in particular are attractive; bushy and fresh light green in colour. Bark silvery-grey, scaly; needles in twos, slender, 5–7 cm long; cones 5–10 cm long, shining reddish-brown, scales thick, without a prickle. Several provenances are available overseas but seed should be collected from good specimens growing locally. 1683.

*P. halepensis* var. *brutia* (P. brutia)     E. Mediterranean
                        —**Calabrian Pine**

20–25 m

Similar to *P. halepensis* but taller, faster growing, and with a better form; more suitable for farm forests and commercial timber production.

*P. nigra* subsp. *laricio* (P. nigra var. laricio, P. laricio)
                        —**Corsican Pine**         Corsica
25–35 m     650 cm     Will grow on         Moderate
                        adverse sites         to fast

Symmetrical, conical, somewhat open tree, with branches arranged in very regular whorls. Best growth on deep soils in high rainfall areas but will grow steadily, although more slowly, on adverse sites including calcareous soils. (*P. contorta* is preferable for poor non-calcareous sites.) Will grow on land-filled areas. Deep-rooted and wind-firm. Drought-resistant once established. Produces good quality timber, very suitable for preservation as posts and poles. Withstands moderate coastal exposure. Bark reddish-brown, flaky; needles in twos, 7–12 cm long, stiff; cones oval, 5–8 cm long, 2–4 cm diam., scales thin with a small prickle; cones falling after the seed has been shed. Several varieties and provenances are available. The subspecies *laricio* as above has proved the most satisfactory in southern Australia. 1779.

*P. patula*—**Mexican Pine**         Mexico
20–25 m     750 mm     Prefers good soils         Fast
Roadsides, avenues, parks, ornamental

Graceful tree. The slender, drooping needles, arranged in clusters give it an attractive and distinctive appearance. Most ornamental of the pines. Bark reddish, flaky; needles in threes, soft, 15–20 cm long; cones 7–10 cm long, semi-conical, scales with a small prickle. 1825.

*Pinus pinaster*

*P. pinaster*—**Maritime Pine**         Mediterranean
15–25 m     500 mm     Prefers sandy soils         Fast
Parks, roadsides

Variable tree; most suitable for sandy areas in 600–800 mm rainfall areas. Good on soils which are marginal for *P. radiata*. Hardy. Wind-firm. Withstands severe coastal exposure. Bark reddish-brown, semi-furrowed; needles in twos, stiff, rigid, 12–20 cm long; cones 12–18 cm long, 4–6 cm diam., semi-conical, scales thick, without a prickle. Several provenances or strains are recognized. The Leiria stand is the fastest growing and most suitable for farm forests. Prone to develop butt sweep or wind lean when young, but firming and straightening (as required) for the first 2–3 years will reduce this tendency. Before 1600.

*Pinus pinea*

237

*Pinus ponderosa*

Below    *Pinus ponderosa—mature tree*

**P. pinea—Stone Pine**                          S. Europe
12–15 m    650 mm    Not suitable for          Moderate
                            calcareous soils

Spreading, flat-topped crown giving the tree an umbrella-like appearance. Good shade tree. Wind-firm. Withstands moderate coastal exposure. Edible seeds. Bark reddish-grey, furrowed; needles in pairs 10–15 cm long; cones nearly globular, 10–15 cm long, 10–12 cm diam., shining nut-brown, scales thick, without a prickle; seeds large 6–8 mm long. Before 1548.

**P. ponderosa—Western Yellow Pine**  W. North America
20–30 m    650 mm    Prefers the more         Moderate
                            fertile soils
Parks, roadsides

Broadly conical, open-crowned, shapely tree when mature. More suitable for the higher rainfall districts. Grows to 60 m in its natural habitat. Frost-resistant. Unusually wind-firm. Commonly planted as a specimen tree. Old trees produce very good timber. Bark reddish-brown, flaky; needles in threes, stiff, 15–25 cm long, densely crowded on the branchlets; cones variable, usually 10–18 cm long, 5–10 cm diam., scales thick, with a small prickle. 1827. Numerous varieties and provenances are known; many unsuitable for local conditions. Seed should be obtained from good local trees or the coast ranges of California and Oregon at elevations up to 800 m.

**P. radiata—Radiata Pine** or                 California
            **Monterey Pine**
20–30 m    650 mm    Requires well-                Fast
                            drained soils

Hardy, reliable, easily grown, versatile tree. Widely planted for windbreaks and shelter for over 100 years. (Introduced to Australia in 1859.) Will exceed 50 m when conditions are favourable (cool, high rainfall districts). Spreading, heavy crown when open grown, but very compact and lightly branched at close spacings. Wind-firm. Frost-resistant. Provided the soil is not poor it will withstand severe coastal exposure. Tolerates smog. Withstands summer drought. Produces good general purpose timber (milling, posts, poles, pulp, etc.). For commercial timber production, moist, well-drained, moderately fertile sites should be selected. Bark grey, furrowed; needles in threes, slender, 10–15 cm long; cones irregular in shape, 10–15 cm long, 6–10 cm diam., asymmetrical, scales thick, without a prickle. 1833.

**P. strobus—Eastern White Pine**          Eastern USA
20–30 m    900 mm    Prefers good soils      Moderate to
                                                          fast

Upright, broadly conical tree, retaining branches to ground level when open grown. One of the more attractive pine trees. Wind-firm. Will grow on most soils. Moderately deep root system. Hardy, vigorous, frost-resistant. Withstands light snow, suitable for sub-alpine areas. Tolerant of shade. Growth slow for the first few years then fast. Bark furrowed; needles

*Pinus strobus*

*Pittosporum eugenioides*

in fives, 7–12 cm long, slender, bluish-green; cones cylindrical, pendulous, 10–15 cm long, 2–3 cm diam., scales thin, without a prickle. 1705.

# PISTACIA

*Pistacia* is derived from the Greek name, *pistake*. About ten species from Europe, Asia and Northern Africa are known.

*P. chinensis*—**Chinese Pistachia**   D.H.          China
Seed
5–10 m        650 mm        Prefers moist,          Fast
                            deep soils
Parks, stream banks, ornamental

Attractive, well-shaped, upright, small tree. Taller in favourable sites. Very similar to *Rhus* in appearance but is non-irritant. Male and female trees. Hardy. Moderately frost-resistant. Vigorous, fibrous root system; large specimens may be transplanted. Can be trimmed. Foliage pinnate, 20–25 cm long, leaflets 10–12, 5–8 cm long, 5–7 mm wide, alternate, turning vivid red in autumn even in mild climates; flowers inconspicuous; fruit a berry, 5–6 mm diam., red, turning blue-black later, on female trees only. 1880.

# PITTOSPORUM   E.H.

Seed
Low windbreaks, parks, roadsides, hedges
The name is from *pitta* meaning pitch, and *sporum*, a seed, alluding to the sticky seeds. There are about 120 species mainly from Australia and New Zealand but a few from tropical Asia and Africa.
They are moderate to dense-crowned good foliage trees which are hardy and easily grown. Will withstand some degree of coastal exposure and can be trimmed to form a dense hedge. Large plants are difficult to transplant. Flowers are scented. Fruit is globular and fleshy, 5–15 mm diameter. Seeds are sticky (spreading by adhering to birds) and germinate readily.

*Pittosporum eugenioides—hedge*

*P. bicolor*—**Banyalla**          Vic., Tas., N.S.W., Qld
5–8 m        650 mm                          Moderate

Small tree more suitable for the cooler districts. Leaves 3–5 cm long, 1 cm wide, dull green, silver underneath; flowers bell-like, 1 cm long, yellow to dark red, yellow stamens; fruit 5–10 mm diam.; seeds reddish.

*P. crassifolium*—**Karo**                          N.Z.
5–6 m        650 mm        Will grow on saline soils    Fast
Streets

Upright small tree. Wind-firm. Tolerates smog. Withstands severe coastal exposure. Drought-resistant. Leaves thick, oval, 5–8 cm long, dark green, silver underneath; flowers brown, in terminal clusters; fruit black.

*P. eugenioides*—**Tarata**                          N.Z.
4–5 m        650 mm                              Fast
Streets, avenues, ornamental

Very attractive, light green, densely-foliaged small tree. Good hedge plant. Leaves light green, wavy margin, soft, 7–12 cm long, 2–3 cm wide; flowers yellow, in terminal clusters; fruit black. Popular garden plant, several horticultural varieties of leaf colourings are available.

*Pittosporum tenuifolium*

**P. phillyreoides—Weeping Pittosporum**   W.A., S.A.,
Vic., N.S.W., Qld, N.T.
5–6 m    250 mm    Prefers sandy soils    Moderate
Streets, shade, fodder, ornamental

Attractive tree, moderately dense crown
with slender, pendulous foliage. Good shade
tree for hot, dry districts. Hardy. Drought-
resistant. Withstands moderate coastal
exposure. Foliage good subsistence fodder.
Leaves 5–10 cm long, 5–10 mm wide, thick; flowers
yellow, in leaf axils; fruit 1 cm diam., orange.

**P. tenuifolium** (P. nigrescens, P. nigricans)    N.Z.
**—Kohuhu**    Moderate
3–6 m    650 mm

Hardy, small tree with light green foliage
and slender trunk. Variable in height,
according to soil conditions. Wind-firm.
Withstands severe coastal exposure.
Tolerates smog. Branchlets almost black;
leaves light silvery-green, 4–5 cm long, 1 cm wide, wavy
margins; flowers dark brown to black, night-scented.
Numerous horticultural varieties are available.

**P. undulatum—Sweet Pittosporum**    Vic., Tas.,
N.S.W., Qld
5–12 m    500 mm    Prefers loams to    Fast
clay loams
Shade, hedges

Bushy, spreading tree, more suitable for
the cooler, moister districts but can be
grown in the warmer, inland areas. Smaller
tree in the lower rainfall zones. Very hardy,
widely planted for shade and hedges; grows

well on clayey soils but not satisfactory on sands or
gravels. Wind-firm. Withstands moderate coastal
exposure. Young plants slightly frost-tender. Branches
exude a sticky resin. Leaves 10–15 cm long, 2–4 cm wide,
slightly wavy margins; flowers white, shiny, sweet-
scented, terminal clusters, 5–7 cm across; fruit 1 cm
diam., bright orange. 1789.

# PLATANUS

*Platanus* is the old Greek name. About ten species from
Europe, Asia and North America are known.

**P. x. acerifolia—London Plane**   D.H.    Hybrid
Similar to *P. orientalis* (see below). Believed
to be a hybrid—*P. orientalis x P. occidentalis*.
Withstands industrial smog better than most
trees hence widely planted in towns and
industrial areas. Will tolerate wet (but not
stagnant) soils. Distinguished from *P. orientalis* by (a) leaf
lobes being more deeply cut (about one-third to the midrib), and (b) fruits single or only in pairs. Before 1700.

**P. orientalis—Plane**   D.H.    S.E. Europe,
Asia Minor
Seed, cuttings
15–25 m    400 mm    Prefers deep soils    Fast
Shade, parks, roadsides, windbreaks

Large, spreading, round crown. Excellent
summer shade and park tree, particularly
if space is adequate to allow full crown and
height development. Best growth on deep
soil in the cooler and moister districts but
will thrive on most soils. Hardy, vigorous, with strong
and extensive root system. Frost-resistant. Wind-firm.
Withstands slight coastal exposure. Very tolerant of
smog. Tolerates some flooding. Can be severely trimmed,
even to form a 2–3 m hedge; crown stands repeated
lopping. May be hard to establish in drier districts but
requires little attention later. Widely planted as a street

*Pittosporum undulatum*

*Platanus orientalis*

tree but because of the maintenance required to keep crown under control (lopping every 3–4 years) it cannot be recommended for street planting. Bark smooth, shed in large flakes; leaves large, 18–25 cm across, roughly triangular, 5 lobes, not deeply cut; fruit globular, rough, 2–3 cm diam., in groups of 3. L.C.

*Podocarpus totara*

# PODOCARPUS

*Podocarpus* is from *podos* meaning a foot and *karpus*, a fruit, referring to the fleshy foot-like stalk on the fruit of some species. There are about 100 species from New Zealand, Australia, southern Africa, Central America, southern and eastern Asia, Indonesia and Philippines.

*P. elatus*—**Plum Pine**   E.C.                   N.S.W., Qld
Seed, cuttings

5–7 m        750 mm      Prefers good        Slow to
                          soils               moderate
Windbreaks, parks, hedges, shade

Upright tree with a dense crown, conical when open grown. Grows to 30 m in its native habitat. Frost-resistant. Can be heavily trimmed to form a compact hedge. Leaves 5–15 cm long, 5–12 mm wide, glossy; fruit globular, 2–3 cm diam. fleshy, purple, bearing the seed at its apex.

*P. lawrencei* (P. alpina)                       Vic., Tas.
   —**Mountain Plum Pine**
Seed, cuttings

2–3 m        650 mm                              Slow
Windbreaks, hedges

Dense-crowned, upright, small tree, retaining branches to ground level. Frost-resistant. Withstands prolonged snow. Can be trimmed to form a compact hedge. Leaves dark green, small, 5–15 mm long; fruit fleshy, dark red, 5–6 mm diam., enclosing the woody cone. 1825.

*P. totara*—**Totara**                            N.Z.
Cuttings

5–8 m        650 mm      Well-drained loams      Slow
Windbreaks, hedges

Upright, dense-crowned tree with horizontal branching and spreading when old. Usually not tall in southern Australia but grows to over 30 m in its native habitat. Better growth in the cooler, mountain districts. Hardy, frost-resistant. Stands severe pruning, forms a very dense hedge. Because of its slow growth it is not suitable for quick shelter or farm forests even though it produces good quality reddish timber. Leaves 1–2 cm long, 2–4 mm wide, sharply pointed (leaves on young shoots longer); flowers not conspicuous; fruit a reddish nut, 5–6 mm diam.

# POPULUS—**Poplar**   D.H.

Cuttings                 Moist, well-drained     Fast
                         soils
Windbreaks, summer shade, farm forests, parks, road-sides, gully erosion and stream bank control, autumn foliage

*Populus* is the old Latin name. About thirty-five species from Europe, Asia and North America are known.

These are attractive, upright trees from Northern Hemisphere countries which retain branches near to ground level when open grown. The leaves tremble in a slight breeze. While preferring cool, moist, well-

*Populus species windbreaks mainly P. deltoides*

occurs. For commercial purposes many thousands of hybrids have been produced, of which a few are outstanding. Two strains of *P. deltoides* var. *angulata*—I 488, a semi-evergreen, and 'Persistente', a Chilean evergreen—have been widely used in commercial plantings but as they are susceptible to 'golden glow' fungi, their future in this field is doubtful. Further information should be sought if commercial planting is proposed.

Because of the extensive hybridization which occurs naturally, the naming of poplars is confused. The names and synonyms given below are more or less as applied locally even though they may not be strictly correct botanically. Wide variations and intermediate forms are likely to be found so that the names given should be regarded as indicating a broad type rather than a rigid species description.

*P. alba* (P. bolleana)—**Silver Poplar**  Europe, Asia
15–20 m  650 mm

Upright when young, spreading with maturity. Silvery foliage. Very strong root system. Suckers prolifically from the roots, more than most trees. Should only be planted where suckers will not be a problem or where they can be kept under control by mowing or cultivation. Very effective for soil stabilization in eroding gullies or as screens around quarries etc. Withstands moderate coastal exposure. Bark grey-white, attractive, particularly when young, with conspicuous lenticels; leaves oval, 5–6 cm long, dark green, white underneath. Terminal and sucker leaves conspicuously lobed, (maple-like), 3–5 lobes. L.C.

*P. alba* var. *pyramidalis*—**Upright Silver Poplar**

A narrow-crowned upright form, with steeply ascending branches. Mature trees usually free of branches for the basal 2–3 m of the trunk.

*P. candicans* (P. macrophylla, P. balsamifera var. candicans)—**Balm of Gilead**  Unknown
10–15 m  650 mm

Spreading branches. Strong-growing. Suckers, but not extensively. Resinous fragrance from the sticky buds in spring. Bees are attracted to the buds in spring and may damage them leading to reduced or deformed growth. Leaves 10–15 cm long, 5–10 cm wide, triangular, covered in fine hairs, bright yellow in autumn. 1755.

*P. canescens*—**Grey Poplar**  W. Europe
20–30 m  600 mm

Similar in appearance to *P. alba*, some authorities regard *P. canescens* as only a geographical form of *P. alba* but there are several botanical differences which probably justify maintaining it as a separate species. Vigorous, suckering vigorously from the roots. Withstands moderate coastal exposure. Tolerates snow. Leaves broadly oval, almost round, 10–12 cm long, more or less evenly toothed, terminal and sucker leaves unevenly toothed (not lobed); mature leaves not hairy underneath.

drained alluvial soils, they will also grow on a wide range of sites including slightly wet sites, and in the hotter and drier districts if irrigation water is available. Commercial plantations should only be planted on sandy loams or deep loams in warmer districts where rainfall and irrigation equivalent to 1000 mm or more per year is available. The timber produced is a light-coloured almost white wood which is in demand for plywood and other high quality uses. Root systems are vigorous and wide spreading; trees should not be planted close to underground earthenware drains, sewerage pipes, etc. or building foundations, hence they are not suitable for street planting. They are hardy, easily established, frost-resistant and wind-firm but not tolerant of shade. Large specimens can be transplanted after leaf fall. They can be heavily lopped and pollarded. Fairly large branches (5–10 cm diam.) can be struck as cuttings in their final planting position. Buds are conical, sticky and resinous. Leaves are broadest below the midpoint, simple, normally serrated and leaf-stalks are long and usually flattened. Flowers and fruits are in catkins and are not particularly conspicuous. Species are recognized mainly by leaves. Leaf shape and size can be very variable on the one tree, depending on position, time of the year, and whether on a sucker, normal or epicormic shoot. Typical leaves occur in mid-summer about half way up the long shoots.

Some poplars, particularly the varieties of *P. deltoides* and *P. fastigata*, are likely to be attacked in mid- to late summer by 'golden glow', a rapid yellowing of the leaves caused by the fungi *Melampsora medusae* and *M. laricipopulina*. Infestations rarely kill the tree although the growth rate may be reduced because of the earlier cessation of photo-synthesis by the leaves. It is widespread but can be controlled by monthly sprayings with 0·075 per cent dithianon or copper oxychloride beginning early summer. *P. simonii* and *P. yunnanensis* seem to be resistant to the disease. Poplars are distinguished from willows (*Salix*) by (a) leaves being about as broad as they are long whereas willows are long and narrow, and (b) buds are covered with overlapping scales whereas willows have a cap-like cover. Male and female flowers are on different trees so that extensive hybridisation

*Populus species windbreaks*
*P. nigra (narrow trees) ard P. deltoides*

*Populus nigra* var. *italica in autumn*

*Populus serotina* var. *aurea*

---

**P. deltoides var. *missouriensis*—Southern Cottonwood**

North America

25–30 m    650 mm

Large, rounded, spreading crown, without a pronounced leading shoot. Produces good timber. Grows in a wide range of soils; one of the most adaptable of the poplars. Does not sucker extensively. Tolerates some flooding. Fresh leaves have some fodder value. Bark grey-brown, furrowed; leaves large, 10–15 cm long, 10–12 cm wide, roughly delta-shaped, margins crenated rather than serrated, bright glossy green, quite hairless, vivid yellow in autumn; liberates masses of seeds with long silky hairs attached, hence the common name of cottonwood. Before 1800.

**P. deltoides var. *angulata*—Carolina Poplar**

North America

Similar to *P. deltoides* var. *missouriensis* but crown slightly more open and not so spreading. Distinguished by its ribbed (angular) stems. Believed to be a natural hybrid. Grows well in irrigation districts and summer rainfall areas. Before 1790.

**P. x generosa**    Hybrid

15–20 m    950 mm

Large-crowned, spreading tree. Vigorous, adaptable to a wide range of sites. Does not sucker extensively. Hybrid between *P. trichocarpa* (male) and *P. deltoides* var. *angulata* (female) with characteristics between these two. Very fast-growing on good sites. Leaves 15–30 cm long, broad, grey underneath, coarsely serrated, margins translucent. 1912.

**P. nigra var. *italica* (P. pyramidalis, P. fastigata, P. dilatata)—Lombardy Poplar**

N. Italy

12–35 m    500 mm

Avenues, ornamental

Very erect, narrow columnar tree; lower branches ascending steeply. Widely planted. Vigorous, hardy, tolerant of drier conditions than most poplars. Wind-firm. Relatively long-lived. Withstands slight coastal exposure. Suckers but not extensively. Good for park and formal avenue planting. Leaves diamond-shaped, 5–10 cm wide, shallow-toothed, flattened petiole. Used extensively as one parent in hybridization work for commercial plantations. L.C.

**P. serotina var. *aurea* (P. canadensis var. aurea)—Golden Poplar**    Hybrid

12–25 m    650 mm

Symmetrical, pyramidal to narrow upright tree, retaining branches to ground level. Strong root system, suckers freely; should only be planted where suckers can be controlled by cultivation, mowing, etc. Foliage yellow to golden, good types are sometimes grafted for ornamental plantings. Leaves triangular, 7–12 cm wide, long tip, serrated margin. 1871.

*P. simonii* c.v. '*fastigiata*'—**Upright Poplar**    N. China
10–15 m    650 mm
Avenues

Upright, narrow, columnar tree, similar in general appearance and use to *P. nigra*. Branches short, small. Produces very few suckers, preferred to *P. nigra* where suckers could be troublesome. Young shoots reddish-brown; leaves small, oval, 5–10 cm long, finely serrated, broadest at or above the middle. 1862.

*P. tremula*—**Aspen**    Europe, Asia, Africa
10–12 m    650 mm

Graceful, spreading, densely-crowned tree. Foliage greyish, leaves quivering on long slender leaf-stalks. Suckers rather freely. Golden colour in autumn. Bark whitish-grey; leaves a rounded, diamond-shape, 5–6 cm wide, slightly lobed, shining, greyish-green; leaf-stalk very long, 8–10 cm, slender and flattened.    L.C.

*P. tremuloides*—**American Aspen** or **Quaking Aspen**    Eastern USA
10–15 m    750 mm

Round crown, slender trunk with ascending branches. Relatively short-lived. Tolerates some degree of wetness. Vigorous surface root system. Suckers rather freely. Leaves quiver. Bark whitish-grey; buds reddish-brown, shining; leaves oval to round, 4–7 cm wide, dull, finely serrated margins, on long 6–10 cm slender, flattened leaf-stalks, turn golden-yellow in autumn. 1812.

*P. trichocarpa*—**Black Cottonwood**    Western USA
20–30 m    750 mm

Large, erect but columnar to slightly spreading tree with upright branches. Straight trunk. Largest of the poplars, exceeds 50 m height and 1 m diam. in its native habitat. Prefers the cooler moist districts but will grow where irrigation water is available. Very fast-growing on good sites. Leaves variable, narrow to broadly oval, 10–15 cm long, 4–6 cm wide, white underneath, finely serrated; leaf-stalks 1–10 cm long, round (most poplars have flattened leaf-stalks). 1892.

*P. yunnanensis*—**Yunnan Poplar**    S.W. China
10–20 m    900 mm

Upright, broadly columnar tree. Relatively new introduction, not well known particularly in regard to its lower rainfall limit; appears to stand warmer and drier conditions satisfactorily. Very fast growth once established. Does not sucker freely. In the drier northern districts bees are attracted to the sticky buds and may damage them, leading to reduced or deformed stems; not a problem in the southern and cooler districts. Branchlets bright red, stems angled; leaves triangular, 10–15 cm long, 5–10 cm wide, bright green, veins slightly reddish. Before 1905.

*Populus tremula in autumn*

# PROSTANTHERA

*Prostanthera* is derived from *prosti* meaning to append and, *anthera* meaning anthers which alludes to the appendages on the anthers of some species. There are about forty species from Australia.

*P. lasianthos*—**Victorian Christmas Bush**    E.H.    Vic., Tas., N.S.W., Qld
Seed
5–10 m    750 mm    Well-drained loams    Fast
Parks, roadsides, streets, ornamental

Attractive, upright, flowering, small tree; develops a slightly spreading bushy crown when open grown. Requires well-drained sites but not satisfactory on sandy soils; most suitable for the cooler, mountain districts. Vigorous, frost-resistant. Not very wind-firm. Large plants cannot be transplanted. May become straggly if not lightly pruned occasionally; pruning improves flowering. Will grow under and close to other trees. Worthy of more extensive planting. Leaves 5–10 cm long, 1–2 cm wide, coarsely serrated; flowers conspicuous, prolific, white tinged with purple, foxglove-like, 2–3 cm long, 1–2 cm wide, flowering at Christmas in terminal clusters. 1807.

*P. nivea*—**Snowy Mintbush**    Vic., N.S.W., Qld
Seed, cuttings
2–3 m    450 mm    Requires well-drained soils    Fast
Low shelter, parks, roadsides, hedges, ornamental

Showy, upright, flowering, bushy shrub with whitish stems. Grows well on granitic sands but not pure sands. Can become straggly and spindly if not pruned occasionally. Frost-resistant. Not easily transplanted. Leaves oval, 1–3 cm long; flowers purple-white, 1–2 cm across, conspicuous, produced prolifically in large terminal clusters. 1864.

*Prostanthera rotundifolia*

**P. rotundifolia—Round Leaf Mintbush**     S.A., Vic.,
                                             Tas., N.S.W.
Seed, cuttings
2–3 m      500 mm      Requires well-        Fast
                       drained soil
Low shelter, parks, streets, hedges, ornamental

Bushy, compact, attractive, flowering shrub. Hardy, vigorous, frost-resistant, easily grown. May require pruning occasionally to retain a good shape; pruning improves flowering. Leaves round, small, 1 cm diam., thick, deep green; flowers deep-blue to purple, 1 cm long, conspicuous, prolific, flowering at 2–3 years, terminal, single and in groups. (*P. ovalifolium* while differing botanically, has similar growth and appearance characteristics.) 1824.

# PRUNUS

*Prunus* is the old Latin name. About 200 species from Europe, Asia and North America with a few from South America are known.
This large group of flowering plants is divided into two main groups: (a) Evergreen, and (b) Deciduous.

*(a) Evergreen—***Cherry Laurels**   E.H.
Cuttings
5–8 m      750 mm      Well-drained soils      Fast
Windbreaks, hedges
These dense, low-growing large shrubs or small trees are hardy, vigorous and frost-resistant. They are more suitable for the cooler, mountain districts and are excellent for low shelter. They can be severely trimmed and fairly large plants can be transplanted. Leaves are large, thick, deep green and glossy, and have a prominent midrib. Flowers are white and arranged in terminal clusters, conspicuous but not showy. Fruit is purplish-black and globular, 1–2 cm in diameter. Although botanically belonging to the genus *Prunus*, they are not like the flowering *Prunus* commonly planted in gardens.

*P. laurocerasus—***Cherry Laurel**     S.E. Europe,
                                         Asia Minor

Leaves 10–20 cm long, 3–5 cm wide, pale green underneath; flowers in clusters 8–12 cm long, erect; fruit like a black cherry. 1576.

*P. lusitanica—***Portugal Laurel**     Portugal

Leaves 8–15 cm long, 3–5 cm wide, toothed, stalk reddish; flowers in clusters 15–20 cm long, pendulous; fruit dark purple, cherry-like but oval in shape. 1648.

*(b) Deciduous—***Flowering Prunus**   D.H.
Cuttings
Parks, roadsides, streets, ornamental
These colourful, showy, flowering shrubs or small trees have flowers appearing before the leaves. If allowed to grow naturally, they form spreading or inverted conical crowns. They are hardy and are suitable for most soils, except poorly-drained sites. Flowers are prolific and usually in the pink to white range. Foliage is colourful in autumn. They are mainly garden plants and numerous hybrids and cultivars are available. The main parent stocks are:

*Prunus amygdalus*—Flowering Almond

*Prunus cerasifera* var. *blieriana*—Flowering Cherry Plum
(natural hybrid)

*Prunus cerasifera* var. *nigra*—Purple Cherry Plum

*Prunus cerasifera* var. *pisardii*—Flowering Cherry Plum

*Prunus mume*—Flowering Apricot

*Prunus persica*—Flowering Peach

*Prunus serrulata*—Flowering Cherry

*Prunus subhirtella* var. *pendula*—Weeping Flowering
Cherry

*P. cerasifera* var. *nigra—***Purple Cherry Plum**  Caucasus
4–7 m      500 mm                                    Fast
Windbreaks, avenues

Upright tree with slender ascending branches, spreading when older. Vigorous, hardy, easily grown. Tolerates poorly-drained sites better than most *Prunus*. Will grow in the warmer and drier districts, particularly if additional water (irrigation) is available. Widely planted. Foliage dark reddish-purple all the year (except winter), one of the richest colourings of the coloured foliage trees. Makes an attractive display when planted with evergreens or green foliaged trees. Leaves oval, 3–6 cm long; flowers pink; fruit a deep red cherry plum, edible. 1880.

*Pseudotsuga menziesii*

will grow in medium rainfall zones. Hardy, frost-resistant, wind-firm. Deep-rooted. Withstands snow. Will survive under shade; young planted trees will 'come through' heavy scrub eventually. Bark dark grey with resin blisters; needles 2–3 cm long, soft, deep green, round at the apex; cones 7–10 cm long, pendulous, scales papery, distinguished by a distinctive 3-lobed bract protruding between the scales. Often sold as a fir (*Abies* sp.) tree, e.g. Abies douglasii or A. menziesii. Several varieties are recognized. 1827.

*P. menziesii* var. *glauca*—**Blue Douglas Fir**

Similar to *P. menziesii* but the foliage is bluish-green and softer. Needles tend to be arranged around the twig rather than more or less in one plane. 1910.

# PYRACANTHA

*Pyracantha* is from *pyr* meaning a fire and *acantha*, a thorn. About ten species from south-eastern Europe and southern Asia are known.

*P. crenulata* (Crataegus crenulata)
    —**Nepal Hawthorn**   E.H.      Himalayas
Seed
3–5 m    650 mm                  Fast
Windbreaks, parks, roadsides, hedges, birds

Upright, bushy shrub producing masses of red berries in winter. Hardy, vigorous, but withstands moderate frosts only. Can be heavily trimmed to form a dense hedge. Tolerates smog. Branches thorny. Leaves 2–5 cm long, narrow; flowers hawthorn-like, white to pink; fruit red berry, 5–10 mm diam., attractive to birds. Distinguished from *Crataegus* by being evergreen whereas the latter is deciduous. 1844.

*P. serrulata*—**Flowering Cherry**      Japan
3–8 m    650 mm     Requires well-    Fast
                          drained soils
Parks, roadsides, ornamental, autumn foliage

Attractive, flowering, small tree, branches ascending when young, spreading crown when older. More suitable for the cooler mountain districts. Hardy, frost-resistant. Should not be pruned. Very showy flowers, produced prolifically in early spring. Leaves oval, large, 12–20 cm long, colouring to rich reds and browns in autumn; flowers in stalked clusters, white in the parent type but numerous horticultural varieties of pink and red tonings and shape are available. L.C.

# PSEUDOTSUGA

The name is from *pseudo* meaning false and *tsuga*, Tsuga. About seven species from North America, China, Japan and Taiwan are known.

*P. menziesii* (P. taxifolia, P. douglasii)
    —**Douglas Fir**   E.C.    W. North America
Seed
20–30 m    750 mm    Well-drained loams    Fast
Farm forests, windbreaks, parks, roadsides

Attractive, upright tree with a dense symmetrical crown. Retains branches to ground level when open grown. Produces Oregon timber of commerce. Best growth on sheltered sites in cool, mountain areas with annual rainfall of over 1200 mm and good soils but

# PYRUS

*Pyrus* is the old Latin name. There are about thirty species from Europe and Asia.

*P. calleryana*—**Chinese Pear**   D.H.      China
Seed
4–5 m    650 mm                  Fast
Parks, roadsides, ornamental

Single-stemmed, bushy-crowned tree. Originally imported as a disease-free root-stock for commercial pears but worth growing in its own right for its rich red and brown autumn colourings. Conditions, growth, etc. same as for ordinary pears. Leaves 5–8 cm long, 3–5 cm wide, serrated; fruit a pear but of poor quality and yield. Birds are fond of it, so few drop to the ground and are not a problem on paths, etc. 1903.

*Pyrus ussuriensis*

**P. ussuriensis—Snow Pear** D.H.　　　　　N.E. Asia
Seed
10–15 m　600 mm　　Requires well-　　　　Fast
　　　　　　　　　　　drained soil
Parks, shade, ornamental

Much-branched tree. Hardy, vigorous, frost-resistant. Attractive in flower, colours well in autumn. Some pruning for shape may be required. Leaves oval, pointed, 5–10 cm long, serrated, bronzy-red and brown in autumn; flowers white, prolific, 2–4 cm across, in dense hemispherical clusters; fruit a typical pear, 3–5 cm diam. 1855.

# QUERCUS—Oak

Seed
Windbreaks, shade, parks

*Quercus* is the old Latin name. About 450 species from Europe, Asia, North Africa, North America and South America are known.
This is a large group of mainly Northern Hemisphere evergreen and deciduous trees. They usually have dense crowns and are erect, single-stemmed and long-lived. Produce durable timber but are not recommended for farm forests as growth is too slow. Prefer cool, mountain areas and moist, well-drained, fertile soils; not satisfactory on deep sands or waterlogged sites. They are very deep rooted, developing a strong tap root; young plants often appear to die off in summer but the root can be actively extending to several feet and top growth resumes the following year. Hardy and frost-resistant, but can only be trimmed for light shaping. They are easily grown and because of the strong tap root young plants are easiest to establish; acorns can also be planted in the final site if this is well prepared. Leaves are green on the upper surface and usually lobed. Fruit is an acorn. Acorns must be collected immediately on falling and either sown or stored in moist sawdust, sand, etc. in a cool place as their natural viability is only a few days. Oaks hybridize readily so it is sometimes difficult to identify a species positively; intermediate forms occur.

**Q. aegilops—Valonia Oak** D.H.　　　S. Europe,
　　　　　　　　　　　　　　　　　　Asia Minor
6–10 m　500 mm　Will grow on　　　Moderate
　　　　　　　　　limestone

Variable tree, but usually upright with a spreading crown. Will grow in drier sites than most oaks. Hardy. Frost-resistant. Branchlets covered in yellow down; leaves oval, 8–12 cm long, lobes 6–8, pointed; acorns very large, up to 5 cm across, about half in the cup. 1731.
There is some doubt whether the name *Q. aegilops* is correct.

**Q. alba—White Oak** D.H.　　　　　　　USA
20–30 m　650 mm　　　　　　　　　Moderate

Large, spreading, dome-shaped, impressive tree, with a relatively short bole. Probably the most majestic of the oaks. Good specimen tree, old specimens can have a spread of 25–30 m. Tall and narrow in forest formation. Will grown on a wide range of soils but growth is fast on good, moist, deep, rich soil. Will tolerate some poor drainage. Hardy. Frost-resistant. Leaves 12–20 cm long, 5–10 cm wide, usually deeply lobed, 7–9 lobes, turn red to red-purple in autumn; acorns oval, 1–2 cm long; about one-quarter in the cup. 1724.

**Q. borealis** (Q. rubra)**—Red Oak** D.H.　　Eastern
　　　　　　　　　　　　　　　　　　　　USA
15–20 m　750 mm　　　　　　　　　Moderate
Ornamental

Upright, broadly columnar rather than spreading. Variable tree, name used rather loosely locally. Foliage red and red-brown in autumn. Branchlets dark red; leaves large, variable, oval, 12–18 cm long, deeply cut into 7–11 lobes, grey underneath, stalks up to 6 cm long; acorns 2–3 cm long, usually globular but maybe slightly elongated, cup varies from saucer-like (enclosing one-fifth to one-sixth of the acorn) to deep (enclosing about one-third of it). 1724.

**Q. canariensis—Algerian Oak** or　　　　Spain,
　　　　　　**Canary Oak** D.H.　　　N. Africa
10–20 m　750 mm　Well-drained soils　Moderate

Densely-crowned, somewhat spreading tree, retaining its leaves well into winter; almost a semi-evergreen in southern districts. (Seems to have been widely planted as Q. lusitanica, which now correctly is *Q. faginea*.) Colourful in autumn, retaining colour until well into winter. Leaves 10–18 cm long, not deeply lobed, teeth rounded, silver-blue underneath; acorns 2–3 cm long, 5–12 mm diam. about one fifth in the cup. 1844.

*Q. castaneifolia*—**Chestnut Oak**   D.H.        Caucasus
20–25 m    750 mm                                Moderate

Attractive, spreading tree with large chestnut-like leaves. Dead autumn leaves remain on the tree for most of the winter. Leaves dark glossy-green, 12–20 cm long, 5–8 cm wide, strongly serrated, with a prominent rib to the top of each serration, not lobed; acorns 2–3 cm long, oval, cup 'mossy', enclosing about one-third of the acorn. 1840.

*Q. cerris*—**Turkey Oak**   D.H.        S. Europe,
                                         Asia Minor
20–30 m    550 mm                        Fast

Handsome, short-boled, pyramidal tree, sometimes with an irregular crown. One of the fastest growing of the oaks. Hardy, withstands dry summers, tolerates dry conditions better than most oaks. Leaves oval, 7–12 cm long, thin, strongly lobed, 5–7 pointed lobes, autumn leaves brown remaining on the tree most of the winter; acorns oval, 2–4 cm long, about half enclosed in the cup, cup large, 'mossy', scales narrow and recurved. 1735.

*Q. coccinea*—**Scarlet Oak**   D.H.        North America
10–15 m    750 mm                           Moderate to slow
Avenues, ornamental, autumn foliage

Erect, pyramidal tree with slender, slightly drooping branches. Leaves turning to scarlet in autumn—then a very striking tree. Leaves 10–20 cm long, 5–10 cm wide, deeply lobed (almost to the midrib) into 5–9 lobes; acorns 1–3 cm long, more or less globular, on a short stalk, about one-third to half enclosed in the cup. Similar in appearance to *Q. palustris* but the latter has less than one-quarter of the acorn enclosed in the cup. 1691.

*Q. faginea* (Q. lusitanica)—**Portugal Oak**   D.H.
                                           Spain, Portugal
10–20 m    750 mm                          Moderate

Spreading, densely-crowned tree. Colourful in autumn. Leaves 4–10 cm long, usually oval but variable in shape, margins serrated, 5–12 pairs of triangular teeth, not deeply lobed, blunt apex, grey underneath; acorns long and slender, 2–4 cm long; 6–12 mm wide. Very similar to *Q. canariensis* in its characteristics except that in this latter species the leaves are larger (10–15 cm long). Plants sold as Q. lusitanica are probably *Q. canariensis*, since the former, correctly *Q. faginea* is uncommon in southern Australia. 1824.

*Q. ilex*—**Holm Oak** or **Holly Oak**   E.H.   S. Europe
15–20 m    550 mm                               Moderate
Roadsides

Large, massive tree with a dense, rounded crown. Wind-firm. Withstands moderate coastal exposure, and drier conditions then most oaks. Small (1-year-old) plants best for planting. Leaves oval, 5–8 cm long, wavy margin, not lobed like a typical oak, greyish underneath; acorns small, 1 cm long, oval, about half in the cup. Before 1580.

*Quercus palustris*

*Quercus robur—golden form*

*Q. lyrata*—**Overcup Oak**   D.H.        Southern USA
15–20 m    650 mm    Will grow on poor soils    Fast

Spreading tree with small often pendulous branches although form can be variable. Tolerates poorly-drained soils better than most oaks. Suitable for heavy clays. Hardy. Frost-resistant. Easily transplanted. Leaves variable, 15–25 cm long, 3–8 cm wide, deeply and irregularly lobed, 5–9 lobes; acorns globular, sessile, 1–2 cm long, almost completely enclosed in the cup. 1786.

*Q. palustris*—**Pin Oak** or        Eastern USA
              **Swamp Oak**   D.H.
10–25 m    750 mm    Will grow on heavy soils    Fast
Avenues, ornamental, autumn foliage

Erect, pyramidal tree with horizontal, regular branching. Will grow in wetter areas than most oaks but still prefers well-drained soils. Hardy. Foliage colours to orange-red and yellow-brown in autumn but the dead leaves remain on the tree for most of the winter. Not as colourful as *Q. coccinea* but often favoured because of its much quicker growth and greater tolerance of less favourable sites. Withstands slight coastal exposure. Leaves 10–15 cm long, 5–10 cm wide, deeply lobed (almost to the midrib), 5–7 lobes; acorns 1 cm diam., more or less globular, less than one-quarter of its length enclosed in the cup, sessile. Before 1770.

*Quercus suber*

*Quercus virginiana in winter*

*Q. palustris* 'Macedon'  D.H.                                    Unknown

Variety of *Q. palustris* with richer autumn colourings. Leaves smaller (7–12 cm long), fewer lobes (3–4), and most of the lobes tipped with a fine soft spine. Origin unknown but common around Mt Macedon, Victoria.

*Q. robur* (Q. pedunculata)                                Europe,
——**English Oak**  D.H.                           N. Africa
15–25 m    650 mm                      Fast then moderate

Spreading, densely-crowned, rounded, heavily-branched tree; often spreading over a distance greater than its height. Good specimen and shade tree. Leaves rounded-oblong, 6–12 cm long, 4–7 rounded lobes, light green underneath; acorns 2–3 cm long, rounded-oblong, about one-third in the cup, on a long stalk (over 2 cm long).  L.C.

*Q. suber*—**Cork Oak**  E.H.              Mediterranean
6–12 m    450 mm    Will grow on poor sites    Slow

Spreading, much-branched, short-trunked tree. Hardy, most drought-resistant of the oaks, suitable for the warmer districts. Grows on a wide range of soils including granitic and gravelly types, but not calcareous or saline. Slightly frost-tender. Bark corky, often several inches thick, yields cork of commerce; leaves oval, small, 4–6 cm long, margins slightly toothed, not divided into lobes, grey and hairy underneath; acorns 2–3 cm long, oval, one-third to half in the cup, cup covered with thick scales. 1600.

*Q. virginiana* (Q. virens)—**Live Oak**  E.H.      Southern
6–10 m    550 mm                                      USA
Roadsides                                              Fast

Attractive, rounded, spreading tree with horizontal branches. One of the evergreen oaks. Withstands severe coastal exposure, will tolerate waves washing up to the roots. Easily transplanted. Leaves not like a typical oak, entire (not divided into lobes), 5–12 cm long, 3–6 cm wide; acorns oval, 2–3 cm long, in clusters of 3–5, each sessile, dark brown to almost black, one-third enclosed in the cup. 1739.

# RHUS

*Rhus* is the old Greek name. About 250 species from Europe, Asia, North America and southern Africa are known.

*R. succedanea*—**Scarlet Rhus** or                    Japan
                **Sumach**  D.H.
Seed
4–5 m    600 mm    Requires well-              Fast
                   drained soils
Parks, roadsides, ornamental, autumn foliage

Erect, single-stemmed, small tree with a spreading crown and vivid scarlet foliage in autumn. Vigorous, frost-resistant. Relatively large specimens can be transplanted. Responds to moderate pruning. Foliage pinnate, leaves 15–25 cm long, leaflets 9–15, 5–8 cm long, 1 cm wide; flowers inconspicuous; fruit globular, 5–8 mm diam., whitish-grey, in clusters 15–20 cm long, a wax-like substance can be pressed from the fruit. 1863. All *Rhus* species contain irritant principles that may cause dermatitis but this species has little effect on most people.

*Rhus succedanea*

*Robinia pseudoacacia and flowers*

# ROBINIA

Genus named after J. Robin, herbalist to Henry IV of France. About 20 species from North America are known.

*R. pseudoacacia*—**Black Locust** or      E. North
             **False Acacia** D.H.      America
Seed
10–15 m    400 mm    Not satisfactory      Fast
                     on light soils
Windbreaks, shade, farm forests, parks, roadsides, gully erosion

Handsome, upright tree with a straight trunk and bright green, pinnate foliage. Produces durable timber. Thorns 2–4 cm long on the branchlets. Suitable for the warmer as well as cooler districts. Hardy, vigorous, frost-resistant. Will grow on limestone soils, but not on compacted clays. Produces prolific crops of root suckers, useful for control of eroding gullies and stream banks. If control is necessary plant only where cultivation or mowing is possible. Because of its nitrifying properties can be planted on 'loose' spoil (mining, etc.) banks. Foliage has useful fodder value. Bark rough, deeply furrowed; leaves pinnate, 15–25 cm long, 10–20 leaflets, 4–5 cm long; flowers whitish, pea-shaped, in showy clusters 15–25 cm long; fruit a pod, 5–8 cm long. Seed requires pre-sowing treatment as for acacias. 1601.

# SALIX    D.H.

Cuttings    650 mm    Require moist, well-      Fast
                     drained soils
Windbreaks, shade, parks, stream bank erosion
*Salix* is the old Latin name. There are over 250 species (some authorities say over 500) from Europe, Asia and North America.

These trees are variously shaped and require moist, well-drained soils. Provided the roots can be kept cool, for example, along streams, channels, dams, lagoons, etc., they can be grown in the hotter and drier districts. Because of their preference for moist situations, the minimum rainfall given is not very meaningful. They can be severely pollarded and pruned and will coppice freely. Large cuttings (several centimetres in diameter) will strike; advanced specimens can be obtained quickly by planting large straight branches directly into their final position. They are frost-resistant and will tolerate some flooding. Develop a very strong and extensive fibrous root system in moist areas so should not be planted near underground earthenware drains as these are blocked rapidly by root growth; they cannot be recommended for planting within about 20 m of them. (They do not appear to block cast iron drains.) Branchlets are tough and flexible. Leaves are narrow, short stalked, several times longer than wide and serrated to varying degrees. Flowers are catkins and usually not conspicuous (except *S. discolor*). Fruits are inconspicuous. Various species hybridize readily so some confusion exists among the names; intermediate forms are likely to be seen. They can be distinguished from poplars by (a) their narrow leaves, and (b) the winter buds which are covered by a cap-like structure, compared with the overlapping scales of poplars.

*S. alba* var. *coerulea*—**Cricket Bat Willow**      Europe
12–15 m

Attractive tree with ascending branches and pendulous branchlets. Twigs dark brown; leaves 5–10 cm long, greyish-white underneath.

*S. alba* subsp. *vitellina* (S. vitellina)      Europe,
         —**Golden Willow**      N. Asia
6–10 m
Ornamental

Round-crowned tree, branches ascending. Long, slender branchlets, bright golden colour, attractive when the tree is leafless. Suitable for wet sites. Relatively long-lived. Leaves pale green to slightly bluish, 7–10 cm long, narrow.

*S. alba* subsp. *vitellina* var. *pendula*—**Weeping Golden**
                             **Willow**      Cultivar

As for *S. alba* subsp. *vitellina* but branchlets pendulous.

*Salix babylonica*

*S. babylonica*—**Weeping Willow**                    China
10–15 m
Fodder

Attractive, bushy-crowned, heavily branched, spreading tree with a short trunk and long pendulous branchlets, up to 3 m long. Best growth is obtained where the roots have access to aerated (running, not stagnant) water, such as along streams or with access to both water and aerated soil e.g. dam banks, lakesides. Branches become brittle on old trees (60–70 years old)—can be restored by heavy lopping. Withstands slight coastal exposure. Foliage good as subsistence fodder. Leaves 5–15 cm long, 1–3 cm wide, serrated.

*S. discolor*—**Pussy Willow**                    Europe
5–10 m
Ornamental
Upright tree with ascending branches. Frequently multi-stemmed. Tolerates wet sites better than most willows. Very strong and extensive root systems. Planted often for its attractive 2–3 cm long, yellow-brown, male catkins. Leaves 7–10 cm long, grey underneath, broader than most willows; buds large and black. 1809.

*S. fragilis*—**Crack Willow**                    Europe, N. Asia
10–15 m
Bushy, round-crowned, somewhat spreading tree with a short thick trunk. Branches ascending, twigs and branchlets brittle at the base. Grows well on stream banks; widely planted for this use. Roots growing in water are red in colour. Leaves large, 12–18 cm long, 2–3 cm wide, serrated. L.C. Local specimens are probably a hybrid between *S. alba* and *S. fragilis*, known in Europe as *S. rubens*.

*S. humboltiana* (S. chilensis)—**Chilean Willow**    South America
8–10 m
Upright, columnar tree. Recent introduction, not yet well known. Leaves similar to *S. babylonica*, but narrower.

*S. matsudana* c.v. *'tortuosa'*—**Corkscrew Willow**    China
5–7 m
Ornamental
Small tree with semi-pendulous foliage. Trunk, branches, twigs contorted, often twisting through 180°, attractive when the tree is leafless. Leaves 8–15 cm long, 1–2 cm wide, serrated. 1913.

## SAMBUCUS

*Sambucus* is the old Latin name. About twenty species occur, with representatives in every continent.

*S. nigra*—**Elder**    D.H.                    Europe, W. Asia, N. Africa
Seed, cuttings
3–5 m        750 mm                    Fast
Low windbreaks, roadsides, parks, hedges

Shrub or small tree growing on a wide range of soils. Relatively long-lived; previously widely planted. Vigorous, hardy, and frost-resistant. Can be heavily trimmed, forms a good hedge. Coppices readily. Bark yellowish, branchlets very pithy; leaves opposite, 7–12 cm long, pinnate, 5–7 leaflets each 2–4 cm long, 1 cm wide, toothed; flowers white, in flat-topped clusters, 10–15 cm across, fragrant; fruit a berry, black, shining, 5–7 mm diam., in clusters. L.C.

## SCHINUS

The genus is named after *Schinus*, the Greek name for a tree which exudes a similar mastic-like substance to some species of *Schinus*. About thirty species from Central and South America are known.

*S. molle*—**Pepper Tree**    E.H.                    Peru
Seed
6–8 m        350 mm    Not suitable for wet soils    Fast
Windbreaks, shade, parks, roadsides
Spreading, rounded, heavily-branched tree with slightly drooping foliage almost to ground level. Branches originate from a few feet above the ground. Widely planted and well known, suitable for the cool as well as the hot and dry districts. Satisfactory on most soils except shallow and swampy sites. Very deep and strong root system. Frost-tender when young. Drought-resistant. Withstands moderate coastal exposure. Can be pollarded or trimmed to form a hedge. Good tree for hard adverse conditions. Leaves pinnate, 15–25 cm long, leaflets 5–8 cm long; flowers yellow-white in clusters 5–8 cm long; fruit red berries, 4–6 mm diam., in clusters.

*Sequoia sempervirens plantation in Victoria*

*S. terebinthifolius*—**Brazilian Pepper Tree**   D.H.   Brazil
Seed

5–7 m        400 mm        Will grow in                Fast
                           alkaline soils
Windbreaks, shade

Attractive, small tree with a rounded, bushy crown and pendulous branchlets. Not widely grown. More suitable for the warmer districts. Easily grown. Hardy, frost-resistant and tolerant of hot, dry summers. Leaves pinnate, 10–18 cm long, leaflets variable in shape and number; fruit masses of bright red globular berries, 3–5 mm diam., not eaten by birds because of their astringent taste.

# SEQUOIA

*Sequoia* is believed to be derived from the American Indian name for the Redwood—*See-qua-yoh*. One species from western USA is known.

*S. sempervirens*—**Redwood**   E.C.        Western USA
Seed

25–30 m    750 mm                    Moderate
Windbreaks, shade, parks, roadsides

Erect, conical, narrow-crowned tree retaining branches to ground level. Long-lived, exceeds 70 m in its native habitat. Will grow on a wide range of soils but best development occurs on deep, fertile soils in sheltered sites within the higher rainfall areas. Only recommended for farm forests where soil is good and rainfall exceeds 1200 mm per year. Difficult to establish in the field but once established is hardy and frost-resistant. Shade-tolerant when young, will 'come through' scrub growth. Cannot be trimmed. One of the few conifers to coppice from cut stumps. Leaves (needles) flat, 6–15 mm long, narrow, arranged more or less in one plane; cones 2–3 cm long, 1–2 cm wide, woody. 1843.

# SEQUOIADENDRON

The name is from *Sequoia* (see *Sequoia*) and *dendron*, a tree. One species occurs in California.

*S. giganteum* (Sequoia gigantea)—**Big Tree**   E.C.
Seed                                    California
25–30 m    900 mm    Rich, well-drained soils    Fast
Windbreaks, shade, parks, roadsides

Erect, pyramidal tree with drooping branches. Retains branches to ground level for some years, but gradually sheds the lower ones with age. Grows to well over 80 m in its native locality. Long-lived. Best growth on fertile soils in sheltered sites where rainfall exceeds 1300 mm per annum, but will grow in

*Sequoiadendron giganteum—young tree*

areas with rainfall down to 750 mm. Difficult to establish and growth slow for the first few years, but then fast. Only suitable for farm forests in high rainfall districts. Hardy, frost-resistant. Shade-tolerant when young, will 'come through' scrub growth. Coppices from cut stumps. Leaves (needles) 3–5 mm long, closely attached to the stem and arranged spirally around it; cones 5–8 cm long, 3–4 cm wide, 35–40 scales, woody. 1853.

# SOPHORA

*Sophora* is from *Sephora*, the Arabian name for a tree bearing pea-shaped flowers. There are about fifty species, mainly from Asia, North America, Pacific islands and New Zealand.

*S. japonica*—**Pagoda Tree**  D.H.                China
Seed
6–20 m      550 mm      Moist, well-drained      Moderate
                        soils
Parks, roadsides, ornamental

Attractive, upright tree suitable for moist, well-drained sites in the warmer as well as the cooler districts. Good specimen tree with pinnate foliage and clusters of colourful yellowish pea-shaped flowers. Round-headed when mature. Frost- and cold-resistant. Leaves

*Sophora tetraptera*

15–25 cm long, leaflets 8–16, 5–6 cm long; flowers creamy-white, 1 cm, in clusters 15–25 cm long; fruit a pod, 7–8 cm long. 1749.

*S. microphylla*

Similar to *S. tetraptera*; some authorities regard it as only a variety of *S. tetraptera*. Differs in that it often remains in the juvenile form for some years as a twiggy shrub; once past this stage growth is rapid. Leaflets also are much smaller, only 5–8 mm long, 3–5 mm wide, 30–40 pairs. Foliage drops off in the spring, showing up the flowers to greater advantage.

*S. tetraptera*—**Kowhai**  E.H.                N.Z.
Seed
5–8 m      650 mm      Will not grow      Fast
                       in wet soils
Parks, roadsides, ornamental

Attractive, lightly-crowned tree with fine fern-like pinnate foliage and thin stems. Hardy, suitable for most soils except poorly-drained. Tolerates smog. large specimens can be transplanted. Withstands moderate pruning. Various geographical forms occur. Leaves 10–15 cm long, leaflets 1–2 cm, numbers variable, from 5 or so pairs on young plants to 20 or more pairs when older; flowers golden, or greenish-yellow (orange colours occur), 2–3 cm long, pea-shaped, abundant, in drooping clusters; fruit 4-winged pod, 15–20 cm long, markedly constricted between the seeds. 1772.

*Sequoiadendron giganteum—mature tree*

*Sorbus aucuparia*

*Sorbus aucuparia—fruit and foliage*

# SORBUS

*Sorbus* is the old Latin name. About one hundred species from Europe, Asia and North America are known.

*S. aucuparia* (Pyrus aucuparia)—**Rowan**   D.H.   Europe, W. Asia

Seed
6–12 m    650 mm                                    Fast
Windbreaks, shade, parks, roadsides, birds, ornamental, autumn foliage

Upright tree, erect when young, often with several closely ascending branches giving a columnar appearance, tending to spread with age. Will grow on poor soils. Hardy, frost-resistant, easily grown. Large specimens can be transplanted. Normally trim only for shaping but in cooler districts trees can be pollarded. Foliage may be scorched by hot winds so should not be planted in exposed position in warmer districts. Birds are very fond of the berries. Leaves pinnate, 15–20 cm long, 6–12 pairs leaflets, 3–5 cm long, serrated, turning vivid red-brown in autumn, particularly in the cooler districts; flowers white, small, hawthorn-like; fruit a berry, 5–7 mm diam., red to orange (some types yellow), in large, conspicuous and decorative clusters. L.C.

*S. decora*   D.H.                              E. North America
5–8 m      650 mm      Suitable for            Moderate
                       adverse sites
Parks, autumn foliage, ornamental

Spreading, small tree with a rounded crown. Hardy, vigorous, withstands cold and some snow. Tolerates smog. Not well known locally but is being increasingly used in Britain and Europe as a street tree, in reclamation plantings, and on similar adverse sites. Foliage pinnate, leaves 10–20 cm long, usually 9–13 leaflets, each 3–7 cm long, serrated, colouring well in autumn; flowers prolific, white, in clusters 5–10 cm across, rather showy, individual flowers 5–10 mm diam.; fruit a bright red, oval to globular berry, 7–10 mm diam., hanging in large clusters.
*S. americana* is similar in appearance but has usually 11–17 leaflets in the leaves.

# STENOCARPUS

The name is from *stenos* meaning narrow and *karpos* meaning a fruit which alludes to the narrow, flat fruit common to many species. About twenty-five species from Australia and New Caledonia are known.

*S. sinuatus*—**Firewheel Tree**   E.H.         N.S.W., Qld
Seed
10–20 m    500 mm    Well-drained,            Slow
                     deep loams
Shade, parks, roadsides, avenues, ornamental

Upright, single-stemmed tree. Slow-growing in southern districts, more suitable for the warmer districts. Withstands only light frosts. Conspicuous colourful flowers—best displays obtained after a hot, dry summer but often not flowering until 10–12 years old. Leaves large, up to 30 cm long, very variable from simple to deeply-lobed (almost pinnate); flowers 5–10 cm long, brilliant red or orange, 10–20 arranged in wheel-like clusters; fruit boat-like pod, 5–10 cm long. 1830.

# STEWARTIA

Genus named after J. Stewart, Earl of Bute (1713–1792). There are about eight species from North America and eastern Asia.

*S. pseudocamellia*   D.H.                              Japan
Seed, cuttings
5–10 m    750 mm                               Moderate
Parks, ornamental

Attractive flowering tree with bright green foliage. Not suitable for alkaline soils, prefers soils high in organic-matter content. Requires some shade during the day rather than full sunshine, otherwise hardy. Leaves, oval, 5–8 cm long, colourful red and yellow in autumn; flowers large, white, cup-shaped, 5–8 cm across, camellia-like; fruit a hairy seed-pod, 2–3 cm long, splitting into 5 valves. 1874.

# STYRAX

*Styrax* is the old Greek name. About one hundred species from Europe, Asia and North America are known.

*S. japonica*—**Japanese Snowdrop Tree**  D.H.  China, Japan

Seed

4–7 m      750 mm                      Moderate

Parks, roadsides, ornamental

Very attractive, slightly spreading and open, small tree. One of the finest of the small trees; more suitable for the cooler districts. Hardy, frost-resistant. Withstands slight to moderate shade. Large specimens can be transplanted when leafless. Leaves oval, 5–12 cm long; shining green; flowers white, 1–2 cm across on long stalks, snowdrop-like, produced in pendulous clusters scattered through the foliage, very showy, summer-flowering; fruit globular, about 2 cm diam., black. 1862.

# SYNCARPIA

The name is from *syn* meaning joined and *karpos*, a fruit, which alludes to several fruits being fused together. There are five species from Australia.

*Stenocarpus sinuatus*

*S. glomulifera* (S. laurifolia)—**Turpentine**  D.H.
N.S.W., Qld

Seed

20–25 m    750 mm     Prefers good, moist,        Fast
well-drained loams

Windbreaks, farm forests, shade

Tall, straight-stemmed, eucalypt-like tree but retaining its lower branches when open grown. Timber durable in salt water. Prefers well-drained, fertile soils but will grow on most soils except light sands. Hardy, but frost-sensitive when young. Coppices freely, can be trimmed to form a hedge. Bark thick, brown, fibrous, flaky; leaves oval, thick, 5–8 cm long, woolly underneath, margins wavy; flowers white, like a eucalypt; fruit globular, formed by a number of capsules fusing together.

# TAMARIX

Cuttings                                      Fast

Windbreaks, parks, roadsides, hedges, gully erosion

*Tamarix* is an old Latin name. Probably about sixty species are known (although some authorities say over one hundred) from southern Europe and Asia.

These short-boled, dense-crowned trees with drooping branchlets are most suitable for the hot, dry districts but can be grown in the cooler areas. They grow well on heavy, alkaline, saline or poorly-drained soils; are hardy, very drought-resistant and easily grown, but are frost-sensitive when young. Wind-firm and can withstand severe coastal exposure, and will tolerate smog. They are not easily transplanted, but coppice freely and may be severely trimmed to form a dense hedge. Can be grown from large cuttings (several centimetres in diameter). Prune after flowering for good floral displays. Leaves are minute and scale-like. Flowers are small and arranged in slender spikes.

*T. aphylla* (T. articulata)—**Athel**  E.H.     N. Africa,
W. Asia

6–10 m     250 mm     Will grow on hot,
deep sands

Erect, narrow-crowned, bole sometimes crooked. Straight-boled types suitable for farm forests. Withstands severe coastal exposure. Not very satisfactory in southern districts. Foliage light green; flowers pink, late summer.

*T. juniperina* (T. japonica)                    China
**Juniper Tamarisk**  D.H.

4–5 m      450 mm

Dense crown, with branches to ground level. Withstands moderate coastal exposure. Branchlets rich brown; foliage light green; flowers pink in racemes, 3–5 cm long, spring-flowering, before the leaves appear. 1877.

*T. parviflora*—**Tamarisk**   D.H.          Mediterranean
5–7 m     400 mm      Will grow in coastal,
                                    sandy soils

Attractive, erect, small tree with slender branches. Can be grown in hot, dry districts if irrigation water is available. Withstands severe coastal exposure. Branchlets purplish; foliage 'feathery'; flowers white, tinged with pink, very small, densely crowded in 4–5 cm long racemes, spring-flowering. 1853.

*T. gallica* (1596) is sometimes cultivated under the name *T. parviflora* and vice versa.

*T. pentandra*—**Late Tamarisk**   D.H.        S.E. Europe,
                                                         W. Asia

4–5 m     450 mm

Shrubby, densely-crowned, small tree. Withstands moderate coastal exposure. Branchlets purplish; foliage bluish; flowers pink to red, in terminal sprays up to 40 cm long, summer-flowering. 1883.

*T. tetrandra*—**Flowering Cypress**   D.H.    S.E. Europe,
                                                           W. Asia

3–4 m     450 mm

Dense shrub with dark coloured almost black bark and arching stems. Withstands moderate coastal exposure. Flowers pink, in 4–5 cm racemes, spring-flowering. Very similar in appearance and characteristics to *T. parviflora*. 1821.

# TAXODIUM

The name is from *taxus*, the Latin name for yew and *eidos*, meaning resemblance. This refers to the similarity of the foliage to the yew (*Taxus*). About three species from southern USA and Mexico are known.

*T. distichum*—**Swamp Cypress** or
                       **Bald Cypress**   D.C.   South-eastern USA
Seed
20–30 m     750 mm     Not on poor or        Moderate
                                    dry soils
Windbreaks, farm forests, parks, roadsides, autumn foliage

Tall, erect, neat, pyramidal tree, spreading when mature. Requires good, moist soil, will grow in the warmer districts if a good supply of ground water is available. In its native habitat grows in swamps. Moderately frost-resistant. Coppices well. Foliage light green, brown in autumn; leaves narrow, 2–3 cm long arranged spirally around the stem; cones globular, 1–2 cm diam., purplish, scales 'square' in cross-section. Tallest tree in the Royal Botanic Gardens, Melbourne (37 m). 1640.

*Taxus baccata*

# TAXUS

*Taxus* is the old Latin name. There are about ten species from Europe, southern Asia and Central America.

*T. baccata*—**Common Yew**   E.C.          Europe
Seed, cuttings
4–5 m     750 mm     Well-drained soils,
                                    but not deep sands      Slow
Low windbreaks, parks, hedges, ornamental

Dense, compact, much-branched small tree. Very long-lived. Tolerates calcareous soils. Frost- and smog-resistant. Can be severely trimmed to form a dense hedge; used extensively in Britain and Europe for topiary work. Bark and leaves poisonous. Leaves linear, 2–3 cm long, arranged spirally around the stem; cone not like a typical conifer, seed is in a woody shell borne in a scarlet fleshy cup. Numerous horticultural varieties are available; smaller strains more commonly seen. L.C.

# TELOPEA

The name is from *telopas*, visible at a distance, which alludes to the conspicuousness of the flowers. Four species from Australia are known.

*T. speciosissima*—**New South Wales Waratah**   E.H.
                                                                       N.S.W.
Seed
3–4 m     650 mm     Will grow on poor soils      Slow
Parks, ornamental

Small tree with large conspicuous flowers. Requires well-drained soil. Will grow on the poorer and sandier types, but also satisfactory on well-drained, heavier soils. Applying fertilizer is not desirable. Not as easy to grow as the Victorian Waratah (*T. oreades*) but is more attractive. Can become straggly, needs trimming to maintain a good shape. Leaves 15–25 cm long, narrow, 1–2 cm wide, serrated, rounded tip; flowers conspicuous, large, globular heads 7–10 cm diam., surrounded by a collar of red bracts. 1789.

# TEMPLETONIA

Genus named after J. Templeton, an Irish botanist. There are about eight species from Australia.

*T. retusa*—**Red Templetonia**   E.H.         W.A., S.A.
Seed
2 m         450 mm       Prefers alkaline soils         Fast
Low shelter, parks, roadsides

Attractive, bushy, winter-flowering shrub. Will grow on deep sands, including calcareous types. Can be grown on acid soils with heavy dressings of lime. Moderately frost-resistant and hardy. More satisfactory in the warmer districts. Not very wind-firm, staking is desirable. Withstands severe coastal exposure. Can be straggly but with an annual light pruning can be formed into a dense, compact shrub. Leaves grey-green, oval, 2–3 cm long; flowers large, pea-shaped, 2–3 cm long, reddish-white; fruit a short legume. 1817.

*Templetonia retusa—flowers—R. Elliot*

# THRYPTOMENE

*Thrypto* means to break or crush; meaning not clear. There are about twenty-five species from Australia.

*T. calycina*—**Grampians Thryptomene**   E.H.      Vic.
Seed
2–3 m       500 mm       Well-drained soils         Fast
Parks, roadsides

Much-branched shrub, very attractive when flowering. Hardy, frost-resistant, easily-grown. Prefers sandy soils but will grow on well-drained loam. May be straggly, requires an occasional pruning for shaping. Can be heavily pruned. Leaves small, 6–12 mm long; flowers small, 3–8 mm across, pink, then white, prolific, late winter-flowering. 'Paynes' hybrid the best form, pink flowers produced abundantly in late winter.

# THUJA

*Thuja* is the old Greek name. About six species occur in Asia and North America.

*T. plicata*—**Arbor Vitae** or                    W. North
         **Western Red Cedar**   E.C.        America
Seed, cuttings
15–20 m     900 mm       Prefers well-            Slow
                         drained loams
Windbreaks, farm forests, roadsides, ornamental

Dense, pyramidal, symmetrical tree retaining its branches to ground level. More suitable for the cool, mountain districts. Fine specimen tree; will grow to over 35 m. Produces reddish durable timber. Hardy, frost-resistant. Tolerates wet conditions better than most conifers. Can be moderately trimmed to form an open hedge. Branchlets arranged in one plane, plate-like; leaves small, 5–6 mm long, closely pressed to the stem; cones small, 1 cm long, erect, 5–6 pairs of scales. 1853.

# THUJOPSIS

The name is from *Thuja*, and *opais* meaning resemblance. Only one species from Japan is known.

*T. dolobrata*—**Mock Thuja**   E.C.              Japan
Seed, cuttings
6–10 m      900 mm       Well-drained,             Slow
                         moist loams
Parks, ornamental

Dense, compact, pyramidal tree; branches to ground level. Attractive specimen tree with thick, 'heavy', plate-like foliage. Hardy, frost-resistant. Stands light trimming. Leaves 5–6 mm long, broad, closely pressed to the stem; cones 1–2 cm long, scales almost leathery, wedge-shaped, prominent triangular tip. 1853.

# TIEGHEMOPANAX

Renamed in 1905 after P. E. L. van Tieghem (1839–1914), a French botanist, and *Panax* the original name given in 1830. There are about thirty-five species from Australia, New Caledonia and New Hebrides.

*T. sambucifolius*—**Elderberry Panax**   E.H. N.S.W., Vic.
Seed
3–8 m       750 mm                                Fast
Parks, roadsides, ornamental

Rather variable small tree; open-grown specimens usually rounded, spreading, with branches to ground level. Attractive the whole year. Hardy, frost-resistant, suitable for most soils except sands. Can be heavily trimmed. Not widely planted but worthy of more attention. Foliage variable, simple or pinnate, 'ash-like', 15–25 cm long, leaflets 3–7 cm long, 1–3 cm wide; flowers white, in terminal clusters, 8–15 cm long; fruit bluish-white berry, 5–8 mm diam., in clusters.

*Tilia* × *vulgaris*

*Tilia* × *vulgaris—fruit*

# TILIA

*Tilia* is the old Latin name. About thirty species occur in Europe, Asia and Central America.

*T. x. vulgaris* (T. europaea)—**Lime** or
                            **Linden** D.H.    Europe
Seed
10–12 m   900 mm   Well-drained loams    Fast
Windbreaks, shade, parks, roadsides, avenues, honey

Dense, round, shapely crown, spreading when mature. Good specimen tree, believed to be a stable natural hybrid between *T. cordata* and *T. platyphyllos*. Most satisfactory in the cool, mountain districts. Easily grown, frost-resistant. Large specimens can be transplanted. Tolerates smog. Should only be trimmed if shaping is necessary. Coppices vigorously from cut stumps. Leaves broadly oval, 7–10 cm long, serrated, golden-brown in autumn; flowers 5–10, in clusters, fragrant, produce copious amounts of nectar; fruit distinctive, 'winged', 4–6 cm long; seed short-lived. L.C.

# TRISTANIA

Genus named after J. M. C. Tristan (1776–1861), a French botanist. There are about fifty species from Australia, New Caledonia, Fiji and Malaysia.

*T. conferta*—**Brush Box**  E.H.        N.S.W., Qld
Seed
5–8 m   600 mm   Prefers well-        Fast
                    drained loams
Windbreaks, parks, roadsides, streets

Dense, symmetrical crown; eucalypt-like, shedding its lower branches. Trunk attractive, reddish-brown but light yellow-green in late summer. Widely planted as a street tree. Suitable for most soils but site must be well-drained and not alkaline. Frost-sensitive when young, otherwise hardy and easily grown. Drought-resistant once established. Tolerates smog. Coppices readily, can be pollarded. Leaves 10–18 cm long, 4–6 cm wide, more or less oval, leathery; flowers white, 1 cm across, stamens conspicuous (like a eucalypt), in terminal clusters, flowering early summer; fruit like a gum nut, 1 cm long, 6–8 cm diam. 1805.

*T. laurina*—**Kanooka**  E.H.       Vic., N.S.W., Qld
Seed
5–7 m   750 mm   Prefers moist, well-   Moderate
                    drained soils
Shade, roadsides, parks

Attractive, upright and compact tree. Slightly frost-sensitive when young otherwise hardy. Will not tolerate wet soils or prolonged flooding. Withstands cold. Leaves glossy, 5–12 cm long, 1–2 cm wide, silky underneath; flowers small, 1 cm diam., eucalypt-like, but with 5 small rounded petals, creamy to orange-yellow, in short clusters; fruit woody, globular, 6–8 mm diam.

# TSUGA

*Tsuga* is the Japanese name. About ten species from North America, Japan, China and the Himalayas are known.

*T. heterophylla*—**Western Hemlock**  E.C.   W. North
Seed                                America
20–30 m   1000 mm   Requires good,   Moderate
                   deep soils     then fast
Farm forests, windbreaks

Attractive, tall, slender and stately tree with a spire-like crown. Reaches 70 m in its natural habitat. Will grow on a wide range of soils, from sandy types to stiff clay, provided rainfall is adequate; best results are in deep, moist loam. Very hardy. Frost-resistant. Withstands cold and prolonged snow, suitable for alpine areas. Not tolerant of shade or smog. Produces good timber, bark rich in tannin but rarely more than 2–3 cm thick. Needles flat, varying markedly in length on the same shoot, 0·5–3 cm long, blunt; cones oval to conical, 1–3 cm long, sessile, terminal. 1851.

*Tristania conferta*

# ULMUS—Elm

Seed, cuttings                                    Moderate
Windbreaks, shade, parks, roadsides, avenues
*Ulmus* is the old Latin name. About thirty-five species from Europe and Asia are known.

These are attractive, large and spreading trees which make good shade and specimen trees. They require well-drained soils, and are not satisfactory in wet sites. Easily grown, hardy, frost-resistant and wind-firm. As the vigorous root system is moisture-seeking, they should not be planted near drains. Coppice freely and can be heavily pollarded. Large specimens can be transplanted. Bark is hard and furrowed. Leaves are oval and serrated. Fruit is papery and disc-shaped, 1–2 cm in diameter. Dutch Elm disease has caused extensive death of elms in North America. It is caused by the fungus *Ceratocystis ulmi*. The sticky spores of the fungus are transferred by elm bark beetles mainly *Scolytus scolytus* and *S. multistriatus* and are deposited on the shoots as the beetles feed. Typical symptoms are the wilting and death of shoots but the focus of attack is disruption of the vascular system caused by toxins produced by the fungus spreading through the system. A fairly resistant variety named Commelin elm has been bred in the Netherlands but is not yet freely available.

*U. glabra* (U. montana)—**Scotch Elm**   D.H.      Britain
10–15 m    650 mm

Compact tree with a short upright bole dividing into several ascending main branches. Spreading crown. Best elm for general planting. Does not produce root suckers, used as stock for grafting other species. Leaves 8–12 cm long, rough, leaf-stalk very short; fruit 2 cm diam., purplish. L.C.

*U. x hollandica*—**Dutch Elm**   D.H.      Britain
15–20 m    550 mm
Gully erosion

Spreading, somewhat open-crowned tree, branch ends pendulous. Variable, believed to be a hybrid between *U. glabra* and *U. carpinifolia*, probably originating in England. (Sometimes known as U. racemosa.) Vigorous, hardy, producing prolific root suckers which have prominent corky wings up to 2 cm wide. Suckers are persistent so plant only where they can be controlled by cultivation or slashing. Useful for the control of gully erosion. Leaves 6–12 cm long, tapering tip, margins not opposite at the base of the leaf; fruit 1–2 cm diam.

*U. parvifolia* (U. chinensis)                     China,
—**Chinese Elm**   E.H.                            Japan
6–10 m    650 mm
Ornamental, autumn foliage

Spreading, rounded crown, with pendulous branchlets. Often evergreen in cooler districts. Very attractive tree, worth planting more frequently. Suckers only slightly from the roots. Withstands slight coastal exposure. Leaves 4–5 cm long, but smaller (5–7 mm long) when young, young foliage 'pinnate' in appearance. 1794.

*U. procera* (U. campestris)—**English Elm**   D.H.   Britain,
                                                      Europe
20–25 m    550 mm
Gully erosion

Fine, impressive tree with a short trunk and heavy, steeply ascending branches giving it a slightly inverted cone shape. Excellent avenue tree where space is not limiting. Formerly widely planted but as it has a vigorous root system which extends over a wide area and suckers prolifically it has fallen into disfavour. Root suckers worse on light than heavy soils. Should only be planted in large open areas (parks) where suckers can be controlled by cultivation or mowing; or plants grafted onto the non-suckering *U. glabra* root-stock should be used. Bark corky; leaves 4–7 cm long, uneven at the base; fruit 1 cm diam. L.C.

*Ulmus procera in autumn*

*U. procera* var. *van houttei*—**Golden Elm**  D.H.  Cultivar
7–10 m    650 mm

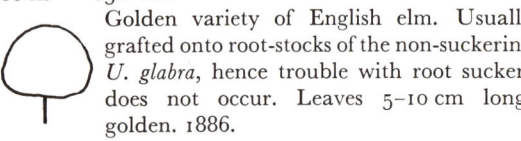

Golden variety of English elm. Usually grafted onto root-stocks of the non-suckering *U. glabra*, hence trouble with root suckers does not occur. Leaves 5–10 cm long, golden. 1886.

*U. pumila*—**Siberian Elm**  D.H.    Siberia, Central Asia
10–15 m    400 mm                                    Fast
Windbreaks, shade, roadsides, parks

Attractive, spreading tree with a bushy rounded crown. Occasionally 25 m in height. Frost-hardy. Withstands long dry periods, severe cold, and snow; good for windswept areas (but deciduous). Bark rough, furrowed; leaves oval, 3–7 cm long; fruit circular, 1 cm diam., seed slightly above centre. 1860.

## VIMINARIA

*Vimen* means a twig which alludes to the twiggy nature of its growth. Only one species from Australia is known.

*V. juncea* (V. denudata)                       W.A., S.A., Vic.,
  —**Golden Spray**  E.H.    Tas., N.S.W., Qld
Seed, cuttings
3–5 m    600 mm    Will grow in          Moderate
                            swampy sites
Parks, roadsides, ornamental

Narrow, upright, rather open, large shrub or small tree with pendulous, apparently leafless, wiry branchlets. Hardy. Frost-resistant. Easily grown. Trimming desirable to maintain shape. 'Leaves' 8–20 cm long, like a piece of wire (actually an elongated leaf-stalk, the true leaves, about 5–7 mm long, are rarely seen); flowers golden with a red-brown base, pea-shaped, in sprays 60–90 cm long, very attractive and conspicuous; fruit a small pod, 2–3 cm long, containing one hard seed. (Seed needs to be treated like acacia seeds before sowing.) 1780.

## VIRGILIA—Virgilia  E.H.        S. Africa

Seed
        650 mm    Prefers moist, well-    Very fast
                        drained soils
Parks, roadsides, ornamental

Genus named after Virgil, the Latin poet. There are two species from South Africa, although some authorities regard the species *V. divaricata* as only a provenance form of *V. oroboides*.

These compact, round-crowned, small trees are amongst the fastest growing of small trees but may be correspondingly short-lived. The flowers are conspicuous, colourful, pink to purplish-pink and pea-shaped. Although slightly frost-sensitive when young, they are otherwise hardy and easily grown. Withstand slight coastal exposure. Can be trimmed, should preferably be cut back fairly heavily every few years to retain shape and vigour. Leaves pinnate; flowers in pendulous clusters 8–15 cm long.

*V. divaricata*—**Pink Virgilia**
3–4 m

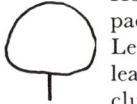

Horizontal branches. Smaller, more compact, and longer-lived than *V. oroboides*. Leaves 10–20 cm long, usually 5–10 pairs leaflets but may be more; flowers pink, in clusters, 7–12 cm long.

*V. oroboides* (V. capensis)—**Cape Virgilia**
3–6 m

Slightly ascending branches and short bole, becoming straggly after a few years if not trimmed. Not wind-firm in exposed positions. Very fast-growing, particularly in light soils. Leaves 10–12 cm long, 10–40 pairs leaflets; flowers pinkish to purplish-white, 7–10 in clusters, 8–15 cm long. 1767.

## ZELKOVA

The name is from *Zelkova*, the Caucasian name. Five species from the eastern Mediterranean, China and Japan are known.

*Ulmus procera avenue*

*Z. carpinifolia*—**Zelkova**   D.H.                    Caucasus
Seed

12–20 m    650 mm                              Moderate
Windbreaks, shade, roadsides, parks

Erect, elm-like tree with a short trunk and several ascending branches, outer branches semi-pendant in older trees. More suitable for the cooler districts. Hardy, easily grown. Large specimens can be transplanted when leafless. Can be pruned heavily and pollarded. Bark smooth grey; leaves oval, 5–10 cm long, 1–3 cm wide, coarsely toothed, similar to English elm but narrower and margins even at the base, autumn foliage distinctive orange-brown; flowers inconspicuous; fruit 5–9 mm across, not winged. 1760.

*Z. serrata* (Z. acuminata)—**Japanese Zelkova**   D.H.
                                                      Japan

Very similar to *Z. carpinifolia* except the leaves are doubly toothed. 1860.
These two Zelkovas are very similar in appearance and growth characteristics to the English Elm but do not sucker from the roots.

# PART III

# REFERENCE LISTS AND

# TABLES

# SYNONYMS

(Corrected to July, 1973 by National Herbarium, Melbourne)

| SYNONYM | CORRECT NAME | SYNONYM | CORRECT NAME |
|---|---|---|---|
| Abies nobilis | *Abies procera* | Crataegus cordata | *Crataegus phaenopyrum* |
| Acacia accola | *Acacia adunca* | crenulata | *Pyracantha crenulata* |
| discolor | *botrycephala* | oxycantha | *Crataegus monogyna* |
| elata | *terminalis* | mexicana | *pubescens* |
| hunteriana | *boormanii* | stipulacea | *pubescens* |
| mollissima | *mearnsii* | Cupressus arizonica | *Cupressus glabra* |
| normalis | *decurrens* | lambertiana var. | *macrocarpa* var. *lambertiana* |
| obliqua | *rotundifolia* | horizontalis | |
| Acer polymorphum | *Acer palmatum* | Cytisus proliferus | *Chamaecytisus proliferus* |
| trifidium | *buergerianum* | Dodonaea attenuata | *Dodonaea angustissima* |
| Acmena smithii | *Eugenia smithii* | Erythrina indica | *Erythrina variegata* |
| Agathis brownii | *Agathis robusta* | Eucalyptus andreana | *Eucalyptus elata* |
| Ailanthus glandulosa | *Ailanthus altissima* | bicolor | *largiflorens* |
| Alnus cordifolia | *Alnus cordata* | bicostata | *st johnii* |
| Angophora intermedia | *Angophora floribunda* | capitellata | *baxteri* |
| lanceolata | *costata* | corymbosa | *gummifera* |
| Araucaria excelsa | *Araucaria heterophylla* | fruticetorum | *polybractea* |
| imbricata | *araucana* | gigantea | *delegatensis* |
| Banksia collina | *Banksia spinulosa* | hemiphloia | *microcarpa* |
| Benthamia fragifera | *Cornus capitata* | leptophylla | *foecunda* |
| Betula alba | *Betula pendula* | lindleyana | *elata* |
| verrucosa | *pendula* | maculosa | *mannifera* |
| Callistemon coccineus | *Callistemon macropunctatus* | numerosa | *elata* |
| lanceolata | *citrinus* | niphophila | *pauciflora, 'Alpine form'* |
| rugulosus | *macropunctatus* | pauciflora var. alpina | *pauciflora, 'Alpine form'* |
| violaceus | *lilacinus* | racemosa | *crebra* |
| Callitris arenosa | *Callitris columellaris* | rostrata | *camaldulensis* |
| calcarata | *endlicheri* | redunca var. elata | *wandoo* |
| cupressiformis | *rhomboidea* | Eugenia myrtifolia | *Eugenia australis* |
| glauca | *columellaris* | paniculata | *australis* |
| gracilis | *preissii* | Fraxinus oxyphylla | *Fraxinus oxycarpa* |
| hugelii | *preissii* | pennsylvanica | *lanceolata* |
| intratropica | *columellaris* | Fugosia patersonii | *Lagunaria patersonii* |
| propinqua | *preissii* | Hakea flexilis | *Hakea muellerana* |
| robusta | *columellaris* | saligna | *salicifolia* |
| robusta | *preissii* | Halesia tetraptera | *Halesia carolina* |
| tasmanica | *rhomboidea* | Heyderia decurrens | *Calocedrus decurrens* |
| Cassia candolleana | *Cassia coluteoides* | Jacaranda mimosaefolia | *Jacaranda acutifolia* |
| eremophila | *nemophila* | Kunzea peduncularis | *Leptospermum phylicoides* |
| floribunda | *corymbosa* | Laburnum vulgare | *Laburnum anagyroides* |
| sturtii | *nemophila* var. *coriacea* | watereri | x *vossii* |
| Casuarina lepidophloia | *Casuarina cristata* | Leptospermum ericoides | *Leptospermum phylicoides* |
| quadrivalis | *stricta* | pubescens | *lanigerum* |
| suberosa | *littoralis* | Libocedrus decurrens | *Calocedrus decurrens* |
| Combretum salicifolium | *Combretum caffrum* | Maclura aurantiaca | *Maclura pomifera* |
| Coprosma baueri | *Coprosma repens* | Melaleuca laxiflora | *Melaleuca crassifolia* |
| retusa | *repens* | leucadendron | *quinquenervia* |
| Cotinus americana | *Cotinus obovatus* | macronycha | *longicoma* |

| SYNONYM | CORRECT NAME |
|---|---|
| pubescens | *lanceolata* |
| Metrosideros tomentosa | *Metrosideros excelsa* |
| Paulownia imperialis | *Paulownia tomentosa* |
| Picea excelsa | *Picea abies* |
| Pinus brutia | *Pinus halepensis* var. *brutia* |
| Pittosporum nigricans | *Pittosporum tenuifolium* |
| nigrescens | *tenuifolium* |
| Podocarpus alpina | *Podocarpus lawrencei* |
| Populus balsamifera var. candicans | *Populus candicans* |
| bolleana | *alba* |
| canadensis var. aurea | *serotina* var. *aurea* |
| dilatata | *Populus nigra* var. *italica* |
| fastigata | *nigra* var. *italica* |
| macrophylla | *candicans* |
| monilifera | *deltoides* |
| pyramidalis | *nigra* var. *italica* |
| Pseudotsuga douglasii | *Pseudotsuga menziesii* |
| taxifolia | ,,        ,, |
| Pyrus aucuparia | *Sorbus aucuparia* |
| Quercus lusitanica | *Quercus faginea* |
| pedunculata | *robur* |
| rubra | *borealis* |
| virens | *virginiana* |
| Rhus cotinoides | *Cotinus obovatus* |
| Salix chilensis | *Salix humboltiana* |
| vitellina | *alba* subsp. *vitellina* |
| Sequoia gigantea | *Sequoiadendron giganteum* |
| Sterculia acerifolia | *Brachychiton acerifolius* |
| diversifolia | *populneus* |
| Syncarpia laurifolia | *Syncarpia glomulifera* |
| Tamarix articulata | *Tamarix aphylla* |
| japonica | *juniperiana* |
| Tilia europaea | *Tilia x vulgaris* |
| Ulmus campestris | *Ulmus procera* |
| chinensis | *parvifolia* |
| montana | *glabra* |
| Viminaria denudata | *Viminaria juncea* |
| Virgilia capensis | *Virgilia orotoides* |
| Zelkova acuminata | *Zelkova serrata* |

# COMMON NAMES BOTANICAL NAMES

# GROUPINGS

## NATIVE—INTRODUCED

### NATIVE

*Acacia*
*Actinostrobus*
*Agathis robusta*
*Agonis*
*Albizia lophantha*
*Alyxia*
*Angophora*
*Araucaria bidwillii*
  *cunninghamii*
*Atriplex*
*Baeckea*
*Banksia*
*Brachychiton*
*Bursaria*
*Callistemon*
*Callitris*

*Calothamnus*
*Capparis*
*Cassia*
*Castanospermum*
*Casuarina*
*Ceratopetalum*
*Chamaelaucium*
*Codonocarpus*
*Crotalaria*
*Dodonaea*
*Dryandra*
*Elaeocarpus*
*Eremophila*
*Eucalyptus*
*Eucryphia lucida*
*Eugenia*

*Ficus*
*Flindersia*
*Geijera*
*Grevillea*
*Hakea*
*Heterodendrum*
*Homalanthus*
*Hymenosporum*
*Leptospermum*
*Leucopogon*
*Melaleuca*
*Melia*
*Myoporum*
*Nothofagus*
  *cunninghamiana*
*Persoonia*

*Phebalium*
*Pittosporum bicolor*
  *phillyreoides*
  *undulatum*
*Podocarpus elatus*
  *lawrencei*
*Prostanthera*
*Stenocarpus*
*Syncarpia*
*Telopea*
*Templetonia*
*Thryptomene*
*Tieghemopanax*
*Tristiania*
*Viminaris juncea*

### INTRODUCED

All genera and species not listed under 'Native' are introduced species.

## CONIFERS—HARDWOODS

### CONIFERS

*Abies*
*Actinostrobus*
*Agathis*
*Araucaria*
*Callitris*
*Calocedrus*
*Cedrus*

*Chamaecyparis*
*Cryptomeria*
*Cupressus*
*Dacrydium*
*Ginkgo*
*Juniperus*
*Larix*

*Metasequoia*
*Picea*
*Pinus*
*Podocarpus*
*Pseudotsuga*
*Sequoia*
*Sequoiadendron*

*Taxodium*
*Taxus*
*Thuja*
*Thujopsis*
*Tsuga*

### HARDWOODS

All genera not listed under 'Conifers' are hardwoods.

# DECIDUOUS—EVERGREEN

DECIDUOUS

(i) Conifers

*Ginkgo*                *Larix*                *Metasequoia*                *Taxodium*

(ii) Hardwoods

| | | | |
|---|---|---|---|
| *Acer* | *Cercidiphyllum* | *Laburnum* | *Pyrus* |
| *Aesculus* | *Combretum* | *Liquidambar* | *Quercus* (except *Q. ilex* |
| *Ailanthus* | *Cornus* | *Liriodendron* | and *Q. virginiana*) |
| *Albizia julibrissin* | (except *C. capitata*) | *Maclura* | *Rhus* |
| *Alnus* | *Cotinus* | *Malus* | *Robinia* |
| (except *A. jorullensis*) | *Crataegus* | *Melia* | *Salix* |
| *Amelanchier* | *Dais* | *Mespilus* | *Sambucus* |
| *Betula* | *Dombeya* | *Morus* | *Sophora japonica* |
| *Brachychiton* | *Erythrina* | *Nyssa* | *Sorbus* |
| (except *B. populneus*) | *Fagus* | *Oxydendrum* | *Stewartia* |
| *Carpinus* | *Fraxinus* | *Parrotia* | *Styrax* |
| *Carya* | *Gleditsia* | *Paulownia* | *Tamarix* |
| *Castanea* | *Halesia* | *Pistacia* | *Tilia* |
| *Catalpa* | *Hovenia* | *Platanus* | *Ulmus* |
| *Cedrela* | *Idesia* | *Populus* | *Zelkova* |
| *Celtis* | *Jacaranda* | *Prunus* | |
| *Cercis* | *Koelreuteria* | (except *P. laurocerasus* | |
| | | and *P. lusitanica*) | |

EVERGREEN

All genera and species not listed under 'Deciduous' are evergreen.

# METRIC CONVERSIONS

## (a) BASIC UNITS

| Metric Name | Metric | Imperial | Imperial | Metric |
|---|---|---|---|---|
| millimetre | 1 mm | 0·039 in | 1 in | 25·4 mm (exactly) |
| centimetre | 1 cm | 0·394 in | 1 in | 2·54 cm (exactly) |
| metre | 1 m | 39·370 in | 1 ft | 30·48 cm (exactly) |
| | 1 m | 3·281 ft | 1 ft | 0·304 8 m (exactly) |
| | 1 m | 1·094 yd | 1 yd | 0·914 4 m (exactly) |
| kilometre | 1 km | 1 094 yd | 1 mile | 1 609 m |
| | 1 km | 0·621 mile | 1 mile | 1·609 km |
| square centimetre | 1 cm² | 0·155 sq in | 1 sq in | 6·451 cm² |
| hectare | 1 ha | 2·471 acres | 1 acre | 0·404 7 ha |
| cubic metre | 1 m³ | 35·315 cu ft | 1 cu ft | 0·025 m³ |
| | 1 m³ | 1·308 cu yd | 1 cu yd | 0·764 m³ |
| millilitre | 1 ml | 0·001 76 pt | 1 pt | 568·3 ml |
| litre | 1 ℓ | 1·76 pt | 1 pt | 0·568 ℓ |
| | 1 ℓ | 0.22 gal | 1 gal | 4·546 ℓ |
| gram | 1 g | 0·035 oz | 1 oz | 28·350 g |
| kilogram | 1 kg | 2·205 lb | 1 lb | 0·454 kg |
| tonne | 1 t | 2 204·6 lb | 1 ton | 1 016 kg |
| | 1 t | 0·984 ton | 1 ton | 1·016 t |
| millimetre (rainfall) | 1 mm | 3·9 points | 1 point | 0·254 mm |
| | | | 1 inch | 254 mm |

$$\text{micron} \quad = 1\ \mu \quad = \frac{1}{1\ 000}\ \text{mm}$$

$$\text{millimicron} = 1\ m\mu = \frac{1}{1\ 000}\ \mu$$

$$= \frac{1}{1\ 000\ 000}\ \text{mm}$$

ppm  = parts per million

## (b) GENERAL

| millimetres (mm) | inches (nearest $\frac{1}{64}$) (in) | centimetres (cm) | inches (nearest $\frac{1}{16}$) (in) | centimetres (cm) | inches (nearest $\frac{1}{8}$) (in) |
|---|---|---|---|---|---|
| 1 | $\frac{3}{64}$ | 1 | $\frac{3}{8}$ | 12 | $4\frac{3}{4}$ |
| 2 | $\frac{5}{64}$ | 2 | $\frac{13}{16}$ | 15 | $5\frac{7}{8}$ |
| 3 | $\frac{1}{8}$ | 3 | $1\frac{3}{16}$ | 18 | $7\frac{1}{8}$ |
| 4 | $\frac{5}{32}$ | 4 | $1\frac{9}{16}$ | 20 | $7\frac{7}{8}$ |
| 5 | $\frac{3}{16}$ | 5 | 2 | 25 | $9\frac{7}{8}$ |
| 6 | $\frac{15}{64}$ | 6 | $2\frac{3}{8}$ | 30 | $11\frac{7}{8}$ |
| 7 | $\frac{9}{32}$ | 7 | $2\frac{3}{4}$ | 35 | $13\frac{3}{4}$ |
| 8 | $\frac{5}{16}$ | 8 | $3\frac{1}{8}$ | 40 | $15\frac{3}{4}$ |
| 9 | $\frac{11}{32}$ | 9 | $3\frac{9}{16}$ | 45 | $17\frac{3}{4}$ |
| 10 | $\frac{25}{64}$ | 10 | $3\frac{15}{16}$ | 50 | $19\frac{5}{8}$ |

| metres (m) | feet (nearest $\frac{1}{4}$ ft) (ft) | metres (m) | feet (nearest ft) (ft) | metres (m) | feet (nearest ft) (ft) |
|---|---|---|---|---|---|
| 1 | $3\frac{1}{4}$ | 12 | 39 | 60 | 197 |
| 2 | $6\frac{1}{2}$ | 15 | 49 | 70 | 230 |
| 3 | $9\frac{3}{4}$ | 18 | 59 | 80 | 262 |
| 4 | 13 | 20 | 66 | 90 | 295 |
| 5 | $16\frac{1}{2}$ | 25 | 82 | 100 | 328 |
| 6 | $19\frac{3}{4}$ | 30 | 98 | | |
| 7 | 23 | 35 | 115 | | |
| 8 | $26\frac{1}{4}$ | 40 | 131 | | |
| 9 | $29\frac{1}{2}$ | 45 | 148 | | |
| 10 | $32\frac{3}{4}$ | 50 | 164 | | |

(c) RAINFALL

| millimetres (mm) | points (pt) | millimetres (mm) | inches (in) |
|---|---|---|---|
| 150 | 590 | 750 | 29·53 |
| 175 | 689 | 800 | 31·50 |
| 200 | 787 | 900 | 35·43 |
| 225 | 886 | 1 000 | 39·37 |
| 250 | 984 | 1 100 | 43·31 |
| 280 | 1 102 | 1 200 | 47·24 |
| 300 | 1 181 | 1 300 | 51·18 |
| 350 | 1 378 | 1 400 | 55·18 |
| 400 | 1 575 | 1 500 | 59·05 |
| 450 | 1 772 | 1 750 | 66·93 |
| 500 | 1 968 | 2 000 | 78·74 |
| 550 | 2 165 | 2 250 | 88·58 |
| 600 | 2 362 | 2 500 | 98·42 |
| 650 | 2 559 | 2 750 | 108·27 |
| 700 | 2 756 | 3 000 | 118·11 |

# CHARACTERISTICS

| SPECIES | Minimum rainfall zone (mm) | | | | | | Height class (m) | | | | | Suitable for | | | | | | | | | |
|---|---|---|---|---|---|---|---|---|---|---|---|---|---|---|---|---|---|---|---|---|---|
| | <300 | 300–400 | 400–500 | 500–650 | 650–900 | >900 | <3 | 3–5 | 5–8 | 8–12 | >12 | Light soils | Heavy soils | Limestone, alkaline | Poorly-drained | Saline | Stony | Coastal | Snow | Smog | Flooding |
| Abies alba | | | | | | • | | | | | • | | | | | | | | • | | |
| A. cephalonica | | | | • | | | | | | | • | | | | | | | | | | |
| A. concolor | | | | | • | • | | | | | • | | | | | | | | • | | |
| A. magnifica | | | | | | • | | | | | • | | | | | | | | • | | |
| A. nordmanniana | | | | | • | • | | | | | • | | | | | | | | | | |
| A. pindrow | | | | | | • | | | | | • | | | | | | | | • | | |
| A. pinsapo | | | | | • | | | | | • | | | | • | | | | | | | |
| A. procera | | | | | | • | | | | | • | | | | | | | | • | | |
| A. sibirica | | | | | | • | | | | | • | | | | | | | | • | | |
| Acacia acinacea | | • | | | | | • | | | | | • | • | | • | | | | | | |
| A. acuminata | • | | | | | | | • | | | | | • | • | • | | | | | | |
| A. adunca | | | • | | | | | | • | | | | | | | | | | | | |
| A. amoena | | | • | | | | • | | | | | | | | | | | | | | |
| A. aneura | • | | | | | | | • | | | | • | • | • | | | | | | | |
| A. baileyana | | | • | | | | | | • | | | • | • | | • | | | • | | | |
| A. boormanii | | | | | • | | | • | | | | | | | | | | | | | |
| A. botrycephala | | | | | • | | | • | | | | | | | | | | | | | |
| A. brachybotrya | • | | | | | | • | | | | | • | • | • | • | | | | | | |
| A. buxifolia | | | | • | | | • | | | | | | | | | | | | | | |
| A. calamifolia | | • | | | | | • | | | | | • | • | • | • | | | | | | |

# CHARACTERISTICS

| SPECIES | Minimum rainfall zone (mm) | | | | | | Height class (m) | | | | | Suitable for | | | | | | | | | |
|---|---|---|---|---|---|---|---|---|---|---|---|---|---|---|---|---|---|---|---|---|---|
| | <300 | 300–400 | 400–500 | 500–650 | 650–900 | >900 | <3 | 3–5 | 5–8 | 8–12 | >12 | Light soils | Heavy soils | Limestone, alkaline | Poorly-drained | Saline | Stony | Coastal | Snow | Smog | Flooding |
| A. cambagei | | • | | | | | | | • | | | • | • | | • | | | | | | |
| A. cardiophylla | | | • | | | | | • | | | | • | • | | | | | | | | |
| A. cognata | | | | | • | | | • | | | | | | | | | | | | | |
| A. colletioides | | • | | | | | • | | | | | | | | | | | | | | |
| A. conferta | | | • | | | | • | | | | | | | | | | | | | | |
| A. cultriformis | | | • | | | | • | | | | | | | | | | | | | | |
| A. cyanophylla | | • | | | | | | • | | | | • | | • | | • | | • | | | |
| A. cyclops | | • | | | | | • | | | | | • | • | • | | • | | • | | | |
| A. dealbata | | | | | • | | | | | | • | | | | | | | | • | | |
| A. deanei | | | • | | | | | | • | | | • | | | | | | | | | |
| A. decora | | | • | | | | • | | | | | • | | | | | • | | | | |
| A. decurrens | | | | | • | | | | | • | | • | | | | | | | | | |
| A. doratoxylon | • | | | | | | | | • | | | | | | | | • | | | | |
| A. drummondii | | | | • | | | • | | | | | • | | | | | | | | | |
| A. falciformis | | | | | • | | | | | • | | • | | • | | | | | | | |
| A. farinosa | | • | | | | | • | | | | | | | | | | | | | | |
| A. farnesiana | | • | | | | | • | | | | | | | | | | | | | | |
| A. fimbriata | | | | | • | | | • | | | | | | | | | | | | | |
| A. floribunda | | | | | • | | | • | | | | • | | | • | | • | • | | • | |
| A. gladiiformis | • | | | | | | • | | | | | | | | | | | | | | |

279

# CHARACTERISTICS

| SPECIES | Minimum rainfall zone (mm) | | | | | | Height class (m) | | | | | Suitable for | | | | | | | | | |
|---|---|---|---|---|---|---|---|---|---|---|---|---|---|---|---|---|---|---|---|---|---|
| | < 300 | 300–400 | 400–500 | 500–650 | 650–900 | > 900 | < 3 | 3–5 | 5–8 | 8–12 | > 12 | Light soils | Heavy soils | Limestone, alkaline | Poorly-drained | Saline | Stony | Coastal | Snow | Smog | Flooding |
| A. glandulicarpa | • | | | | | | • | | | | | • | | • | | | | | | | |
| A. glaucescens | | | | | • | | | | | • | | | | | | | | • | | | |
| A. gracifolia | • | | | | | | • | | | | | | | | | | | | | | |
| A. hakeoides | • | | | | | | | • | | | | • | | • | | | | | | | |
| A. harpophylla | | • | | | | | | | | • | | | • | | | | | | | | |
| A. homalophylla | • | | | | | | | | | • | | | • | | | | | | | | |
| A. howittii | | | | | • | | | | • | | | | • | | | | | | | | |
| A. implexa | | | | • | | | | | | • | | | | | | | | | | | |
| A. inophloia | | • | | | | | • | | | | | | | | | | | | | | |
| A. iteaphylla | | | • | | | | | • | | | | • | • | | | • | | | | | |
| A. jonesii | | | • | | | | | | • | | | | • | | | | | | | | |
| A. leprosa | | | • | | | | | • | | | | | | | | | | | | | |
| A. ligulata | • | | | | | | • | | | | | • | • | • | | • | | • | | | |
| A. linifolia | | | | | • | | | • | | | | | | | | | | • | | | |
| A. loderi | | • | | | | | | | • | | | • | | • | | • | | | | | |
| A. longifolia | | | | | • | | | • | | | | • | | | • | | | • | | • | |
| A. longifolia var. sophorae | | | • | | | | | • | | | | • | | | • | | | • | | • | |
| A. mearnsii | | | | • | | | | | | | • | • | | | | | | | | | |
| A. melanoxylon | | | | | • | | | | | | • | • | • | | | | | | • | • | • |
| A. microbotrya | | • | | | | | | • | | | | • | | • | | | | | | | |

# CHARACTERISTICS

| SPECIES | Minimum rainfall zone (mm) | | | | | | Height class (m) | | | | | Suitable for | | | | | | | | | |
|---|---|---|---|---|---|---|---|---|---|---|---|---|---|---|---|---|---|---|---|---|---|
| | <300 | 300–400 | 400–500 | 500–650 | 650–900 | >900 | <3 | 3–5 | 5–8 | 8–12 | >12 | Light soils | Heavy soils | Limestone, alkaline | Poorly-drained | Saline | Stony | Coastal | Snow | Smog | Flooding |
| A. montana | | • | | | | | • | | | | | • | | • | | | | | | | |
| A. myrtifolia | | | • | | | | • | | | | | | | | | | | | | | |
| A. notabilis | • | | | | | | • | | | | | • | • | • | | | | | | | |
| A. obliquinervia | | | | | • | | | | • | | | | | | | | | | • | | |
| A. oswaldii | | • | | | | | | | • | | | • | | • | | • | | | | | |
| A. pendula | | • | | | | | | | • | | | | • | • | | | | | | | |
| A. penninervis | | | | | • | | | | | • | | | | | | | | | • | | |
| A. podalyriifolia | | | | | • | | | • | | • | | | • | | | | | | | | |
| A. pravissima | | | | | • | | | | • | | | • | | | | | | • | • | | • |
| A. prominens | | | | • | | | | | • | | | | | | | | | • | | | |
| A. pruinosa | | | | | • | | | | • | | | | | | | | | • | | • | |
| A. pubescens | | | | • | | | | • | | | | | | | | | | | | | |
| A. pycnantha | | | • | | | | | • | | | | • | • | | | | • | • | | • | |
| A. retinodes | | | • | | | | | | • | | | • | • | | • | • | • | • | | | • |
| A. riceana | | | | | | | • | | | | | • | • | | | | | | | | |
| A. rigens | | • | | | | | • | | | | | | | | | | | | | | |
| A. rotundifolia | | • | | | | | • | | | | | | | | | | | | | | |
| A. rubida | | | | | • | | | • | • | • | | | | | | | | | | | |
| A. salicina | | • | | | | | | | • | | | • | • | • | | | | • | | | |
| A. saligna | | • | | | | | | • | • | | | • | • | • | • | | • | • | | • | |

# CHARACTERISTICS

| SPECIES | Minimum rainfall zone (mm) | | | | | | Height class (m) | | | | | Suitable for | | | | | | | | | |
|---|---|---|---|---|---|---|---|---|---|---|---|---|---|---|---|---|---|---|---|---|---|
| | <300 | 300–400 | 400–500 | 500–650 | 650–900 | >900 | <3 | 3–5 | 5–8 | 8–12 | >12 | Light soils | Heavy soils | Limestone, alkaline | Poorly-drained | Saline | Stony | Coastal | Snow | Smog | Flooding |
| A. sclerosperma | • | | | | | | | | • | | | | | | | | | • | | | |
| A. silvestris | | | | | • | | | | | | • | | | | | | • | | • | | |
| A. sowdenii | | • | | | | | | | • | | | • | | • | | | | | | | |
| A. spectabilis | | | • | | | | | • | | | | | | | | | | | | | |
| A. stenophylla | • | | | | | | | | • | | | | • | | | • | | | | | |
| A. stricta | | | | | • | | • | • | | | | | • | | • | | | | | | |
| A. suaveolens | | | | | • | | • | | | | | • | | | | | | | | | |
| A. subporosa | | | | | • | | | | | • | | | • | | | | | | | | • |
| A. subulata | | • | | | | | • | | | | | • | | | | | | | | | |
| A. terminalis | | | | | • | | | | | | • | • | | • | | | | • | | | |
| A. trineura | | • | | | | | • | • | | | | • | | | | | | | | | |
| A. triptera | | | • | | | | • | • | | | | | | | | | | | | • | |
| A. undulifolia | | | • | | | | | • | | | | | • | | | | • | | | | |
| A. verniciflua | | | • | | | | | • | | | | | • | | | | • | | | | |
| A. verticillata | | | | | • | | | • | | | | | | | • | | | | | | |
| A. vestita | | | • | | | | • | | | | | | | | | | | | | | |
| A. victoriae | | • | | | | | • | • | | | | • | | • | | | • | | | | |
| A. wattsiana | | • | | | | | • | | | | | | • | • | | | | | | | |
| Acer buergerianum | | | | | • | | | • | | | | | | | | | | | | | |
| A. campestre | | | | | • | | | | | • | | | | | | | | | | | |

# CHARACTERISTICS

| SPECIES | Minimum rainfall zone (mm) | | | | | | Height class (m) | | | | | Suitable for | | | | | | | | | |
|---|---|---|---|---|---|---|---|---|---|---|---|---|---|---|---|---|---|---|---|---|---|
| | <300 | 300–400 | 400–500 | 500–650 | 650–900 | >900 | <3 | 3–5 | 5–8 | 8–12 | >12 | Light soils | Heavy soils | Limestone, alkaline | Poorly-drained | Saline | Stony | Coastal | Snow | Smog | Flooding |
| A. cappadocicum | | | | | | ● | | | | ● | ● | | | | | | | | | | |
| A. ginnala | | | | | ● | | | | ● | | | | | | | | | | ● | | |
| A. grosseri | | | | | ● | | | ● | ● | | | | | | | | | | | | |
| A. negundo | | | | ● | | | | | | ● | | | | | | | | | ● | ● | |
| A. palmatum | | | | | ● | | | ● | ● | ● | | | | | | | | | | | |
| A. platanoides | | | | | ● | | | | | | ● | | | | | | | ● | ● | ● | |
| A. pseudoplatanus | | | | | ● | | | | | | ● | | ● | | | | | | ● | ● | |
| A. rubrum | | | | | ● | | | | | | ● | | ● | | ● | | | ● | ● | ● | ● |
| A. saccharinum | | | | | ● | | | | | | ● | | | | | | | | ● | | |
| A. saccharum | | | | | | ● | | | | | ● | | | | | | | | ● | | |
| A. tegmentosum | | | | | | ● | | | ● | | | | | | | | | | ● | | |
| Actinostrobus pyramidalis | | | ● | | | | ● | | | | | | | | ● | ● | | ● | | | |
| Aesculus hippocastanum | | | | | ● | | | | | ● | | | | | | | | | | | |
| A. × carnea | | | | | ● | | | | | ● | | | | | | | | | | | |
| A. indica | | | | | ● | | | | | | ● | | | | | | | | | | |
| Agathis australis | | | | | ● | | | | | | ● | ● | | | ● | | | | | | |
| A. robusta | | | | | ● | | | | | | ● | | | | | | | | | | |
| Agonis flexuosa | | | | ● | | | | | ● | | | ● | ● | ● | ● | | | | | | |
| A. juniperina | | | ● | | | | | ● | | | | | | | | | | ● | | | |
| Ailanthus altissima | | | | | ● | | | | | | ● | ● | | | | | | | | ● | ● |

# CHARACTERISTICS

| SPECIES | Minimum rainfall zone (mm) | | | | | | Height class (m) | | | | | Suitable for | | | | | | | | | |
|---|---|---|---|---|---|---|---|---|---|---|---|---|---|---|---|---|---|---|---|---|---|
| | <300 | 300–400 | 400–500 | 500–650 | 650–900 | >900 | <3 | 3–5 | 5–8 | 8–12 | >12 | Light soils | Heavy soils | Limestone, alkaline | Poorly-drained | Saline | Stony | Coastal | Snow | Smog | Flooding |
| Albizia julibrissin | | | | • | | | | | • | | | • | | | | | | | | | |
| A. lophantha | | | | • | | | | | • | | | • | | | | | | • | | | |
| Alnus cordata | | | | • | | | | | | • | | | | | • | | | | • | | • |
| A. glutinosa | | | | | • | | | | | • | | | | | • | | | | • | | • |
| A. jorullensis | | | | | • | | | | | • | | | | | • | | | | | | |
| Alyxia buxifolia | | | | • | | | • | | | | | • | | | | | | • | | | |
| Amelanchier canadensis | | | | | • | | | | • | | | | | | | | | | | | |
| Angophora costata | | | | | • | | | | | | • | • | • | | | | • | • | | • | |
| A. floribunda | | | | | • | | | | | • | | • | • | | | | • | • | | | |
| Araucaria araucana | | | | | • | | | | | | • | | | | | | | | | | |
| A. bidwillii | | | | | • | | | | | | • | | | | | | | • | | | |
| A. cunninghamii | | | | • | | | | | | | • | | | | | | | • | | • | |
| A. heterophylla | | | | • | | | | | | | • | • | • | | | | | • | | | |
| Arbutus menziesii | | | | | • | | | | | • | | | | | | | | | | | |
| A. unedo | | | | | • | | | | • | | | | | • | | | | • | | | |
| Atriplex nummularia | | • | | | | | • | | | | | • | | | | • | | | | | |
| Azara microphylla | | | | | • | | | | • | | | | | | | | | | | | |
| Baeckea virgata | | | | | • | | • | | | | | • | | | | | | | | | |
| Banksia ashbyii | • | | | | | | • | | | | | • | | | | | | | | | |
| B. attenuata | | | | | • | | | • | | | | | | | | | | | | | |

# CHARACTERISTICS

| Species | Minimum rainfall zone (mm) | | | | | | Height class (m) | | | | | Suitable for | | | | | | | | | |
|---|---|---|---|---|---|---|---|---|---|---|---|---|---|---|---|---|---|---|---|---|---|
| | <300 | 300–400 | 400–500 | 500–650 | 650–900 | >900 | <3 | 3–5 | 5–8 | 8–12 | >12 | Light soils | Heavy soils | Limestone, alkaline | Saline | Poorly-drained | Stony | Coastal | Snow | Smog | Flooding |
| B. ericifolia | | | • | | | | • | | | | | • | | | | • | | • | | | |
| B. grandis | | | • | | | | | | • | | | • | | | | | | • | | | |
| B. ilicifolia | | | | | • | | | | • | | | • | | | | | | • | | | |
| B. integrifolia | | | | • | | | | | | • | | • | | | | | | • | | | |
| B. littoralis | | | | • | | | | | | • | | | | | | • | | | | | • |
| B. marginata | | | | • | | | | | • | | | • | | | | | | • | | | |
| B. menziesii | | | • | | | | | | • | | | • | | | | | | • | | | |
| B. ornata | | • | | | | | • | | | | | • | | | | | | | | | |
| B. prionotes | | | • | | | | | | • | | | | • | | | | | | | | |
| B. serrata | | | • | | | | | | • | | | • | | | | | | • | | | |
| B. spinulosa | | | | | • | | | | | | • | • | | | | • | | | | | • |
| Betula lutea | | | | | • | | | | | | • | | | | | • | | | | | • |
| B. maximowiczii | | | | | • | | | | | • | | | | | | • | • | | | | |
| B. papyrifera | | | | | • | | | | | • | | | | | | • | | | | | • |
| B. pendula | | | | | • | | | | | • | | | | | | • | | | | | |
| Brachychiton acerifolius | | | | | | | | | | • | | | • | | | | | | | | |
| B. discolor | | | | • | | | | | | • | • | • | • | | | | | | | | |
| B. populneus | | • | | | | | | | | • | • | • | | • | | | | | | • | |
| Bursaria spinosa | | | | • | | | | • | • | | | • | • | • | | | • | • | | | • |
| Callistemon acuminatus | | | | | • | | • | • | | | | | | | | | | • | | | |

# CHARACTERISTICS

| SPECIES | Min. rainfall zone (mm) <300 | 300–400 | 400–500 | 500–650 | 650–900 | >900 | Height class (m) <3 | 3–5 | 5–8 | 8–12 | >12 | Suitable for: Light soils | Heavy soils | Limestone, alkaline | Poorly-drained | Saline | Stony | Coastal | Snow | Smog | Flooding |
|---|---|---|---|---|---|---|---|---|---|---|---|---|---|---|---|---|---|---|---|---|---|
| C. citrinus | | | | • | | | • | | | | | • | • | | • | | | • | | | |
| C. lilacinus | | | • | | | | • | | | | | | • | | | | | • | | | |
| C. linearis | | | • | | | | • | | | | | • | | | | | | | | | |
| C. macropunctatus | | | • | | | | | • | | | | • | • | • | • | | | | | | • |
| C. pallidus | | | | | • | | | • | | | | | | | | | | | | | |
| C. paludosus | | | | • | | | | • | | | | | | | | | | | | | • |
| C. phoeniceus | | | • | | | | | • | | | | | • | | • | | | | | | |
| C. pinifolius | | | • | | | | | • | | | | • | | | | | | | | | |
| C. rigidus | | | • | | | | | • | | | | • | • | • | | | | | | | |
| C. salignus | | | | • | | | | | • | | | | • | | • | • | | • | | | • |
| C. shiresii | | | | | • | | | • | | | | | | | | | | | | | |
| C. sieberi | | | | | • | | • | | | | | | | | | | | | | | |
| C. speciosus | | | | • | | | • | | | | | • | | | • | | | • | | | |
| C. teretifolius | | • | | | | | • | | | | | • | | • | | | | | | | |
| C. viminalis | | | | • | | | | | • | • | • | | • | | • | | | • | | | • |
| Callitris columellaris | | • | | | | | | | | • | • | • | | • | | | | • | | | |
| C. endlicheri | | • | | | | | | | | | • | • | | • | | | | | | | |
| C. oblonga | | | | • | | | | • | | | | • | | | | | | | | | |
| C. preissii | | | • | | | | | | | | • | • | | • | | | | | | | |
| C. rhomboidea | | | • | | | | | | • | | | • | | | | | | • | | | |

# CHARACTERISTICS

| SPECIES | \< 300 | 300–400 | 400–500 | 500–650 | 650–900 | \> 900 | \< 3 | 3–5 | 5–8 | 8–12 | \> 12 | Light soils | Heavy soils | Limestone, alkaline | Poorly-drained | Saline | Stony | Coastal | Snow | Smog | Flooding |
|---|---|---|---|---|---|---|---|---|---|---|---|---|---|---|---|---|---|---|---|---|---|
| | Minimum rainfall zone (mm) | | | | | | Height class (m) | | | | | Suitable for | | | | | | | | | |
| C. verrucosa | • | | | | | | | | • | | | • | | | | | | | | | |
| Calocedrus decurrens | | | | | • | | | | | | • | | • | | | | | | • | | |
| Calodendrum capense | | | | • | | | | | | • | | | | | | | | • | | | |
| Calothamnus asper | | | • | | | | • | | | | | • | • | | | | | • | | | |
| C. chrysantherus | | | • | | | | • | | | | | • | • | | | | | • | | | |
| C. gilesii | | | • | | | | • | | | | | • | • | | | | | • | | | |
| C. quadrifidus | | | • | | | | • | | | | | • | • | | • | | | • | | | |
| C. rupestris | | | • | | | | • | | | | | • | • | | | | | • | | | |
| C. sanguineus | | | • | | | | • | | | | | • | • | | | | | • | | | |
| C. schaueri | | | • | | | | • | | | | | • | • | | | | | • | | | |
| C. torulosa | | | • | | | | • | | | | | • | • | | | | | • | | | |
| C. villosus | | | • | | | | • | | | | | • | • | | | | | • | | | |
| Capparis mitchellii | • | | | | | | | • | • | | | | • | | • | | | | | | |
| Carpinus betulus | | | | | | • | | | | | • | | | | | | | | • | | |
| Carya illinoensis | | | | | • | | | | | | • | | | | | | | | | | |
| C. ovata | | | | | • | | | | | | • | | | | | | | | | | |
| Cassia artemisioides | • | | | | | | • | | | | | • | • | • | | | | • | | | |
| C. coluteoides | | | • | | | | • | | | | | • | | • | | | | • | | | |
| C. corymbosa | | | | • | | | • | | | | | • | | • | | | | • | | | |
| C. nemophila | • | | | | | | • | | | | | • | | • | | | | • | | | |

# CHARACTERISTICS

| SPECIES | \<300 | 300–400 | 400–500 | 500–650 | 650–900 | >900 | \<3 | 3–5 | 5–8 | 8–12 | >12 | Light soils | Heavy soils | Limestone, alkaline | Poorly-drained | Saline | Stony | Coastal | Snow | Smog | Flooding |
|---|---|---|---|---|---|---|---|---|---|---|---|---|---|---|---|---|---|---|---|---|---|
| | | | Minimum rainfall zone (mm) | | | | | | Height class (m) | | | | | | | Suitable for | | | | | |
| C. nemophila *var.* coriacea | • | | | | | | • | | | | | • | | • | | | | • | | | |
| Castanea dentata | | | | | • | | | | | | • | • | • | | | | | | | | |
| C. sativa | | | | | • | | | | | | • | | | | | | | | | | |
| Castanospermum australe | | | | | • | | | | | | • | | | | | | | | | | |
| Casuarina cristata | | • | | | | | | | | • | • | • | • | • | | • | | • | | • | |
| C. cunninghamiana | | | • | | | | | | | • | | • | • | • | | • | | • | | • | • |
| C. glauca | | | • | | | | | | | • | | • | • | • | • | • | | • | | • | • |
| C. leuhmannii | | • | | | | | | | | • | | • | • | • | | • | | • | | | |
| C. littoralis | | | | • | | | | | • | | | • | • | • | • | • | • | • | | | |
| C. muellerana | • | | | | | | • | | | | | • | | | | | | | | | |
| C. obesa | | | | • | | | | | | • | | | | | • | • | | • | | | • |
| C. paludosa | | | • | | | | • | • | | | | | | | • | | | | | | • |
| C. stricta | | | • | | | | | | | • | • | • | • | • | • | • | • | • | | • | • |
| C. torulosa | | | | • | | | | | | • | • | • | | | | • | | • | | • | |
| Catalpa bignonioides | | | | | • | | | | | • | • | | | | | | | | | • | |
| C. speciosa | | | | | | • | | | | | | | | | | | | | | | |
| Cedrela sinensis | | | | | • | | | | | • | • | | | | | | | | | | |
| Cedrus atlantica | | | | | • | | | | | • | • | | | • | | | | | | | |
| C. deodara | | | | | • | | | | | • | • | | | | | | | | | | |
| C. libanii | | | | | • | | | | | • | • | | | | | | | | | | |

288

# CHARACTERISTICS

| S P E C I E S | Minimum rainfall zone (mm) | | | | | | Height class (m) | | | | | Suitable for | | | | | | | | | |
|---|---|---|---|---|---|---|---|---|---|---|---|---|---|---|---|---|---|---|---|---|---|
| | < 300 | 300–400 | 400–500 | 500–650 | 650–900 | > 900 | < 3 | 3–5 | 5–8 | 8–12 | > 12 | Light soils | Heavy soils | Limestone, alkaline | Poorly-drained | Saline | Stony | Coastal | Snow | Smog | Flooding |
| Celtis australis | | | | • | | | | | | • | | | | • | • | | | | | | |
| C. occidentalis | | | • | • | | | | | | | • | | | • | | | | | | • | |
| Ceratonia siliqua | | | • | | | | | | | • | • | • | | • | | | | | | • | |
| Ceratopetalum apetalum | | | | | • | | | | • | • | | | • | | | | | • | | | |
| C. gummiferum | | | | | • | | | | • | • | | | | | | | | | | | |
| Cercidiphyllum japonicum | | | | | • | • | | | | • | | | | | | | | | | | |
| C. japonicum var. sinensis | | | | | • | • | | | | • | | | | | | | | | | | |
| Cercis siliquastrum | | | | • | | | | | • | | | | | • | | | | • | | | |
| Chamaecyparis lawsoniana | | | | | • | | | | | | • | | | | | | | | | | |
| C. nootkatensis | | | | | | • | | | | | • | | | | | | | | | | |
| C. obtusa | | | | | • | | | | | | • | | | | | | | | | | |
| C. pisifera | | | | | • | | | | | | • | | | | | | | | | | |
| C. thyoides | | | | | • | | | | | • | | | | | | | | | | | |
| Chamaecytisus proliferus | | | • | | | | | | • | | | | | • | | | | • | | | |
| Chamaelaucium uncinatum | | | • | | | | • | | | | | | | | | | | | | | |
| Cinnamomum camphora | | | | • | | | | | • | • | | | | | | | | | | • | |
| Codonocarpus cotinifolius | • | | | | | | | • | • | | | • | | | | | | | | | |
| Combretum caffrum | | | | • | | | | | • | | | | | | | | | | | | |
| Coprosma repens | | | | • | | | | • | • | | | | | | | | | • | | • | • |
| Cornus capitata | | | | | • | | | | • | • | | | | | | | | | | | |

289

# CHARACTERISTICS

| SPECIES | \<300 | 300–400 | 400–500 | 500–650 | 650–900 | \>900 | \<3 | 3–5 | 5–8 | 8–12 | \>12 | Light soils | Heavy soils | Limestone, alkaline | Poorly-drained | Saline | Stony | Coastal | Snow | Smog | Flooding |
|---|---|---|---|---|---|---|---|---|---|---|---|---|---|---|---|---|---|---|---|---|---|
| | **Minimum rainfall zone (mm)** | | | | | | **Height class (m)** | | | | | **Suitable for** | | | | | | | | | |
| C. florida | | | | | • | | | • | • | | | | | | | | | | | | |
| C. kousa | | | | | • | | | | • | | | | | | | | | | | | |
| C. kousa var. chinensis | | | | | • | | | | • | | | | | | | | | | | | |
| Corynocarpus laevigatus | | | | | • | | | | • | • | | | | | | | | • | | | |
| Cotinus obovatus | | | | | • | | | | • | • | | | | | | | | | | | |
| Crataegus crus-galli | | | | | • | | | • | | | | | | | | | | | | | |
| C. monogyna | | | | | • | | | | • | | | | | | | | | • | • | • | |
| C. phaenopyrum | | | | | • | | | | • | | | | | | | | | • | | | |
| C. pubescens | | | | | • | | | | • | | | | | | | | | • | | | |
| C. tanacetifolia | | | | | • | | | • | • | | | | | | | | | | | | |
| Crotalaria laburnifolia | | | | | • | | • | | | | | | | | | | | | | | |
| Cryptomeria japonica var. elegans | | | | | • | | | | | • | | | | | | | | | | | |
| Cupressus benthamii | | | | | • | | | | | • | • | • | | | | | | | | • | |
| C. funebris | | | | | • | | | | | | • | | • | | | | | | | | |
| C. glabra | | | | • | | | | | | • | • | | | • | | | | • | | | |
| C. lusitanica | | | | • | | | | | | • | | | • | | | | | | | | |
| C. macrocarpa | | | | | • | | | | | • | • | • | • | • | | | | • | • | | |
| C. macrocarpa var. lambertiana | | | | • | | | | | | • | • | • | • | • | | | | • | | | |
| C. sempervirens | | | | • | | | | | | • | • | | | • | | | | | | • | |
| C. sempervirens var. stricta | | | • | | | | | | | | • | | | • | | | | | | • | |

# CHARACTERISTICS

| SPECIES | Minimum rainfall zone (mm) |||||| Height class (m) ||||| Suitable for ||||||||||
|---|---|---|---|---|---|---|---|---|---|---|---|---|---|---|---|---|---|---|---|---|---|
| | <300 | 300–400 | 400–500 | 500–650 | 650–900 | >900 | <3 | 3–5 | 5–8 | 8–12 | >12 | Light soils | Heavy soils | Limestone, alkaline | Poorly-drained | Saline | Stony | Coastal | Snow | Smog | Flooding |
| C. torulosa | | | | | • | | | | | • | | | | | | | | | | | |
| Dacrydium cupressinum | | | | | • | | | | | • | • | | | • | | | | | | | |
| Dais cotinifolia | | | | | • | | | • | | | | | | | | | | | | | |
| Dodonaea angustissima | • | | | | | | • | | | | | • | • | | | | | • | | | |
| D. cuneata | • | | | | | | • | | | | | • | • | | | | | • | | | |
| D. microzyga | • | | | | | | • | | | | | | • | | | | | • | | | |
| D. viscosa | | | • | | | | | • | • | | | | • | | | | | • | | • | |
| D. viscosa var. purpurea | | | • | | | | • | | | | | | • | | | | | • | | • | |
| Dombeya natalensis | | | | • | | | | • | • | | | | | | | | | | | | |
| Dryandra formosa | | | | • | | | • | | | | | • | | • | | | | • | | | |
| D. nobilis | | | • | | | | • | | | | | • | | • | | | | | | | |
| D. sessilis | | | • | | | | | • | | | | • | | • | | | | | | | |
| Elaeagnus angustifolia | | | | • | | | | | • | • | | | | • | | | | | • | | |
| Elaeocarpus dentatus | | | | • | | | | | • | | | | | | | | | | | | |
| E. reticulatus | | | | | • | | | | • | | | | | | | | | | | | |
| Eremophila alternifolia | | • | | | | | • | | | | | • | | • | | | | | | | |
| E. longifolia | | • | | | | | | | • | | | • | • | • | | | | | | | |
| E. maculata | | • | | • | | | | • | | | | • | • | • | | | | | | | |
| Erythrina crista-gallii | | | | | | | | | • | • | | | | | | | | | | | |
| E. variegata | | | | | • | | | | | • | | | | | | | | | | | |

291

# CHARACTERISTICS

| SPECIES | Min. rainfall zone (mm) | | | | | | Height class (m) | | | | | Suitable for | | | | | | | | | |
|---|---|---|---|---|---|---|---|---|---|---|---|---|---|---|---|---|---|---|---|---|---|
| | <300 | 300–400 | 400–500 | 500–650 | 650–900 | >900 | <3 | 3–5 | 5–8 | 8–12 | >12 | Light soils | Heavy soils | Limestone, alkaline | Poorly-drained | Saline | Stony | Coastal | Snow | Smog | Flooding |
| Eucalyptus acaciaeformis | | | | | • | | | | | • | | | • | | | | | | • | | |
| E. accedens | | • | | | | | | | | • | | | | | | | • | | | | |
| E. aggregata | | | | | • | | | | | | • | | | | • | | | | • | | • |
| E. alpina | | | | • | | | | | • | | | | | | • | | | | • | | |
| E. amplifolia | | | | | • | | | | | • | • | | • | | • | | | | | | |
| E. annulata | | | • | | | | | | • | | | | • | | | | | | | | |
| E. approximans | | | | | • | | • | | | | | | | | | | | | | | |
| E. archeri | | | | | • | | | • | | | | | | | • | | | | • | | |
| E. astringens | | | • | | | | | | | | • | | • | • | | • | | | | | |
| E. baxteri | | | | • | | | | | | | • | • | | | | | | | | | |
| E. bosistoana | | | | | • | | | | | | • | | • | | | | | | | | |
| E. botryoides | | | | • | | | | | | | • | • | • | | • | • | | • | | • | • |
| E. brockwayii | | • | | | | | | | | | • | | • | • | | • | | • | | | |
| E. burdettiana | | | | • | | | • | | | | | | | | | | | • | | | |
| E. burracoppinensis | | • | | | | | | | | | | • | | | | | | | | | |
| E. caesia | | | • | | | | | • | | | | • | | | | | | | | | |
| E. calophylla | | | | • | | | | • | • | • | | • | | | | | | • | | • | |
| E. calophylla var. rosea | | | | • | | | | | • | • | | • | | | | | | • | | | |
| E. calycogona | | • | | | | | | | • | | | • | • | • | | | | | | | |
| E. camaldulensis | | | | | | | | | | | • | | • | | • | | | | | | • |

# CHARACTERISTICS

| SPECIES | Minimum rainfall zone (mm) | | | | | | Height class (m) | | | | | Suitable for | | | | | | | | | |
|---|---|---|---|---|---|---|---|---|---|---|---|---|---|---|---|---|---|---|---|---|---|
| | < 300 | 300–400 | 400–500 | 500–650 | 650–900 | > 900 | < 3 | 3–5 | 5–8 | 8–12 | > 12 | Light soils | Heavy soils | Limestone, alkaline | Poorly-drained | Saline | Stony | Coastal | Snow | Smog | Flooding |
| E. campaspe | • | | | | | | | • | • | | | | • | • | | • | | | | | |
| E. chapmaniana | | | | | | • | | | | • | • | | | | | | | | • | | |
| E. cinerea | | | | • | | | | | • | • | | • | | | • | | | | | • | • |
| E. citriodora | | | | • | | | | | • | • | • | | | | | | | • | | • | |
| E. cladocalyx | | | • | | | | | | | | • | • | | • | | | | | | • | |
| E. cladocalyx var. nana | | | • | | | | | • | | | | • | | • | | | | | | • | |
| E. coccifera | | | | | • | | | | | | • | | | | | | | | • | | |
| E. cornuta | | | | • | | | | | • | | | | • | • | | • | | • | | | |
| E. corrugata | | • | | | | | | | | | | • | | | | | • | | | | |
| E. cosmophylla | | | | • | | | | • | | • | • | | • | | • | | | | | | • |
| E. crebra | | | | • | | | | | • | | | | • | | | | | | | | |
| E. crenulata | | | | • | | | | | • | | | • | • | | • | | | • | | • | |
| E. crucis | • | | | | | | | • | | | | | | | | | | • | | | |
| E. dawsonii | | | | • | | | | | • | • | • | • | • | | | | • | | | | |
| E. decipiens | | | | | • | | | | | | | | | • | | | | | • | | |
| E. delegatensis | | | | | | • | | | | | • | | | | | | | | | | |
| E. desmondensis | | | • | | | | | • | | | | • | | | | | | | | | |
| E. diclsii | | • | | | | | | • | | | | • | • | • | | • | | | | | |
| E. diptera | • | | | | | | | | • | | | • | • | • | | • | | • | | | |
| E. diversifolia | | | • | | | | | • | | | | • | | • | | | • | • | | | |

# CHARACTERISTICS

| SPECIES | Min. rainfall <300 | 300–400 | 400–500 | 500–650 | 650–900 | >900 | Height <3 | 3–5 | 5–8 | 8–12 | >12 | Light soils | Heavy soils | Limestone, alkaline | Poorly-drained | Saline | Stony | Coastal | Snow | Smog | Flooding |
|---|---|---|---|---|---|---|---|---|---|---|---|---|---|---|---|---|---|---|---|---|---|
| E. doratoxylon | | | • | | | | | • | | | | • | | | | • | | | | | |
| E. dumosa | | • | | | | | | • | • | | | • | | • | | • | | | | | |
| E. dundasii | | • | | | | | | | | • | • | • | • | • | | | | | | | |
| E. ebbanoensis | • | | | | | | | | | | | • | | | | | | | | | |
| E. elata | | | | | • | | | • | | | • | | | | | | | | | | • |
| E. eremophila | • | | | | | | | | • | | | • | • | • | | • | | • | | | |
| E. erythrocorys | | | | • | | | | | • | | | • | • | • | | • | | • | | | |
| E. erythronema | | • | | | | | | | • | | | • | • | • | | | | | | | |
| E. fasciculosa | | | • | | | | | | • | • | | • | | | | | | | | | |
| E. ficifolia | | | | • | | | | | | • | | • | | | | | | • | | • | |
| E. flocktoniae | • | | | | | | | • | | • | | | | • | | • | • | | | | |
| E. foecunda | • | | | | | | | • | | | | | | | | | | | | | |
| E. forrestiana | | | • | | | | | | • | • | | • | • | • | | | • | • | | | |
| E. gardneri | | • | | | | | | | | | | • | • | • | | | | | | | |
| E. gillii | • | | | | | | | | | | | | • | • | | | | | | | |
| E. glaucescens | | | | | | • | | | | • | | | | | | | | | • | | |
| E. globulus | | | | | • | | | | | | • | | | | • | | | | | • | |
| E. globulus var. compacta | | | | • | | | | • | | | | | | | | | | | | | |
| E. gomphocephala | | | • | | | | | | | | | • | | • | | | | • | | | |
| E. gracilis | • | | | | | | | | • | • | | • | | • | | • | | | | | |

# CHARACTERISTICS

| SPECIES | <300 | 300–400 | 400–500 | 500–650 | 650–900 | >900 | <3 | 3–5 | 5–8 | 8–12 | >12 | Light soils | Heavy soils | Limestone, alkaline | Poorly-drained | Saline | Stony | Coastal | Snow | Smog | Flooding |
|---|---|---|---|---|---|---|---|---|---|---|---|---|---|---|---|---|---|---|---|---|---|
| | **Minimum rainfall zone (mm)** | | | | | | **Height class (m)** | | | | | **Suitable for** | | | | | | | | | |
| E. griffithsii | • | | | | | | | | • | | | | | | | • | | | | | |
| E. grossa | • | | | | | | | • | | | | | • | • | | | | | | | |
| E. gummifera | | | | | • | | | | | | • | | | | | | | | | • | |
| E. gunnii | | | | | • | | | | | | • | | • | | | | | | • | | |
| E. incrassata | • | | | | | | | • | | | | • | | | | • | | • | | | |
| E. johnstonii | | | | | | • | | | | | • | | | | | | | | • | | |
| E. kingsmillii | | • | | | | | • | • | | | | • | | | | | | | | | |
| E. kitsoniana | | | | | • | | • | • | • | | | | | | • | | | | • | | • |
| E. kondininensis | | | • | | | | | | • | • | | | | | • | • | | | | | |
| E. kruseana | | • | | | | | • | | | | | • | | | | | | | | | |
| E. lacliae | | | | • | | | | | | | • | | | | | | | | | | |
| E. lansdowneana | | | • | | | | | • | • | | | | | | | | | | | | |
| E. largiflorens | | | • | | | | | | | | • | | • | • | | • | | | | | • |
| E. lehmannii | | | • | | | | | | • | | | • | • | | | | | • | | | |
| E. leptopoda | | • | | | | | | • | | • | | • | | | | | | | | | |
| E. le souefii | | • | | | | | | | | | | • | • | | | • | | | | | |
| E. leucoxylon | | | • | | | | | | | | • | | • | • | | | | • | | | |
| E. leucoxylon var. rosea | | | • | | | | | | | | • | | • | • | | | | • | | | |
| E. leucoxylon var. macrocarpa | | | • | | | | | | • | | | | • | • | | | | • | | | |
| E. linearis | | | | | • | | | | • | • | | • | | | | | | | • | | |

295

# CHARACTERISTICS

| SPECIES | Minimum rainfall zone (mm) | | | | | | Height class (m) | | | | | Suitable for | | | | | | | | | |
|---|---|---|---|---|---|---|---|---|---|---|---|---|---|---|---|---|---|---|---|---|---|
| | < 300 | 300–400 | 400–500 | 500–650 | 650–900 | > 900 | < 3 | 3–5 | 5–8 | 8–12 | > 12 | Light soils | Heavy soils | Limestone, alkaline | Poorly-drained | Saline | Stony | Coastal | Snow | Smog | Flooding |
| E. longicornis | | • | | | | | | | | | • | | • | • | | • | | | | | |
| E. loxophleba | | • | | | | | | | | • | | • | | | | | | | | | |
| E. macrandra | | | | • | | | | | • | | | • | | | • | • | | • | • | | |
| E. macrocarpa | | | • | | | | • | | | | | • | | | | | | | | | |
| E. macrorhyncha | | | | • | | | | | | | • | | | | | | • | | | | |
| E. maculata | | | • | | | | | | | | • | • | • | | • | | | • | | • | • |
| E. mannifera subsp. maculosa | | | | | • | | | | | | • | | | | | | • | | | | |
| E. melliodora | | | | • | | | | | | | • | | | | | | | | • | | |
| E. microcarpa | | | • | | | | | | | | • | | • | | | | | | | | • |
| E. miniata | | | | • | | | | | • | | | | | | | | | | | | |
| E. mitchelliana | | | | | • | | | | | • | | | | | | | | | | | |
| E. mitrata | | • | | | | | • | | | | | • | | | | | | | • | | |
| E. moorei | | | | • | | | | | | • | | | | | • | | | • | | | |
| E. muellerana | | | | | • | | | | | | • | | • | | • | | | | • | | • |
| E. neglecta | | | | | • | | | | • | | | | • | | | | | | | | • |
| E. nicholii | | | | | • | | | | | • | | | | | | | • | | | | |
| E. nitens | | | | | | • | | | | | • | | | | | | | | • | | |
| E. nutans | | | • | | | | | • | | | | • | • | • | | | • | | | | |
| E. obliqua | | | | | • | | | | | | • | | | | | | | | • | | |
| E. occidentalis | | | • | | | | | | | | • | | • | | • | • | | • | | | • |

296

# CHARACTERISTICS

| SPECIES | <300 | 300–400 | 400–500 | 500–650 | 650–900 | >900 | <3 | 3–5 | 5–8 | 8–12 | >12 | Light soils | Heavy soils | Limestone, alkaline | Poorly-drained | Saline | Stony | Coastal | Snow | Smog | Flooding |
|---|---|---|---|---|---|---|---|---|---|---|---|---|---|---|---|---|---|---|---|---|---|
| | Minimum rainfall zone (mm) | | | | | | Height class (m) | | | | | Suitable for | | | | | | | | | |
| E. oleosa | | • | | | | | | | | | | • | | • | | • | | | | | |
| E. orbifolia | | • | | | | | | • | • | | | | • | | | | | | | | |
| E. oreades | | | | | | • | | | | | • | | | | | | | | • | | |
| E. ovata | | | | | • | | | | | | • | | | | • | | | | | | • |
| E. paniculata | | | | • | | | | • | • | • | • | | • | | | | • | | | | |
| E. pauciflora | | | | • | | | | | | | • | | | | • | | | • | • | | • |
| E. pauciflora 'Alpine form' | | | | | • | | | | • | | | | | | • | | | | • | | |
| E. pauciflora var. nana | | | | | • | | • | | | | | | | | • | | | | • | | |
| E. pellita | | | | • | | | | | | | | | • | | • | | | | | | |
| E. perriniana | | | | | • | | | | • | • | • | | | | | | | | • | | |
| E. pileata | | • | | | | | | • | | | | • | | • | | • | | | | | |
| E. platycorys | | | • | | | | | | | | | | | | | • | | | | | |
| E. platypus | | | • | | | | | • | • | | | • | • | • | • | | | • | | | |
| E. platypus var. heterophylla | | | • | | | | | • | • | | | • | • | | • | | | • | | | |
| E. polyanthemos | | | | • | | | | | | • | • | | • | • | | | • | | | | • |
| E. polybractea | | • | | | | | | • | • | | | | | | | | | | | | |
| E. porosa | • | | | | | | | • | | | | • | | • | | | | | | | |
| E. prcissiana | | | | • | | | | • | | | | • | | | | | | | | | |
| E. pulverulenta | | | | • | | | | • | • | | | | • | | | | • | | | | |
| E. punctata | | | | | • | | | | | | • | • | • | | | | | | | | |

297

# CHARACTERISTICS

| SPECIES | <300 | 300–400 | 400–500 | 500–650 | 650–900 | >900 | <3 | 3–5 | 5–8 | 8–12 | >12 | Light soils | Heavy soils | Limestone, alkaline | Poorly-drained | Saline | Stony | Coastal | Snow | Smog | Flooding |
|---|---|---|---|---|---|---|---|---|---|---|---|---|---|---|---|---|---|---|---|---|---|
| | | | | | | | | | | | | | | | | | | | | | |
| E. pyriformis | | | | • | | | | • | | | | | | | | | | | | | |
| E. radiata | | | | | • | | | | | | • | | | | | | | | • | | |
| E. regnans | | | | | | • | | | | | | | | | | | | | • | | |
| E. rhodantha | | | • | | | | • | | | | | | | | | | | | | | |
| E. risdonii | | | | | • | | | | • | | | • | | | | | | | • | | |
| E. robusta | | | | | • | | | | | | • | | • | • | • | • | | • | | • | • |
| E. rubida | | | | | • | | | | | | • | | | | | | | | | | |
| E. saligna | | | | | • | | | | | | • | | • | | | | | | | | |
| E. salmonophloia | | • | | | | | | | | | • | • | • | • | • | • | | | | | |
| E. salubris | • | | | | | | | | | • | | | • | • | | | | | | | |
| E. sargentii | | • | | | | | | | • | | | • | • | | | • | | | | | |
| E. scoparia | | | | | • | | | | | • | | | | | | | | | • | | |
| E. sepulcralis | | | | • | | | | | • | | | • | | | | | • | • | | | |
| E. sideroxylon | | | | • | | | | | | | • | | • | | | | • | | | | |
| E. sideroxylon var. pallens | | | | • | | | | | | | • | • | • | | | | • | | | | |
| E. sieberi | | | | | • | | | | | | | | | | | | | | | | |
| E. socialis | | • | | | | | • | | | | | | | | | | | | | | |
| E. spathulata | | | • | | | | | • | • | | | | • | • | • | • | | | | | • |
| E. steedmanii | | • | | | | | | • | | | | • | • | • | • | | | • | | | |
| E. stellulata | | | | | • | | | | | • | | | • | • | • | | | | • | | |

# CHARACTERISTICS

| SPECIES | Min. rainfall zone (mm) | | | | | | Height class (m) | | | | | Suitable for | | | | | | | | | |
|---|---|---|---|---|---|---|---|---|---|---|---|---|---|---|---|---|---|---|---|---|---|
| | <300 | 300–400 | 400–500 | 500–650 | 650–900 | >900 | <3 | 3–5 | 5–8 | 8–12 | >12 | Light soils | Heavy soils | Limestone, alkaline | Poorly-drained | Saline | Stony | Coastal | Snow | Smog | Flooding |
| E. st johnii | | | | | • | | | | | | • | | • | | • | | | | • | | |
| E. stoatei | | | • | | | | | • | • | | | | | • | | | | | | | |
| E. striaticalyx | | • | | | | | | | | • | | | | • | | • | | | | | |
| E. stricklandii | • | | | | | | | | | • | | • | | • | | | • | • | | | |
| E. tasmanica | | | | | • | | | | | • | • | | | | | | | | • | | |
| E. tetragona | | | • | | | | | • | | • | | • | | | | | | • | | | |
| E. tetraptera | | | • | | | | • | • | | | | • | | | | | | | | | |
| E. torquata | • | | | | | | | • | • | | | • | • | • | | | | | | | |
| E. urnigera | | | | | • | | | | • | • | | | | | | | | | | | |
| E. viminalis | | | | | • | | | | | | • | • | • | | | | | | • | | • |
| E. viridis | | | • | | | | | | • | | | | • | | | | • | | | | |
| E. wandoo | | | • | | | | | | | • | • | • | • | | | | | | | | |
| E. websterana | • | | | | | | • | • | | • | | | | | | | | | | | |
| E. woodwardii | | • | | | | | | • | | | | | • | | | • | | | | | |
| E. youngiana | | • | | | | | | • | | | | | | • | | | | | | | |
| Eucryphia cordifolia | | | | | • | | | | • | • | | | | | | | | | | | |
| E. lucida | | | | | • | | | | • | • | • | • | | | | | | | • | | |
| Eugenia australis | | | | | • | | | | • | • | • | • | | | | | | • | | | |
| E. smithii | | | | | • | | | | • | • | • | • | • | | | | | • | | • | |
| E. ventenatii | | | | | • | | | | • | • | | • | | | | | | • | | | |

299

# CHARACTERISTICS

| SPECIES | Minimum rainfall zone (mm) | | | | | | Height class (m) | | | | | Suitable for | | | | | | | | | |
|---|---|---|---|---|---|---|---|---|---|---|---|---|---|---|---|---|---|---|---|---|---|
| | <300 | 300–400 | 400–500 | 500–650 | 650–900 | >900 | <3 | 3–5 | 5–8 | 8–12 | >12 | Light soils | Heavy soils | Limestone, alkaline | Poorly-drained | Saline | Stony | Coastal | Snow | Smog | Flooding |
| Fagus sylvatica | | | | | | | | | | | • | | | • | | | | | • | | |
| Ficus macrophylla | | | | | • | | | | | | • | | • | | • | • | | | | • | |
| Flindersia maculosa | | • | | | | | | • | | | | | | | | | | | | | |
| Fraxinus americana | | | | | • | | | | | | • | | | • | | | | | | | • |
| F. angustifolia | | | | • | | | | | | | • | | | | | | | | | | |
| F. excelsior | | | | • | | | | | | | • | | • | | | | | | • | • | • |
| F. excelsior var. aurea | | | | • | | | | | | • | | | | | | | | | • | • | |
| F. lanceolata | | | • | | | | | | | • | | | | | | | | | | | |
| F. ornus | | | | • | | | | | | • | | | | | | | | | • | • | |
| F. oxycarpa | | | • | | | | | | | • | | | • | • | | | | | | | |
| F. x 'raywoodii' | | | | • | | | | | | | • | | • | • | | | | | | • | |
| F. syriaca | | | | • | | | | | • | | | | • | | | | | | | | |
| F. velutina | | | • | | | | | | • | | | | | • | | • | | | | • | |
| Geijera parviflora | | • | | | | | | • | | | | • | | • | | | | | | • | |
| Ginkgo biloba | | | | | | | | | | • | • | | | • | | | | | | | |
| Gleditsia triacanthos | | | | | • | | | | | • | • | | • | • | | • | | • | • | • | |
| G. triacanthos var. inermis | | | | | • | | | | | | • | | • | • | | • | | • | • | | |
| Grevillea asplenifolia | | | | | • | | • | • | | | | • | • | | | | | | • | | |
| G. banksii | | | | • | | | • | | | | | • | | | | | | | | | |
| G. barklyana | | | | | • | | • | • | • | • | | | | | | | | | | | |

# CHARACTERISTICS

| SPECIES | Min. rainfall <300 | 300–400 | 400–500 | 500–650 | 650–900 | >900 | Height <3 | 3–5 | 5–8 | 8–12 | >12 | Light soils | Heavy soils | Limestone, alkaline | Poorly-drained | Saline | Stony | Coastal | Snow | Smog | Flooding |
|---|---|---|---|---|---|---|---|---|---|---|---|---|---|---|---|---|---|---|---|---|---|
| G. confertifolia | | | | • | | | • | | | | | | | | | | | | | | |
| G. glabrata | | | | • | | | • | | | | | • | | | | | | | | | |
| G. hookeriana | | | | • | | | • | | | | | | | | | | | | | | |
| G. ilicifolia | | | • | | | | • | | | | | • | | • | | | | | | | |
| G. lavandulacea | | | • | | | | • | | | | | | | | | | | | • | | |
| G. longifolia | | | | | • | | | • | | | | • | • | | | | | | • | | |
| G. nematophylla | | • | | | | | • | • | | | | • | | • | | | | • | | | |
| G. robusta | | | | • | | | | | | | • | | | | • | | | | | | |
| G. rosmarinifolia | | | | • | | | • | | | | • | | | | | | | | • | | |
| G. striata | | | • | | | | | | • | • | | | • | • | | | | | | | |
| G. thelemanniana | | | • | | | | • | | | | | | | | | | | | | | |
| G. vestita | | | • | | • | | • | | | | | | | • | | | | | | | |
| G. victoriae | | | | | • | | • | | | | | | | | | | | | | | |
| Griselina littoralis | | | | | | | | | • | | | | | | | | | • | | | |
| Hakea bucculenta | | | • | | | | | • | | | | | | | | | | | | | |
| H. elliptica | | | | • | | | | • | | | | • | • | | | | | • | | | |
| H. eriantha | | | | | • | | | • | • | | | | | | | | | | | | |
| H. francisiana | | | | • | | | | • | | | | | | | | | | | | | |
| H. laurina | | | • | | | | | • | | | | | • | | | | | • | | • | |
| H. leucoptera | | | • | | | | | • | | | | • | • | • | • | | | | | | |

# CHARACTERISTICS

| SPECIES | Minimum rainfall zone (mm) | | | | | | Height class (m) | | | | | Suitable for | | | | | | | | | |
|---|---|---|---|---|---|---|---|---|---|---|---|---|---|---|---|---|---|---|---|---|---|
| | <300 | 300–400 | 400–500 | 500–650 | 650–900 | >900 | <3 | 3–5 | 5–8 | 8–12 | >12 | Light soils | Heavy soils | Limestone, alkaline | Poorly-drained | Saline | Stony | Coastal | Snow | Smog | Flooding |
| H. muellerana | | | | • | | | | • | | | | | | | | | | | | | |
| H. multilineata | | • | | | | | | • | | | | | | • | | | | | | | |
| H. nodosa | | | | • | | | | • | | | | | | | • | | | | | | • |
| H. petiolaris | | | • | | | | | • | | | | | • | • | | | | | | | |
| H. preissii | | • | | | | | • | | | | | • | | | | | | | | | |
| H. pubescens | | | | • | | | • | | | | | | | | | | | | | | |
| H. salicifolia | | | | • | | | | • | • | | | | | | • | | | | | • | |
| H. sericea | | | | | • | | • | | | | | | | | | | | | | | |
| H. suaveolens | | | • | | | | • | • | | | | • | • | | | | | • | • | • | |
| H. ulicina | | | | • | | | | • | | | | | | | | | | | | | |
| H. victoriae | | | • | | | | • | • | | | | | | | | | • | | | | |
| H. vittata | | • | | | | | • | | | | | • | | | | | | | | | |
| Halesia carolina | | | | | • | | | | • | • | | | | | | | | | | | |
| Heterodendrum oleifolium | | • | | | | | | | • | | | | | | | | | | | | |
| Hoheria populnea | | | | | • | | | • | | | | • | | | | | | • | • | | |
| Homalanthus populifolius | | | | | • | | | • | | | | | | | | | | | | | |
| Hovenia dulcis | | | | | • | | | | • | | | | | | | | | | | | |
| Hymensporum flavum | | | | | • | | | | • | | | | | | | | | • | | | |
| Idesia polycarpa | | | | | • | | | | | • | | | | | | | | | | | |
| Ilex aquifolium | | | | | • | | | • | • | • | • | | | | | | | • | • | • | |

# CHARACTERISTICS

| SPECIES | Minimum rainfall zone (mm) | | | | | | Height class (m) | | | | | Suitable for | | | | | | | | | |
|---|---|---|---|---|---|---|---|---|---|---|---|---|---|---|---|---|---|---|---|---|---|
| | <300 | 300–400 | 400–500 | 500–650 | 650–900 | >900 | <3 | 3–5 | 5–8 | 8–12 | >12 | Light soils | Heavy soils | Limestone, alkaline | Poorly-drained | Saline | Stony | Coastal | Snow | Smog | Flooding |
| Jacaranda acutifolia | | | | | • | | | • | • | • | • | | | | | | | | | | |
| Juglans nigra | | | | | | • | | | | | • | | | • | | | | | | | • |
| Juniperus communis | | | | | • | | | | • | • | | | • | • | | | | | • | | |
| Koelreuteria paniculata | | | | • | | | | • | • | | | | | • | | | | • | | • | |
| Kunzea ambigua | | | | • | | | | | | | | | | | | | | • | | | |
| Laburnum anagyroides | | | | | | • | | | • | | | • | • | • | | | • | | | | |
| L. vossii | | | | | | • | | | • | | | • | • | • | | | • | | | | |
| Lagunaria patersonii | | | • | | | | | | • | | | | • | • | • | • | | • | | • | • |
| Larix decidua | | | | | • | | | | | | • | | | | | | | | • | | |
| L. eurolepis | | | | | • | | | | | • | | | | | | | | • | • | | |
| Laurus nobilis | | | | | • | | | • | • | • | | • | | | | | | • | | | |
| Leptospermum laevigatum | | | | • | | | • | | | | | • | | | | | | • | | | |
| L. laevigatum *var.* minus | | • | | | | | • | | | | | | | | | | | | | | |
| L. lanigerum | | | | • | | | | • | | | | | • | • | • | | | | | | • |
| L. phylicoides | | | | • | | | | | • | | | • | • | | • | | | • | • | | • |
| L. scoparium | | | | • | • | | | • | | | | | • | | • | | | | | | • |
| Leucadendron argenteum | | | | | • | | | • | | • | | | | | | | | | | | |
| Leucopogon parviflora | | | | • | | | | • | | | | • | | | | | | • | | | |
| Liquidambar formosana | | | | | • | | | | | • | • | | | | | | | | | | |
| L. styraciflua | | | | | • | | | | | • | • | | | | • | | | | | | • |

303

# CHARACTERISTICS

| SPECIES | Min. rainfall <300 | 300–400 | 400–500 | 500–650 | 650–900 | >900 | Height <3 | 3–5 | 5–8 | 8–12 | >12 | Light soils | Heavy soils | Limestone, alkaline | Poorly-drained | Saline | Stony | Coastal | Snow | Smog | Flooding |
|---|---|---|---|---|---|---|---|---|---|---|---|---|---|---|---|---|---|---|---|---|---|
| Liriodendron tulipifera | | | | | • | | | | | | • | | | | | | | | | | |
| Maclura pomifera | | | • | | | | | | • | • | • | | | • | | | | | | | |
| Magnolia grandiflora | | | | | • | | | | | • | • | | | | | | | | | | |
| Malus | | | | | | | | • | | | | | | | | | | | | | |
| Melaleuca acuminata | • | | | | | | • | | | | | • | • | | | | | | | | |
| M. armillaris | | | • | | | | | • | | | | • | • | • | • | • | | • | | • | • |
| M. crassifolia | | | • | | | | | | | | | • | | | | | | | | | |
| M. decussata | | | • | | | | | • | | | | • | • | | • | • | | | | • | • |
| M. diosmifolia | | | | • | | | • | | | | | | | | • | | | | | | |
| M. elliptica | | | | • | | | • | | | | | • | | | | | | | | | |
| M. ericifolia | | | | • | | | | • | | | | | • | • | • | • | | • | | | • |
| M. fulgens | | | • | | | | • | | | | | | | | | | | | | | |
| M. gibbosa | | | | • | | | • | | | | | | | | | | | | | | |
| M. glomerata | • | | | • | | | | • | | | | | | | • | | | | | | |
| M. halmaturorum | | | • | | | | | | • | | | | • | • | • | • | | • | | • | • |
| M. huegelii | | | • | | | | • | | | | | | | • | | • | | • | | • | |
| M. hypericifolia | | | • | | | | • | | | | | | • | • | • | • | | • | | • | |
| M. incana | | | • | | | | • | | | | | | • | | • | | | | | • | • |
| M. lanceolata | | | | • | | | | • | • | | | • | • | • | • | • | • | | | • | • |
| M. lateritia | | | • | | | | • | | | | | | | | | | | | | | |

# CHARACTERISTICS

| SPECIES | Minimum rainfall zone (mm) | | | | | | Height class (m) | | | | | Suitable for | | | | | | | | | |
|---|---|---|---|---|---|---|---|---|---|---|---|---|---|---|---|---|---|---|---|---|---|
| | < 300 | 300–400 | 400–500 | 500–650 | 650–900 | > 900 | < 3 | 3–5 | 5–8 | 8–12 | > 12 | Light soils | Heavy soils | Limestone, alkaline | Poorly-drained | Saline | Stony | Coastal | Snow | Smog | Flooding |
| M. linariifolia | | | | • | | | | | • | • | | | • | | | • | | • | | • | • |
| M. longicoma | | • | | | | | • | | | | | • | | | | | • | | | | |
| M. neglecta | | | • | | | | • | | | | | | | | | | | | | | |
| M. nesophila | | | • | | | | • | | | | | • | • | • | | • | | • | | | |
| M. pentagona | | | • | | | | • | | | | | • | | | | | | | | | |
| M. quinquenervia | | | | • | | | | | | • | • | | | | • | • | | • | | | • |
| M. radula | | | | • | | | • | | | | | | | | | | | | | | |
| M. squarrosa | | | | | • | | | | • | | | | • | | • | • | | • | | | • |
| M. styphelioides | | | | • | | | | | • | • | | | • | | • | • | | • | | • | • |
| M. wilsonii | | | • | | | | • | • | | | | | | • | | | | | | | • |
| Melia azedarach | | | • | | | | | | • | • | | | | | | | | • | | | |
| Mespilus germanica | | | | | • | | | | • | | • | | | | | | | | | | |
| Metasequoia glyptostroboides | | | | | • | | | | • | | | | | | | | | | | | |
| Metrosideros excelsa | | | | • | | | | | • | | | | | | | | | • | | • | |
| Morus alba | | | | | • | | | | • | | | | | | | | | | | | |
| Myoporum floribundum | | | | • | | | • | • | | | | | | | | | | | | | |
| M. insulare | | | | • | | | | • | • | | | • | • | • | • | • | | • | | • | |
| M. platycarpum | | • | | | | | | | | | | | | • | | | | • | | | |
| Nothofagus cunninghamiana | | | | | | • | | | | | • | | | | | | | • | | | |
| N. fusca | | | | | • | | | | • | | | | | | | | | • | | | |

# CHARACTERISTICS

| SPECIES | Minimum rainfall zone (mm) | | | | | | Height class (m) | | | | | Suitable for | | | | | | | | | |
|---|---|---|---|---|---|---|---|---|---|---|---|---|---|---|---|---|---|---|---|---|---|
| | <300 | 300–400 | 400–500 | 500–650 | 650–900 | >900 | <3 | 3–5 | 5–8 | 8–12 | >12 | Light soils | Heavy soils | Limestone, alkaline | Poorly-drained | Saline | Stony | Coastal | Snow | Smog | Flooding |
| N. obliqua | | | | | | • | | | | | • | | | | | | | | • | | |
| Nyssa sylvatica | | | | | • | | | | | | • | | | | • | | | | | | • |
| Olea europaea | | • | | | | | | | • | | | • | | • | | | | | | • | |
| Oxydendrum arboreum | | | | | • | | | • | | | | | | | | | | | | | |
| Parrotia persica | | | | | • | | | | • | | | | | | | | | | | | |
| Paulownia tomentosa | | | | | • | | | | | • | | | | | | | | • | | | |
| Persoonia pinifolia | | | | | • | | | • | | | | | | | | | | | | | |
| Phebalium squameum | | | | | • | | | | | • | | | | | | | | • | | | |
| Photinia serrulata | | | | • | | | | | • | | | | | | | | | | | | |
| Picea abies | | | | | • | | | | | | • | | | | | | | | • | • | |
| P. engelmannii | | | | | • | | | | | | • | | | | | | | | • | | |
| P. omorika | | | | | • | | | | | | • | | | • | | | | | | | |
| P. pungens | | | | | • | | | | | | • | | | | | | | | • | • | |
| P. sitchensis | | | | | | • | | | | | • | | | | | | | | • | | |
| Pinus canariensis | | | • | | | | | | | | • | | | • | | | | | | | |
| P. contorta | | | | | • | | | | | | • | • | | | | | • | • | • | | |
| P. halepensis | | | • | | | | | | | | • | • | • | • | | • | • | • | | | |
| P. halepensis var. brutia | | | • | | | | | | | | • | • | • | • | | • | • | • | | | |
| P. nigra subsp. laricio | | | | | • | | | | | | • | • | | • | | | • | • | • | | |
| P. patula | | | | | • | | | | | | • | | | | | | | | | | |

# CHARACTERISTICS

| SPECIES | Min. rainfall (mm) <300 | 300–400 | 400–500 | 500–650 | 650–900 | >900 | Height (m) <3 | 3–5 | 5–8 | 8–12 | >12 | Light soils | Heavy soils | Limestone, alkaline | Poorly-drained | Saline | Stony | Coastal | Snow | Smog | Flooding |
|---|---|---|---|---|---|---|---|---|---|---|---|---|---|---|---|---|---|---|---|---|---|
| P. pinaster | | | | • | | | | | | | • | • | | | | | | • | | | |
| P. pinea | | | | | • | | | | | | • | | | • | | | | • | | | |
| P. ponderosa | | | | | • | | | | | | • | | | | | | | | • | | |
| P. radiata | | | | | • | | | | | | • | • | | | | | | • | • | • | |
| P. strobus | | | | | • | | | | | | • | | | | | | | | | | |
| Pistacia chinensis | | | | | • | | | | • | • | | | | | | | | | | | |
| Pittosporum bicolor | | | | | • | | | | • | | | | | | | | | | | | |
| P. crassifolium | | | | | • | | | | • | | | | | • | | • | | • | | • | |
| P. eugenioides | | | | | • | | | • | | | | | | | | | | • | | | |
| P. phillyreoides | • | | | | | | | | • | | | • | • | • | | • | | • | | | |
| P. tenuifolium | | | | | • | | | | • | | | | | | | • | | | | • | |
| P. undulatum | | | | • | | | | | | • | | | • | | | | | • | | | • |
| Platanus x acerifolia | | | • | | | | | | | | • | | • | | | | | | • | • | |
| P. orientalis | | | • | | | | | | | | • | | • | | • | | | • | • | • | • |
| Podocarpus elatus | | | | | • | | | | • | | | | | | | | | | | | |
| P. lawrencei | | | | | • | | • | | | | | | | | | | | | • | | |
| P. totara | | | | | • | | | | • | | | | | | | | | | | | |
| Populus alba | | | | | • | | | | | | • | | | | • | | | • | | | • |
| P. alba var. pyramidalis | | | | | • | | | | | | • | | | | • | | | • | | | • |
| P. candicans | | | | | • | | | | | • | | | | | | | | | | | |

# CHARACTERISTICS

| SPECIES | Minimum rainfall zone (mm) | | | | | | Height class (m) | | | | | Suitable for | | | | | | | | | |
|---|---|---|---|---|---|---|---|---|---|---|---|---|---|---|---|---|---|---|---|---|---|
| | <300 | 300–400 | 400–500 | 500–650 | 650–900 | >900 | <3 | 3–5 | 5–8 | 8–12 | >12 | Light soils | Heavy soils | Limestone, alkaline | Poorly-drained | Saline | Stony | Coastal | Snow | Smog | Flooding |
| P. canescens | | | | ● | | | | | | | ● | | | | | | | ● | ● | | |
| P. deltoides var. missouriensis | | | | | ● | | | | | | ● | | | | | | | | | | ● |
| P. deltoides var. angulata | | | | | ● | | | | | | ● | | | | | | | | | | ● |
| P. x generosa | | | | | ● | | | | | | ● | | | | | | | | | | ● |
| P. nigra var. italica | | | | ● | | | | | | | ● | | ● | | ● | | | ● | | | |
| P. serotina var. aurea | | | | | ● | | | | | | ● | | | | | | | | | | ● |
| P. simonii c.v. 'fastigiata' | | | | | ● | | | | | ● | | | | | | | | | | | |
| P. tremula | | | | | ● | | | | | ● | | | | | ● | | | | ● | | ● |
| P. tremuloides | | | | | ● | | | | | ● | | | | | | | | | ● | | ● |
| P. trichocarpa | | | | | ● | | | | | | ● | | | | | | | | ● | | ● |
| P. yunnanensis | | | | | ● | | | | | | ● | | | | | | | | | | ● |
| Prostanthera lasianthos | | | | | ● | | | | ● | ● | | | | | | | | | ● | | |
| P. nivea | | | ● | | | | ● | | | | | | | | | | | | | | |
| P. rotundifolia | | | | ● | | | ● | | | | | | | | | | | | | | |
| Prunus cerasifera var. nigra | | | ● | | | | | | | | | | | | | | | | | | |
| P. laurocerasus | | | | | ● | | | | ● | | | | ● | | | | | ● | ● | ● | |
| P. lusitanica | | | | | ● | | | | ● | | | | | ● | | | | ● | ● | ● | |
| P. serrulata | | | | | ● | | | | ● | | | | | | | | | | | | |
| Pseudotsuga menziesii | | | | | | | | | | | ● | | | | | | | | ● | | |
| P. menziesii var. glauca | | | | | ● | | | | | | ● | | | | | | | | ● | | |

# CHARACTERISTICS

| SPECIES | Minimum rainfall zone (mm) | | | | | | Height class (m) | | | | | Suitable for | | | | | | | | | |
|---|---|---|---|---|---|---|---|---|---|---|---|---|---|---|---|---|---|---|---|---|---|
| | <300 | 300–400 | 400–500 | 500–650 | 650–900 | >900 | <3 | 3–5 | 5–8 | 8–12 | >12 | Light soils | Heavy soils | Limestone, alkaline | Poorly-drained | Saline | Stony | Coastal | Snow | Smog | Flooding |
| Pyracantha crenulata | | | | | • | | | • | | | | | | | | | | | • | • | |
| Pyrus calleryana | | | | | • | | | • | | | | | | | | | | | | | |
| P. ussuriensis | | | | • | | | | | | • | • | | | | | | | | | | |
| Quercus aegilops | | | | • | | | | | • | • | | | | • | | | | | | | |
| Q. alba | | | | | • | | | | | | • | | | | | | | | | | |
| Q. borealis | | | | | • | | | | | | • | | | | | | | | | | |
| Q. canariensis | | | | | • | | | | | • | • | | | | | | | | | | |
| Q. castaneifolia | | | | | • | | | | | | • | | | | | | | | | | |
| Q. cerris | | | | • | | | | | | | • | | | | | | | | | | |
| Q. coccinea | | | | | • | | | | | | • | | | | | | | | | | |
| Q. faginea | | | | | • | | | | | • | • | | | | | | | | | | |
| Q. ilex | | | | • | | | | | | • | • | | | • | | | | • | • | | |
| Q. lyrata | | | | | • | | | | | | • | | | | | | | | | | • |
| Q. palustris | | | | | • | | | | | | • | | • | | • | | | • | | | • |
| Q. palustris 'Macedon' | | | | | • | | | | | | • | | | | | | | | | | |
| Q. robur | | | | | • | | | | | | • | | | | | | | | | | |
| Q. suber | | | • | | | | | | | • | | • | | | | | • | | | | |
| Q. virginiana | | | | • | | | | | • | • | | | | | | | | | | | |
| Rhus succedanea | | | | • | | | | • | | • | | | | | | | | | | | |
| Robinia pseudoacacia | | | • | | | | | | | • | • | | • | • | | | | | | • | • |

# CHARACTERISTICS

| SPECIES | \<300 | 300–400 | 400–500 | 500–650 | 650–900 | \>900 | \<3 | 3–5 | 5–8 | 8–12 | \>12 | Light soils | Heavy soils | Limestone, alkaline | Poorly-drained | Saline | Stony | Coastal | Snow | Smog | Flooding |
|---|---|---|---|---|---|---|---|---|---|---|---|---|---|---|---|---|---|---|---|---|---|
| | | Minimum rainfall zone (mm) | | | | | | Height class (m) | | | | | | | | Suitable for | | | | | |
| Salix alba *var.* coerulea | | | | | • | | | | | | • | | | | | | | | | | • |
| S. alba *subsp.* vitellina | | | | | • | | | | • | • | | | | | • | | | | | | • |
| S. alba *subsp.* vitellina *var.* pendula | | | | | • | | | | • | • | | | | | • | | | | | | • |
| S. babylonica | | | | | • | | | | | | • | | | | | | | • | | | • |
| S. discolor | | | | | • | | | | • | • | | | | | • | | | | | | • |
| 'S. fragilis' | | | | | • | | | | | • | • | | | | | | | | | | • |
| S. humboltiana | | | | | • | | | | | • | | | | | | | | | | | |
| S. matsudana *c.v.* 'tortuosa' | | | | | • | | | | • | | | | | | | | | | | | • |
| Sambucus nigra | | | | | • | | | | • | | | | | | | | | | | | |
| Schinus molle | | • | | | | | | • | | | | • | • | • | | • | | • | | • | |
| S. terebinthifolius | | | • | | | | | | • | | | | | • | | | | | | | |
| Sequoia sempervirens | | | | | • | | | | • | | • | | | | | | | | | | |
| Sequoiadendron giganteum | | | | | • | | | | • | | • | | | | | | | | | | |
| Sophora japonica | | | | | • | | | | • | • | • | | | | | | | | | | |
| S. microphylla | | | | | • | | | | • | | | | | | | | | | • | | |
| S. tetraptera | | | | | • | | | | • | | | | | | | | | | • | | |
| Sorbus aucuparia | | | | | • | | | | | • | • | | | | | | | | • | • | |
| S. decora | | | | | • | | | | • | | | | | | | | | | • | | |
| Stenocarpus sinuatus | | | | • | | | | | | • | • | | | | | | | | • | • | |
| Stewartia pseudocamellia | | | | • | | | | | • | • | | | | | | | | | | | |

# CHARACTERISTICS

| SPECIES | <300 | 300–400 | 400–500 | 500–650 | 650–900 | >900 | <3 | 3–5 | 5–8 | 8–12 | >12 | Light soils | Heavy soils | Limestone, alkaline | Poorly-drained | Saline | Stony | Coastal | Snow | Smog | Flooding |
|---|---|---|---|---|---|---|---|---|---|---|---|---|---|---|---|---|---|---|---|---|---|
| | Minimum rainfall zone (mm) | | | | | | Height class (m) | | | | | Suitable for | | | | | | | | | |
| Styrax japonica | | | | | ● | | | | ● | | | | | | | | | | | | |
| Syncarpia glomulifera | | | | | ● | | | | | | ● | | | | | | | | | | |
| Tamarix aphylla | ● | | | | | | | | ● | ● | | ● | ● | ● | ● | ● | | ● | | ● | |
| T. juniperina | | | ● | | | | | ● | | | | | ● | ● | ● | ● | | ● | | | |
| T. parviflora | | | ● | | | | | | | | | ● | ● | ● | ● | ● | | ● | | ● | |
| T. pentandra | | | ● | | | | | ● | | | | ● | ● | ● | ● | ● | | ● | | | |
| T. tetrandra | | | ● | | | | | ● | | | | | ● | ● | ● | ● | | ● | | | |
| Taxodium distichum | | | | | ● | | | | | | ● | | | | ● | | | | | | ● |
| Taxus baccata | | | | | ● | | | ● | | | | | ● | ● | | | | | ● | | |
| Telopea speciosissima | | | | | ● | | | ● | | | | ● | | | | | ● | | | | |
| Templetonia retusa | | | ● | | | | ● | | | | | ● | | ● | | | | ● | | | |
| Thryptomene calycina | | | | ● | | | ● | | | | | ● | | | | | | | | | |
| Thuja plicata | | | | | ● | | | | | | ● | | | | | | | | ● | | |
| Thujopsis dolobrata | | | | | ● | | | | ● | ● | | | | | | | | | ● | | |
| Tieghemopanax sambucifolius | | | | | ● | | | ● | ● | | | | | | | | | | ● | | |
| Tilia x vulgaris | | | | | ● | | | | | ● | | | | | | | | | ● | ● | |
| Tristania conferta | | | | ● | | | | | ● | | | | | | | | | | | ● | |
| T. laurina | | | | | ● | | | | ● | | | | | | | | | | | | |
| Tsuga heterophylla | | | | | | ● | | | | | ● | ● | ● | | | | | | ● | | |
| Ulmus glabra | | | | | ● | | | | | | ● | | | | | | | | ● | | |

# CHARACTERISTICS

| SPECIES | Minimum rainfall zone (mm) | | | | | | Height class (m) | | | | | Suitable for | | | | | | | | | |
|---|---|---|---|---|---|---|---|---|---|---|---|---|---|---|---|---|---|---|---|---|---|
| | < 300 | 300–400 | 400–500 | 500–650 | 650–900 | > 900 | < 3 | 3–5 | 5–8 | 8–12 | > 12 | Light soils | Heavy soils | Limestone, alkaline | Poorly-drained | Saline | Stony | Coastal | Snow | Smog | Flooding |
| U. x hollandica | | | | ● | | | | | | | ● | | | | | | | | | ● | |
| U. parvifolia | | | | | ● | | | | | ● | | | | | | | | ● | ● | | |
| U. procera | | | | ● | ● | | | | | | ● | | ● | | | | | | ● | ● | |
| U. procera *var.* van houttei | | | | | | | | | | ● | | | | | | | | | | | |
| U. pumila | | | ● | | | | | | | ● | ● | | | | | | | | ● | | |
| Viminaria juncea | | | | ● | | | | ● | | | | | ● | | ● | | | | | | ● |
| Virgilia divaricata | | | | | ● | | | ● | | | | | | | | | | ● | | | |
| V. oroboides | | | | | ● | | | ● | ● | | | | | | | | | ● | | | |
| Zelkova carpinifolia | | | | | ● | | | | | | ● | | | | | | | | ● | | |
| Z. serrata | | | | | ● | | | | | | ● | | | | | | | | ● | | |

# USES

| SPECIES | Farm forests | Windbreaks Low | Windbreaks High | Shade | Parks | Road sides | Streets | Avenues | Control of — Sand drift | Control of — Gully erosion | H'dges | Fodder | Ornamental | Honey | Birds | Autumn foliage |
|---|---|---|---|---|---|---|---|---|---|---|---|---|---|---|---|---|
| Abies alba | • | | • | • | • | | | | | | | | | | | |
| A. cephalonica | • | | • | • | • | | | | | | | | | | | |
| A. concolor | • | | • | • | • | | | • | | | | | | | | |
| A. magnifica | • | | • | • | • | | | | | | | | | | | |
| A. nordmanniana | • | | • | • | • | | | | | | | | | | | |
| A. pindrow | • | | • | • | • | | | | | | | | | | | |
| A. pinsapo | • | | • | • | • | | | | | | | | | | | |
| A. procera | • | | • | • | • | | | | | | | | | | | |
| A. sibirica | • | | • | • | • | | | | | | | | | | | |
| Acacia acinacea | • | • | | | | | | | | • | | | | | | |
| A. acuminata | • | • | | • | • | | • | | | | | | • | | | |
| A. adunca | | | | | • | • | | | | | | | • | | | |
| A. amoena | | • | | | • | | | | | | | | • | | | |
| A. aneura | | • | | • | | | • | | | | | • | • | | | |
| A. baileyana | | • | | | • | | | • | | | | | • | | | |
| A. boormanii | | • | | | • | • | | | | | | | • | | | |
| A. botrycephala | | | | | • | • | | | | | | | • | | | |
| A. brachybotrya | | • | | | | | • | | | | | | • | | | |
| A. buxifolia | | | | | • | | | | | | | | • | | | |
| A. calamifolia | | • | | | • | | | | | | | | • | | | |

# USES

| SPECIES | Farm forests | Windbreaks Low | Windbreaks High | Shade | Parks | Road sides | Streets | Avenues | Control of Sand drift | Control of Gully erosion | H'dges | Fodder | Ornamental | Honey | Birds | Autumn foliage |
|---|---|---|---|---|---|---|---|---|---|---|---|---|---|---|---|---|
| A. cambagei | | | | • | | | | | | | | | | | | |
| A. cardiophylla | | • | | | • | • | • | | | | | | • | | | |
| A. cognata | | • | | | • | | • | | | | | | • | | | |
| A. colletioides | | • | | | | | | | | • | | | | | | |
| A. conferta | | • | | | • | | • | | | | | | • | | | |
| A. cultriformis | | • | | | • | | | | | | | | • | | | |
| A. cyanophylla | | • | | | • | | | | • | | | | • | | | |
| A. cyclops | | • | | | | | | | • | | | | | | | |
| A. deanei | • | • | • | | | | | | | • | | | | • | | |
| A. decora | | • | | | • | | • | | | | | | • | | | |
| A. decurrens | • | | • | • | | | | | | • | | | | | | |
| A. doratoxylon | | • | | | • | • | | | | | | | | | | |
| A. drummondii | | • | | | • | | | | | | | | • | | | |
| A. falciformis | • | | • | | | | | | | | | | | | | |
| A. farinosa | | • | | | | | | | | | | | • | | | |
| A. farnesiana | | • | | | • | | | | | | • | | • | | | |
| A. fimbriata | | | | | • | | • | | | | | | • | | | |
| A. floribunda | | • | | | | | | | • | | | | • | | | |
| A. gladiiformis | | • | | | | | | | | | | | • | | | |

# USES

| SPECIES | Farm forests | Windbreaks Low | Windbreaks High | Shade | Parks | Road sides | Streets | Avenues | Control of Sand drift | Control of Gully erosion | H'dges | Fodder | Ornamental | Honey | Birds | Autumn foliage |
|---|---|---|---|---|---|---|---|---|---|---|---|---|---|---|---|---|
| A. glandulicarpa | | | | | | | | | | | | | • | | | |
| A. glaucescens | | | | | • | • | | | | | | | • | | | |
| A. gracifolia | | | | | • | | | | | | | | | | | |
| A. hakeoides | | • | | | • | | | | | • | | | • | | | |
| A. harpophylla | | • | | | | | | | | • | | • | | | | |
| A. homalophylla | | | • | • | | | | | | | | • | | | | |
| A. howittii | | • | • | | • | | • | • | | • | | | • | | | |
| A. implexa | | • | • | • | • | | | | | | | | | | | |
| A. inophloia | | • | | | | | | | | | | | | | | |
| A. iteaphylla | | • | | | • | | • | | | | | | | | | |
| A. jonesii | | • | | | | | | | | | | | • | | | |
| A. leprosa | | • | | | | | | | | | | | • | | | |
| A. ligulata | | • | | | | | | | | | | | • | | | |
| A. linifolia | | • | | | | | | | | | | | | | | |
| A. loderi | | • | | • | • | • | | | | | | | • | | | |
| A. longifolia | | • | | | • | | | | • | | • | | • | | | |
| A. longifolia *var.* sophorae | | • | | | | | | | • | • | | | | | | |
| A. mearnsii | • | | • | • | | | | | | • | | | | | | |
| A. melanoxylon | • | | • | • | • | • | | • | | • | | | | | | |
| A. microbotrya | | • | | | | | | | | | | | • | | | |

315

## USES

| SPECIES | Farm forests | Windbreaks Low | Windbreaks High | Shade | Parks | Road sides | Streets | Avenues | Control of — Sand drift | Control of — Gully erosion | H'dges | Fodder | Ornamental | Honey | Birds | Autumn foliage |
|---|---|---|---|---|---|---|---|---|---|---|---|---|---|---|---|---|
| A. montana | | ● | | | ● | | | | | | | | ● | | | |
| A. myrtifolia | | ● | | | ● | | | | | | | | ● | | | |
| A. notabilis | | ● | | | | ● | ● | | | | | | ● | | | |
| A. obliquinervia | ● | ● | | | | | | | | | | | | | | |
| A. oswaldii | | ● | | ● | | | | | | | | ● | | | | |
| A. pendula | | | | ● | | ● | ● | ● | | | | ● | ● | | | |
| A. penninervis | ● | | ● | | | | | | | ● | | | | | | |
| A. podalyriifolia | | | | | | | | | | | | | | | | |
| A. pravissima | | ● | | | ● | | ● | | | ● | ● | | ● | | | |
| A. prominens | | ● | | | ● | ● | | | | | | | ● | | | |
| A. pruinosa | | | | | ● | | | | | | | | ● | | | |
| A. pubescens | | | ● | | ● | | | | | | | | ● | | | |
| A. pycnantha | | ● | | | | | | | | | | | ● | | | |
| A. retinodes | | ● | | | | | ● | | | ● | | | ● | | | |
| A. riceana | | ● | | | | | ● | | | | | | ● | | | |
| A. rigens | | ● | | | ● | | | | | | | | | | | |
| A. rotundifolia | | ● | | | ● | | | | | | | | | | | |
| A. rubida | | ● | | | ● | | | | | ● | | | ● | | | |
| A. salicina | | ● | | | ● | | | | | ● | ● | ● | ● | | | |
| A. saligna | | ● | | | ● | ● | ● | | | | | | ● | | | |

# USES

| SPECIES | Farm forests | Windbreaks Low | Windbreaks High | Shade | Parks | Road sides | Streets | Avenues | Control of – Sand drift | Control of – Gully erosion | H'dges | Fodder | Ornamental | Honey | Birds | Autumn foliage |
|---|---|---|---|---|---|---|---|---|---|---|---|---|---|---|---|---|
| A. sclerosperma | | • | | | | • | | | | | | | | | | |
| A. silvestris | • | | • | | | | | | | • | | | | | | |
| A. sowdenii | | • | | | | | | | | | | • | | | | |
| A. spectabilis | | • | | • | • | • | • | | | | | | • | | | |
| A. stenophylla | | • | | | | | • | | | | | | • | | | |
| A. stricta | | • | | | • | | | | | | | | | | | |
| A. suaveolens | | • | | | | | | | | | • | | • | | | |
| A. subporosa | | • | | | • | | | | | | | | • | | | |
| A. subulata | | • | | | | | | | | | | | • | | | |
| A. terminalis | | | • | • | • | • | | • | | | | | • | | | |
| A. trineura | | • | | | | | | | | | | | • | | | |
| A. triptera | | • | | | | | | | | | | | • | | | |
| A. undulifolia | | • | | | | | | | | | | | • | | | |
| A. verniciflua | | • | | | | | | | | | | | • | | | |
| A. verticillata | | • | | | | | | | | | | | • | | | |
| A. vestita | | • | | | | | | | | | | | • | | | |
| A. victoriae | | • | | | | | | | | | | | | | | |
| A. wattsiana | | • | | | | | | | | | | | | | | |
| Acer buergerianum | | | | | • | | • | | | | | | • | | | • |
| A. campestre | | | • | • | • | • | | | | | | | • | | | • |

## USES

| SPECIES | Autumn foliage | Birds | Honey | Ornamental | Fodder | H'dges | Control of — Gully erosion | Control of — Sand drift | Avenues | Streets | Road sides | Parks | Shade | Windbreaks High | Windbreaks Low | Farm forests |
|---|---|---|---|---|---|---|---|---|---|---|---|---|---|---|---|---|
| A. cappadocicum | • | | | | | | | | | | • | • | • | | | |
| A. ginnala | • | | | • | | | | | | | | • | | | | |
| A. grosseri | • | | | • | | | | | | | | • | | • | | |
| A. negundo | • | | | • | | | | | | | • | • | • | | | |
| A. palmatum | • | | | • | | | | | | • | • | • | • | | • | |
| A. platanoides | • | | | • | | | | | | | • | • | • | • | | |
| A. pseudoplatanus | • | | | • | | | | | | | • | • | • | • | | • |
| A. rubrum | • | | | • | | | | | | | • | • | • | • | | |
| A. saccharinum | • | | | • | | | | | | | • | • | • | • | | • |
| A. saccharum | • | | | • | | | | | | | • | • | • | • | | • |
| A. tegmentosum | • | | | • | | | | | | | • | • | • | • | | |
| Actinostrobus pyramidalis | | | | | | • | | | | • | | • | | | • | |
| Aesculus hippocastanum | | | | | | | | | • | | | | • | | | |
| A. × carnea | • | | | • | | | | | • | | | • | • | | | |
| A. indica | • | | | • | | | | | • | | • | • | | | | |
| Agathis australis | • | | | | | | | | | | | • | | | | • |
| A. robusta | | | | | | | | | • | | | • | | | | • |
| Agonis flexuosa | | | | | | | | | | • | | • | | | | |
| A. juniperina | | | | | | | | | | • | | • | | | | |
| Ailanthus altissima | • | | | | | | • | | | | | • | • | | | |

## USES

| SPECIES | Farm forests | Windbreaks Low | Windbreaks High | Shade | Parks | Road sides | Streets | Avenues | Control of — Sand drift | Control of — Gully erosion | H'dges | Fodder | Ornamental | Honey | Birds | Autumn foliage |
|---|---|---|---|---|---|---|---|---|---|---|---|---|---|---|---|---|
| Albizia julibrissin | | | | | • | | | • | | | | | | | | • |
| A. lophantha | | • | | | • | | | | | | | | | | | |
| Alnus cordata | | | • | | • | • | | | | • | | | | | | • |
| A. glutinosa | | | • | | • | • | | | | • | | | | | | • |
| A. jorullensis | | | • | | • | • | | | | • | | | | | | |
| Alyxia buxifolia | | • | | | | | | | | | • | | | | | |
| Amelanchier canadensis | | • | | | • | • | | | | | | | • | | • | • |
| Angophora costata | | | • | | • | • | | | | | | | | | | |
| A. floribunda | | | • | | • | • | | | | | | | | • | | |
| Araucaria araucana | • | | | | • | • | | | | | | | | | | |
| A. bidwillii | • | | • | | • | • | | | | | | | | | | |
| A. cunninghamii | • | | | | • | • | | • | | | | | | | | |
| A. heterophylla | • | | | | • | • | | • | | | | | • | | | |
| Arbutus menziesii | | • | | | • | | | | | | • | | | | | |
| A. unedo | | • | | | • | | • | | | | • | | | | | |
| Atriplex nummularia | | • | | | | | | | | | • | • | | | | |
| Azara microphylla | | • | | | • | • | • | | | | • | | • | | | |
| Baeckea virgata | | • | | | | | | | | | • | | | | | |
| Banksia ashbyii | | | | | • | | | | | | | | | • | • | |
| B. attenuata | | • | | | • | | | | | | | | | • | • | |

## USES

| SPECIES | Farm forests | Windbreaks Low | Windbreaks High | Shade | Parks | Road sides | Streets | Av-enues | Sand drift | Gully erosion | H'dges | Fodder | Ornamental | Honey | Birds | Autumn foliage |
|---|---|---|---|---|---|---|---|---|---|---|---|---|---|---|---|---|
| B. ericifolia | | • | | | | • | | | | | | | | • | • | |
| B. grandis | | | | • | • | • | • | | | | | | | • | • | |
| B. ilicifolia | | • | | • | | | | | | | | | | • | • | |
| B. integrifolia | | | • | | | | | | • | | | | | • | • | |
| B. littoralis | | • | | | • | • | | | | | | | | • | • | |
| B. marginata | | | • | | • | | | | • | | | | | • | • | |
| B. menziesii | | | • | | • | | | | | | | | | • | • | |
| B. ornata | | • | | | • | • | | | | | | | | • | • | |
| B. prionotes | | | • | | | • | | | | | | | • | • | • | |
| B. serrata | | | • | | • | | | | • | | | | | • | • | |
| B. spinulosa | | • | | | | | | | | | | | | • | • | |
| Betula lutea | • | | | | • | • | | | | | | | • | | | • |
| B. maximowiczii | | | | | • | • | | | | | | | • | | | • |
| B. papyrifera | | | | | • | • | | | | | | | • | | | • |
| B. pendula | | | | | • | • | | | | | | | • | | | • |
| Brachychiton acerifolius | | | | • | • | | | • | | | | | • | | | • |
| B. discolor | | | | | • | | | • | | | | | | | | • |
| B. populneus | | | | • | • | | | • | | | | • | | • | | |
| Bursaria spinosa | | • | | | | | | | | • | • | | | | | |
| Callistemon acuminatus | | • | | | • | | • | | | | | | • | | • | |

# USES

| SPECIES | Autumn foliage | Birds | Honey | Ornamental | Fodder | H'dges | Control of — Gully erosion | Control of — Sand drift | Avenues | Streets | Road sides | Parks | Shade | Windbreaks High | Windbreaks Low | Farm forests |
|---|---|---|---|---|---|---|---|---|---|---|---|---|---|---|---|---|
| C. citrinus | | • | | • | | • | | | | • | | • | | | • | |
| C. lilacinus | | • | | • | | | | | | • | | • | | | • | |
| C. linearis | | • | | • | | • | | | | • | | • | | | • | |
| C. macropunctatus | | • | | • | | • | | | | • | | • | | | • | |
| C. pallidus | | • | | • | | | | | | • | | • | | | • | |
| C. paludosus | | • | | • | | | | | | • | | • | | | • | |
| C. phoeniceus | | • | | • | | • | | | | • | | • | | | • | |
| C. pinifolius | | • | | • | | | | | | • | | • | | | • | |
| C. rigidus | | • | | • | | | | | | • | | • | | | • | |
| C. salignus | | • | | • | | | | | • | • | | • | | | • | |
| C. shiresii | | • | | • | | • | | | | • | | • | | | • | |
| C. sieberi | | • | | • | | | | | | • | | • | | | • | |
| C. speciosus | | • | | • | | | | | | • | | • | | | • | |
| C. teretifolius | | • | | • | | | | | | • | | • | | | • | |
| C. viminalis | | • | | • | | | | | • | • | | • | | | • | |
| Callitris columellaris | | | | | | | | | • | | • | • | • | • | | • |
| C. endlicheri | | | | | | | | | • | | • | • | • | • | | • |
| C. oblonga | | | | • | | | | | | | | | | | • | |
| C. preissii | | | | | | | | | • | | • | | • | • | | • |
| C. rhomboidea | | | | | | | | | • | | • | • | | | • | |

## USES

| SPECIES | Farm forests | Windbreaks Low | Windbreaks High | Shade | Parks | Road sides | Streets | Av-enues | Sand drift | Gully erosion | H'dges | Fodder | Orna-mental | Honey | Birds | Autumn foliage |
|---|---|---|---|---|---|---|---|---|---|---|---|---|---|---|---|---|
| C. verrucosa | | • | | | | | | | | | • | | | | | |
| Calocedrus decurrens | | | | | • | • | | • | | | | | • | | | |
| Calodendrum capense | | | | | • | • | | • | | | | | • | | | |
| Calothamnus asper | | • | | | | | | | | | | | • | | • | |
| C. chrysantherus | | • | | | | | | | | | | | • | | • | |
| C. gilesii | | • | | | | | | | | | • | | • | | • | |
| C. quadrifidus | | • | | | | | | | | | • | | • | | • | |
| C. rupestris | | • | | | | | | | | | | | • | | • | |
| C. sanguineus | | • | | | | | | | | | | | • | | • | |
| C. schaueri | | • | | | | | | | | | | | • | | • | |
| C. torulosa | | • | | | | | | | | | | | • | | • | |
| C. villosus | | • | | | | | | | | | | | • | | • | |
| Capparis mitchellii | | • | | • | • | | | | | | | • | | | | |
| Carpinus betulus | | • | | | • | • | | • | | | • | | • | | | • |
| Carya illinoensis | | | • | • | | • | | | | | | | • | | | • |
| C. ovata | | | • | • | | • | | | | | | | • | | | • |
| Cassia artemisioides | | • | | | • | | | | | | • | | • | | | |
| C. coluteoides | | • | | | • | | | | | | | | • | | | |
| C. corymbosa | | • | | | • | | | | | | • | | • | | | |
| C. nemophila | | • | | | • | | | | | | • | • | • | | | |

# USES

| SPECIES | Farm forests | Windbreaks Low | Windbreaks High | Shade | Parks | Road sides | Streets | Avenues | Control of Sand drift | Control of Gully erosion | H'dges | Fodder | Ornamental | Honey | Birds | Autumn foliage |
|---|---|---|---|---|---|---|---|---|---|---|---|---|---|---|---|---|
| C. nemophila *var.* coriacea | | • | | | • | | | | | | | | • | | | |
| Castanea dentata | | | • | • | • | | | | | | | | • | | | • |
| C. sativa | | | | • | • | | | | | | | | • | | | • |
| Castanospermum australe | | | • | • | • | • | | | | | | | • | | | |
| Casuarina cristata | • | | • | • | • | • | | | | | | • | | | | |
| C. cunninghamiana | • | | • | | • | • | | • | | • | | • | | | | |
| C. glauca | • | | • | • | • | • | | | | • | • | • | | | | |
| C. leuhmannii | • | | • | • | • | • | | | | | | | | | | |
| C. littoralis | • | | • | | • | • | | | | | | | | | | |
| C. muellerana | | • | | | | | | | | | | | | | | |
| C. obesa | | | • | | | | | | | | | | • | | | |
| C. paludosa | | • | | | • | • | | | | | | | • | | | |
| C. stricta | • | | • | • | • | • | | • | | | | • | | | | |
| C. torulosa | • | | • | | • | • | | • | | | | | | | | |
| Catalpa bignonioides | | | | • | • | • | | | | | | | • | | | |
| C. speciosa | | | | | • | | | | | | | | • | | | • |
| Cedrela sinensis | | | | • | | | | | | | • | | • | | | |
| Cedrus atlantica | | | • | | • | • | | • | | | | | | | | |
| C. deodara | | | • | | • | • | | • | | | | | | | | |
| C. libanii | | | • | | • | • | | • | | | | | | | | |

## USES

| Species | Farm forests | Windbreaks – Low | Windbreaks – High | Shade | Parks | Road sides | Streets | Avenues | Control of – Sand drift | Control of – Gully erosion | H'dges | Fodder | Ornamental | Honey | Birds | Autumn foliage |
|---|---|---|---|---|---|---|---|---|---|---|---|---|---|---|---|---|
| Celtis australis | | | | • | | | • | | | | | | | | | • |
| C. occidentalis | | | • | • | • | • | • | | | | | | • | | | • |
| Ceratonia siliqua | | | • | • | • | • | | | | | • | • | | | | |
| Ceratopetalum apetalum | | | | | • | • | | | | | • | | | | | |
| C. gummiferum | | | | | • | • | | | | | | | • | | | |
| Cercidiphyllum japonicum | | | | • | • | • | | • | | | | | • | | | • |
| C. japonicum var. sinensis | | | | • | • | • | | | | | | | • | | | • |
| Cercis siliquastrum | | • | | | • | | • | | | | | | • | | | • |
| Chamaecyparis lawsoniana | | | • | | • | | • | • | | | | | • | | | |
| C. nootkatensis | | | • | | • | | | • | | | | | • | | | |
| C. obtusa | | | • | | • | | • | • | | | | | • | | | |
| C. pisifera | | | • | | • | | • | | | | | | • | | | |
| C. thyoides | | | • | | • | | • | • | | | | | • | | | |
| Chamaecytisus proliferus | | • | | | | | | | | | • | • | | | | |
| Chamaelaucium uncinatum | | • | | | | | | | | | • | | • | | • | |
| Cinnamomum camphora | | | | | | | • | • | | | • | | • | | | |
| Codonocarpus cotinifolius | | | | | | | | | | | | | • | | | |
| Combretum caffrum | | • | | | • | • | | | | | • | | | | | • |
| Coprosma repens | | • | | | • | • | | | | | • | | | | | |
| Cornus capitata | | • | | | • | | | • | | | | | | | | • |

324

# USES

| SPECIES | Farm forests | Windbreaks Low | Windbreaks High | Shade | Parks | Road sides | Streets | Avenues | Control of Sand drift | Control of Gully erosion | H'dges | Fodder | Ornamental | Honey | Birds | Autumn foliage |
|---|---|---|---|---|---|---|---|---|---|---|---|---|---|---|---|---|
| C. florida | | | | | ● | ● | | ● | | | | | | | | ● |
| C. kousa | | | | | ● | ● | | | | | | | ● | | | ● |
| C. kousa *var.* chinensis | | | | | ● | ● | | | | | | | | | | ● |
| Corynocarpus laevigatus | | | | ● | ● | ● | | | | | ● | | ● | | | |
| Cotinus obovatus | | | | | ● | | | | | | | | | | | ● |
| Crataegus crus-gallii | | | | | ● | ● | | | | | ● | | ● | | ● | ● |
| C. monogyna | | | | | ● | ● | | | | | ● | | ● | | ● | ● |
| C. phaenopyrum | | | | | ● | ● | ● | ● | | | ● | | ● | | ● | ● |
| C. pubescens | | | | | ● | ● | | | | | ● | | ● | | ● | ● |
| C. tanacetifolia | | | | | ● | ● | | | | | ● | | ● | | ● | ● |
| Crotalaria laburnifolia | | | | | ● | | | | | | | | ● | | | |
| Cryptomeria japonica *var.* elegans | | | | | ● | | ● | ● | | | | | ● | | | ● |
| Cupressus benthamii | | | ● | | ● | ● | | | | | ● | | | | | |
| C. funebris | | | ● | | ● | ● | | | | | ● | | ● | | | |
| C. glabra | | | ● | | ● | ● | | | | | | | | | | |
| C. lusitanica | ● | | ● | | ● | ● | | | | | ● | | | | | |
| C. macrocarpa | | | ● | | ● | | | | | | ● | | | | | |
| C. macrocarpa *var.* lambertiana | | | ● | | ● | | | | | | ● | | | | | |
| C. sempervirens | | | ● | | ● | | | ● | | | ● | | ● | | | |
| C. sempervirens *var.* stricta | | | ● | | ● | | | ● | | | ● | | ● | | | |

## USES

| SPECIES | Farm forests | Windbreaks Low | Windbreaks High | Shade | Parks | Road sides | Streets | Avenues | Sand drift | Gully erosion | H'dges | Fodder | Ornamental | Honey | Birds | Autumn foliage |
|---|---|---|---|---|---|---|---|---|---|---|---|---|---|---|---|---|
| C. torulosa | | | • | | • | | | • | | | • | | • | | | |
| Dacrydium cupressinum | | | | | • | | | | | | | | • | | | |
| Dais cotinifolia | | | | | • | • | | | | | | | • | | | |
| Dodonaea angustissima | | • | | | | | | | | | • | • | • | | | |
| D. cuneata | | • | | | | | | | | | • | | • | | | |
| D. microzyga | | • | | | | | | | | | • | | • | | | |
| D. viscosa | | • | | | | | | | | | • | • | • | | | |
| D. viscosa var. purpurea | | • | | | | | • | | | | • | | • | | | |
| Dombeya natalensis | | | | | • | • | | | | | | | • | | | • |
| Dryandra formosa | | | | | • | • | | | | | | | • | | | |
| D. nobilis | | | | | • | • | | | | | | | • | | | |
| D. sessilis | | | | | • | • | | | | | | | • | | | |
| Elaeagnus angustifolia | | • | | | • | • | | | | | • | | | | | |
| Elaeocarpus dentatus | | | | | • | • | | | | | | | • | | | |
| E. reticulatus | | • | | | • | • | | | | | | | | | | |
| Eremophila alternifolia | | • | | • | • | • | | | | | | | • | | | |
| E. longifolia | | • | | • | • | • | | | | • | | • | • | | | |
| E. maculata | | • | | • | • | • | | | | | | • | • | | | |
| Erythrina crista-gallii | | | | | • | • | | | | | | | • | | | • |
| E. variegata | | | | | • | • | | | | | | | • | | | • |

# USES

| SPECIES | Farm forests | Windbreaks Low | Windbreaks High | Shade | Parks | Road sides | Streets | Av-enues | Control of Sand drift | Control of Gully erosion | H'dges | Fodder | Orna-mental | Honey | Birds | Autumn foliage |
|---|---|---|---|---|---|---|---|---|---|---|---|---|---|---|---|---|
| Eucalyptus acaciaeformis | | | • | • | • | | | | | | | | | | | |
| E. accedens | | | | | • | • | | | | | | | | | | |
| E. aggregata | | | • | • | | • | | | | | | | | | | |
| E. alpina | | • | | | • | • | • | | | | | | | | • | |
| E. amplifolia | | | • | | | • | | | | | | | | | | |
| E. annulata | | • | | | • | • | | | | | | | • | • | | |
| E. approximans | | • | | | • | | • | | | | | | | | | |
| E. archeri | | • | | | | | | | | | | | | | | |
| E. astringens | • | | • | • | | | | | | | | | | | | |
| E. baxteri | • | | • | | | | | | | | | | | | | |
| E. bosistoana | • | | • | • | • | | | | | | | | | • | | |
| E. botryoides | | | | • | • | | • | | | | | | | • | | |
| E. brockwayii | • | | • | • | • | • | | | | | | | | | | |
| E. burdettiana | | • | | | • | | | | | | | | • | | | |
| E. burracoppinensis | | • | | | • | | | | | | | | | | | |
| E. caesia | | • | | | • | | • | | | | | | • | | • | |
| E. calophylla | | | | | • | • | • | | | | | | • | • | • | |
| E. calophylla var. rosea | | | | | • | • | • | | | | | | • | | • | |
| E. calycogona | | • | | | • | | | | | | | | | | | |
| E. camaldulensis | • | | • | • | • | | | | | | | | | • | | |

## USES

| SPECIES | Farm forests | Windbreaks Low | Windbreaks High | Shade | Parks | Road sides | Streets | Avenues | Control of — Sand drift | Control of — Gully erosion | H'dges | Fodder | Ornamental | Honey | Birds | Autumn foliage |
|---|---|---|---|---|---|---|---|---|---|---|---|---|---|---|---|---|
| E. campaspe | | • | | • | • | | • | | | | | | • | | | |
| E. chapmaniana | | | • | • | • | | | | | | | | | | | |
| E. cinerea | | • | | • | • | | • | | | | | | • | • | | |
| E. citriodora | • | | | | • | • | | • | | | | | • | | | |
| E. cladocalyx | • | | • | | | | | | | | | | | • | | |
| E. cladocalyx var. nana | | | | | | | | | | | | | | | | |
| E. coccifera | | • | • | | | • | | | | | | | | | | |
| E. cornuta | | • | | | • | • | | | | | | | • | | • | |
| E. corrugata | | • | | | • | • | | | | | | | • | | | |
| E. cosmophylla | • | • | | | | • | • | | | | | | | • | • | |
| E. crebra | • | • | • | • | • | | • | | | | | | | • | | |
| E. crenulata | | • | | | • | | | | | | | | • | | | |
| E. crucis | | • | | | | | | | | | | | • | | | |
| E. dawsonii | • | • | • | | | • | | | | | | | | | | |
| E. decipiens | | | | • | | | | | | | | | | | | |
| E. delegatensis | • | | • | | | | | | | | | | | | | |
| E. desmondensis | | • | | | | | | | | | | | • | | • | |
| E. dielsii | | • | | | | • | | | | | | | • | | | |
| E. diptera | | • | | • | • | | | | | | | | | | | |
| E. diversifolia | • | | | | • | • | | | | | | | | • | | |

# USES

| SPECIES | Farm forests | Windbreaks Low | Windbreaks High | Shade | Parks | Road sides | Streets | Avenues | Control of Sand drift | Control of Gully erosion | H'dges | Fodder | Ornamental | Honey | Birds | Autumn foliage |
|---|---|---|---|---|---|---|---|---|---|---|---|---|---|---|---|---|
| E. doratoxylon | | | | | • | | • | | | | | | • | • | | |
| E. dumosa | | • | | | | | | | | | | | | • | | |
| E. dundasii | • | | • | | | | | | | | | | | | | |
| E. ebbanoensis | | • | | | | • | | • | | | | | | | | |
| E. elata | • | | • | | • | | | • | | | | | • | | | |
| E. eremophila | | • | | | • | • | • | | | | | | • | • | • | |
| E. erythrocorys | | • | | | • | | • | | | | | | • | | • | |
| E. erythronema | | | | | • | | | • | | | | | • | | • | |
| E. fasciculosa | | | | • | • | • | | | | | | | • | • | • | |
| E. ficifolia | | | | • | • | | • | • | | | | | • | • | • | |
| E. flocktoniae | • | | • | | | | | | | | | | | | | |
| E. foecunda | | • | | | | | • | | | | | | | | | |
| E. forrestiana | • | • | | | • | • | • | • | | | | | • | • | | |
| E. gardneri | • | | | • | | | | | | | | | • | | • | |
| E. gillii | | • | | | • | • | | | | | | | | • | • | |
| E. glaucescens | • | | | | • | | | • | | | | | • | | | |
| E. globulus | • | • | • | • | • | | • | • | | | | | • | • | | |
| E. globulus var. compacta | | | | | | | | | | | | | | | | |
| E. gomphocephala | | | | • | • | • | | | | | | | | • | • | |
| E. gracilis | | • | | • | | • | | | | | | | | • | | |

329

## USES

| SPECIES | Farm forests | Windbreaks Low | Windbreaks High | Shade | Parks | Road sides | Streets | Av-enues | Sand drift | Gully erosion | H'dges | Fodder | Ornamental | Honey | Birds | Autumn foliage |
|---|---|---|---|---|---|---|---|---|---|---|---|---|---|---|---|---|
| E. griffithsii | | • | | | | • | | | | | | | • | | • | |
| E. grossa | | • | | | | | | | | | | | • | • | • | |
| E. gummifera | • | | • | • | • | • | | | | | | | | • | | |
| E. gunnii | • | | • | • | | | | | | | | | | | | |
| E. incrassata | | • | | | • | • | | | • | | | | • | • | • | |
| E. johnstonii | • | | • | | | | | | | | | | | | | |
| E. kingsmillii | | • | | | • | | | | | | | | • | | • | |
| E. kitsoniana | | • | • | | • | • | • | | | | | | • | | | |
| E. kondininensis | | | | | • | • | | • | | | | | • | | • | |
| E. kruseana | | | | | • | | | | | | | | • | | | |
| E. lacliae | | | | • | | • | | | | | | | • | | | |
| E. lansdowneana | • | • | | | | • | | | | | | | | • | | |
| E. largiflorens | • | | • | | • | | | | | | | | | • | | |
| E. lehmannii | | • | | | • | • | | | | | | | • | | • | |
| E. leptopoda | | • | | | • | • | • | | | | | | | | | |
| E. le souefii | • | • | | | • | • | | | | | | | | | | |
| E. leucoxylon | • | | • | • | • | • | | • | | | | | • | • | • | |
| E. leucoxylon *var.* rosea | • | • | • | • | • | • | | • | | | | | • | • | • | |
| E. leucoxylon *var.* macrocarpa | • | | | • | • | • | | • | | | | | • | • | • | |
| E. linearis | | • | | | | • | | • | | | | | • | | | |

# USES

| SPECIES | Farm forests | Windbreaks Low | Windbreaks High | Shade | Parks | Road sides | Streets | Av-enues | Sand drift | Gully erosion | H'dges | Fodder | Orna-mental | Honey | Birds | Autumn foliage |
|---|---|---|---|---|---|---|---|---|---|---|---|---|---|---|---|---|
| E. longicornis | ● | | ● | ● | ● | | | | | | | | | | | |
| E. loxophleba | | ● | | ● | ● | | | | | | | | | ● | | |
| E. macrandra | | ● | | | ● | ● | ● | | | | | | ● | ● | ● | |
| E. macrocarpa | | | | | | | | | | | | | ● | | | |
| E. macrorhyncha | ● | | ● | ● | ● | | | | | | | | | ● | | |
| E. maculata | ● | | ● | ● | ● | ● | | | | | | | ● | ● | | |
| E. mannifera subsp. maculosa | | | | | | ● | | | | | | | ● | | | |
| E. melliodora | ● | | ● | ● | ● | ● | | | | | | | | ● | ● | |
| E. microcarpa | ● | | ● | ● | ● | | | | | | | | | ● | | |
| E. miniata | | | | | ● | ● | ● | | | | | | ● | | ● | |
| E. mitchelliana | | ● | | | ● | | | ● | | | | | ● | | ● | |
| E. mitrata | | ● | | | ● | | | | | | | | ● | | ● | |
| E. moorei | | ● | | | ● | | | | | | | | | ● | | |
| E. muellerana | ● | | ● | ● | | | | | | | | | | ● | | |
| E. neglecta | | ● | | | ● | ● | | | | | | | | | | |
| E. nicholii | | | | ● | ● | ● | ● | | | | | | ● | | | |
| E. nitens | ● | | | | | | | | | | | | | | | |
| E. nutans | | ● | | | ● | ● | ● | | | | | | ● | | ● | |
| E. obliqua | ● | | ● | ● | | | | | | | | | | ● | | |
| E. occidentalis | ● | | ● | ● | | | | | | | | | | ● | | |

331

## USES

| SPECIES | Autumn foliage | Birds | Honey | Ornamental | Fodder | H'dges | Gully erosion | Sand drift | Avenues | Streets | Road sides | Parks | Shade | High | Low | Farm forests |
|---|---|---|---|---|---|---|---|---|---|---|---|---|---|---|---|---|
| E. oleosa | | | • | | | | | | | | • | | • | | | |
| E. orbifolia | | • | | • | | | | | | | | • | | | | |
| E. oreades | | | | | | | | | | | | | | | | • |
| E. ovata | | | • | | | | | | | | | • | | • | | |
| E. paniculata | | | • | | | | | | | | | • | • | • | | • |
| E. pauciflora | | • | | | | | | | | | | • | • | • | | |
| E. pauciflora 'Alpine form' | | • | • | | | | | | | | | • | | | • | |
| E. pauciflora var. nana | | • | | | | | | | | | | • | | | • | |
| E. pellita | | | | | | | | | | | • | • | • | • | | • |
| E. perriniana | | | | • | | | | | | | | | | | • | |
| E. pileata | | | | | | | | | | | • | • | | | • | |
| E. platycorys | | | | • | | | | | | | • | • | | | • | |
| E. platypus | | • | • | | | | | | | • | • | • | • | | • | |
| E. platypus var. heterophylla | | | | | | | | | | • | • | • | • | • | • | |
| E. polyanthemos | | | • | | | | | | | | • | • | | | • | • |
| E. polybractea | | | • | | | | | | | | | • | | | • | |
| E. porosa | | | • | | | | | | | | | • | | | • | |
| E. preissiana | | • | | • | | | | | | | | • | | | • | |
| E. pulverulenta | | | | • | | | | | | | | • | | | • | |
| E. punctata | | | | | | | | | | | | • | • | | | • |

# USES

| SPECIES | Farm forests | Windbreaks Low | Windbreaks High | Shade | Parks | Road sides | Streets | Av-enues | Control of Sand drift | Control of Gully erosion | H'dges | Fodder | Ornamental | Honey | Birds | Autumn foliage |
|---|---|---|---|---|---|---|---|---|---|---|---|---|---|---|---|---|
| E. pyriformis | | • | | | | | | | | | | | • | | • | |
| E. radiata | • | | • | | • | • | | | | | | | | | | |
| E. regnans | • | | | • | | | | | | | | | | | | |
| E. rhodantha | | • | | | • | • | | | | | | | • | | | |
| E. risdonii | | • | | | • | | | | | | | | • | | | |
| E. robusta | • | | • | | • | • | | | | | | | | • | | |
| E. rubida | • | | • | | • | • | | | | | | | • | • | | |
| E. saligna | • | | • | | • | • | | | | | | | | • | | |
| E. salmonophloia | • | | • | • | • | | • | | | | | | | • | | |
| E. salubris | | • | | | • | | • | | | | | | • | | | |
| E. sargentii | | • | | • | • | | • | | | | | | | | | |
| E. scoparia | | • | | | • | • | | | | | | | • | | | |
| E. sepulcralis | • | | • | • | • | | • | | | | | | • | | • | |
| E. sideroxylon | | | | | | | | | | | | | • | • | • | |
| E. sideroxylon *var.* pallens | | | | | | | | | | | | | • | | • | |
| E. sieberi | • | | | • | | | | | | | | | | • | | |
| E. socialis | | • | | | | | | | | | | | | • | | |
| E. spathulata | | • | | | • | • | • | | | | | | • | | | |
| E. steedmanii | | • | | | • | • | • | | | | | | • | | • | |
| E. stellulata | | • | | • | | • | | | | | | | • | • | • | |

333

## USES

| SPECIES | Autumn foliage | Birds | Honey | Ornamental | Fodder | H'dges | Gully erosion | Sand drift | Avenues | Streets | Road sides | Parks | Shade | Windbreaks High | Windbreaks Low | Farm forests |
|---|---|---|---|---|---|---|---|---|---|---|---|---|---|---|---|---|
| E. st johnii | | | • | | | | | | | | | | • | • | | • |
| E. stoatei | | | | • | | | | | | • | • | • | | | • | |
| E. striaticalyx | | | | | | | | | | | • | • | | | • | |
| E. stricklandii | | • | | • | | | | | | • | • | • | • | | | |
| E. tasmanica | | | | | | | | | | | | • | | | • | • |
| E. tetragona | | • | | • | | | | | | | | • | | | • | |
| E. tetraptera | | • | | • | | | | | | | | • | | | | |
| E. torquata | | • | | • | | | | | | • | • | • | | | | |
| E. urnigera | | | | • | | | | | | | | • | • | | • | |
| E. viminalis | | | • | • | | | | | • | | • | • | | • | | • |
| E. viridis | | | • | • | | | | | | | • | • | | | • | |
| E. wandoo | | | • | | | | | | | | | | • | • | | • |
| E. websterana | | | | • | | | | | | | | • | | | • | |
| E. woodwardii | | • | | • | | | | | | • | • | • | • | | | |
| E. youngiana | | • | | • | | | | | | | | • | | | • | |
| Eucryphia cordifolia | | | | • | | | | | | | | • | | | | |
| E. lucida | | | • | | | | | | | | • | • | • | • | | |
| Eugenia australis | | | | | | | | | | | • | • | | | | |
| E. smithii | | | | | | • | | | | • | | • | | | • | |
| E. ventenatii | | | | • | | • | | | | | | • | | | | |

# USES

| SPECIES | Farm forests | Windbreaks Low | Windbreaks High | Shade | Parks | Road sides | Streets | Avenues | Sand drift | Gully erosion | H'dges | Fodder | Ornamental | Honey | Birds | Autumn foliage |
|---|---|---|---|---|---|---|---|---|---|---|---|---|---|---|---|---|
| Fagus sylvatica | | | | | | | | • | | | | | | | | • |
| Ficus macrophylla | | | • | • | • | | | | | | | | | | | |
| Flindersia maculosa | | | | • | • | | | | | | | • | • | | | |
| Fraxinus americana | • | | • | • | • | • | | | | | | | • | | | • |
| F. angustifolia | | | • | • | • | • | | | | | | | • | | | • |
| F. excelsior | • | | | | | | • | • | | | | | • | | | • |
| F. excelsior var. aurea | • | | | | | | • | • | | | | | • | | | • |
| F. lanceolata | | | • | | • | • | | | | | | | | | | • |
| F. ornus | | | | | | | • | • | | | | | | | | • |
| F. oxycarpa | | | | | | | • | | | | | | | | | • |
| F. x 'raywoodii' | | | | | | | • | • | | | | | • | | | • |
| F. syriaca | | • | | | • | • | | | | | | | | | | • |
| F. velutina | | | • | | • | • | | | | | | | | | | • |
| Geijera parviflora | | • | | • | • | • | | | | | | • | | • | | |
| Ginkgo biloba | | | | | • | | • | | | | | | • | | | • |
| Gleditsia triacanthos | | | • | • | • | • | • | • | | • | • | • | | | | • |
| G. triacanthos var. inermis | | | • | • | • | • | • | | | • | | • | | | | |
| Grevillea asplenifolia | | • | | | • | | • | | | • | • | | | | • | |
| G. banksii | | • | | | • | | | | | | | | | | • | |
| G. barklyana | | | | | • | • | | | | | | | | | • | |

335

## USES

| SPECIES | Farm forests | Windbreaks Low | Windbreaks High | Shade | Parks | Road sides | Streets | Avenues | Control of Sand drift | Control of Gully erosion | H'dges | Fodder | Ornamental | Honey | Birds | Autumn foliage |
|---|---|---|---|---|---|---|---|---|---|---|---|---|---|---|---|---|
| G. confertifolia | | • | | | | | | | | | | | | | • | |
| G. glabrata | | • | | | • | • | | | | | | | | | • | |
| G. hookeriana | | • | | | • | • | | | | | | | | | • | |
| G. ilicifolia | | | | | • | | | | | | | | | | • | |
| G. lavandulacea | | • | | | • | | | | | | • | | | | • | |
| G. longifolia | | • | | | • | | • | • | | • | • | | | | • | |
| G. nematophylla | | • | | | • | • | | | | | | | | | • | |
| G. robusta | | | • | • | • | • | | | | | | | | • | • | |
| G. rosmarinifolia | | • | | | | | | | | | • | | | | • | |
| G. striata | | • | | | • | | | | | | | • | | | • | |
| G. thelemanniana | | | | | | | | | | | | | • | | • | |
| G. vestita | | | | | | | | | | | | | | | • | |
| G. victoriae | | | | | | | | | | | | | | | • | |
| Griselina littoralis | | • | | | | | | | | | • | | | | | |
| Hakea bucculenta | | • | | | • | • | • | | | | | | | • | • | |
| H. elliptica | | • | | | • | • | • | | | | | | | • | • | |
| H. eriantha | | • | | | • | | | | | | | | | • | • | |
| H. francisiana | | | | | • | • | • | | | | | | | • | • | |
| H. laurina | | • | | | • | • | • | | | | | | | • | • | |
| H. leucoptera | | • | | | • | | • | | | | • | | | • | • | |

# USES

| SPECIES | Farm forests | Windbreaks Low | Windbreaks High | Shade | Parks | Road sides | Streets | Avenues | Sand drift | Gully erosion | H'dges | Fodder | Ornamental | Honey | Birds | Autumn foliage |
|---|---|---|---|---|---|---|---|---|---|---|---|---|---|---|---|---|
| H. muellerana | | • | | | • | • | • | | | | | | | • | • | |
| H. multilineata | | • | | | • | • | • | | | | • | | | • | • | |
| H. nodosa | | • | | | • | • | • | | | | | | | • | • | |
| H. petiolaris | | • | | | • | • | • | | | | | | | • | • | |
| H. preissii | | • | | | • | • | • | | | | | | • | • | • | |
| H. pubescens | | • | | | • | • | • | | | | | | | • | • | |
| H. salicifolia | | • | | | • | • | • | | | | | | | • | • | |
| H. sericea | | | | | • | | | | | | | | | | | |
| H. suaveolens | | • | | | • | • | • | | | | • | | | • | • | |
| H. ulicina | | • | | | • | • | • | | | | | | | • | • | |
| H. victoriae | | • | | | • | • | • | | | | | | | • | • | |
| H. vittata | | • | | | • | • | • | | | | | | | • | • | |
| Halesia carolina | | • | | • | • | • | | • | | | | | • | | | • |
| Heterodendrum oleifolium | | • | | | • | • | • | | | | | • | | | | |
| Hoheria populnea | | • | | • | • | • | | | | | | | • | | | |
| Homalanthus populifolius | | • | | • | • | • | | | | | | | • | | | |
| Hovenia dulcis | | | | • | • | | | | | | | | | | | |
| Hymensporum flavum | | | | • | • | • | • | | | | | | • | | | • |
| Idesia polycarpa | | | | | • | | • | • | | | | | • | | | • |
| Ilex aquifolium | | | | | • | | • | • | | | • | | • | | | |

337

# USES

| SPECIES | Farm forests | Windbreaks Low | Windbreaks High | Shade | Parks | Road sides | Streets | Avenues | Control of Sand drift | Control of Gully erosion | H'dges | Fodder | Ornamental | Honey | Birds | Autumn foliage |
|---|---|---|---|---|---|---|---|---|---|---|---|---|---|---|---|---|
| Jacaranda acutifolia | | | | • | • | • | | | | | | | | | | • |
| Juglans nigra | • | | • | • | • | | | | | | | | • | | | • |
| Juniperus communis | | | • | | • | • | | • | | | • | | • | | | |
| Koelreuteria paniculata | | • | | | • | • | • | | | | | | • | | | • |
| Kunzea ambigua | | | | | • | • | | | | | • | | • | | | |
| Laburnum anagyroides | | | | | • | • | | | | | | | • | | | |
| L. vossii | | | | | • | • | | | | | | | • | | | |
| Lagunaria patersonii | | • | | • | • | | | • | | | | | | | | |
| Larix decidua | • | | | • | • | • | | • | | | | | • | | | • |
| L. eurolepis | • | | | • | • | • | | • | | | | | • | | | • |
| Laurus nobilis | | | • | | • | | • | • | | | | | | | | |
| Leptospermum laevigatum | | • | | • | | | | | | | • | | | • | | |
| L. laevigatum var. minus | | | | | | | | | | | • | | | | | |
| L. lanigerum | | • | | | • | | | | | | • | | | • | | |
| L. phylicoides | | • | | | • | | | | | | • | | | • | | |
| L. scoparium | | • | | | • | | | | | | • | | | • | | |
| Leucadendron argenteum | | • | | | • | • | | • | | | | | • | | | |
| Leucopogon parviflora | | | | • | | | | | | | | | | | | |
| Liquidambar formosana | | | | • | • | | | | | | | | • | | | • |
| L. styraciflua | | | | • | • | | | • | | | | | • | | | • |

# USES

| SPECIES | Farm forests | Windbreaks High | Windbreaks Low | Shade | Parks | Road sides | Streets | Avenues | Sand drift | Gully erosion | H'dges | Fodder | Ornamental | Honey | Birds | Autumn foliage |
|---|---|---|---|---|---|---|---|---|---|---|---|---|---|---|---|---|
| Liriodendron tulipifera | | | | • | • | • | | | | | | | | | | • |
| Maclura pomifera | | • | | | | | | | | | • | | • | | | • |
| Magnolia grandiflora | | | | | • | | | • | | | | | | | | |
| Malus | | | | | • | • | | | | | | | | | | • |
| Melaleuca acuminata | | | • | | • | | | | | | | | • | | • | |
| M. armillaris | | | • | | • | | • | • | | | • | | • | | • | |
| M. crassifolia | | | • | | • | | | | | | | | • | | • | |
| M. decussata | | | • | | • | • | • | | | | • | | • | | • | |
| M. diosmifolia | | | • | | • | • | • | | | | • | | • | | • | |
| M. elliptica | | | • | | • | | | | | | | | • | | • | |
| M. ericifolia | | | • | | • | • | • | | | | • | | • | | • | |
| M. fulgens | | | • | | • | | | | | | | | • | | • | |
| M. gibbosa | | | • | | • | | | | | | | | • | | • | |
| M. glomerata | | | • | | • | • | | | | | | | • | | • | |
| M. halmaturorum | | | • | | • | | | | | | | | • | | • | |
| M. huegelii | | | • | | • | | | | | | • | | • | | • | |
| M. hypericifolia | | | • | | • | • | • | • | | | • | | • | | • | |
| M. incana | | | • | | • | • | • | | | | • | | • | | • | |
| M. lanceolata | | | • | | • | • | • | | | | | | • | | • | |
| M. lateritia | | | • | | • | | | | | | | | • | | • | |

## USES

| SPECIES | Farm forests | Windbreaks — Low | Windbreaks — High | Shade | Parks | Road sides | Streets | Avenues | Control of — Sand drift | Control of — Gully erosion | H'dges | Fodder | Ornamental | Honey | Birds | Autumn foliage |
|---|---|---|---|---|---|---|---|---|---|---|---|---|---|---|---|---|
| M. linariifolia | | • | | • | • | • | • | • | | | | | • | | • | |
| M. longicoma | | • | | | • | | | | | | | | • | | • | |
| M. neglecta | | • | | | • | | | | | | | | • | | • | |
| M. nesophila | | • | | | • | • | • | • | | | • | | • | | • | |
| M. pentagona | | • | | | • | | | | | | | | • | | • | |
| M. quinquenervia | | | • | • | • | • | | | | | | | • | | • | |
| M. radula | | • | | | • | | | | | | | | • | | • | |
| M. squarrosa | | • | | • | • | | • | | | | • | | • | | • | |
| M. styphelioides | | • | | • | • | | • | | | | | | • | | • | |
| M. wilsonii | | • | | | • | | | | | | • | | • | | • | |
| Melia azedarach | | | | • | • | • | | • | | | | | | | | • |
| Mespilus germanica | | | | | • | | | • | | | | | | | | • |
| Metasequoia glyptostroboides | • | • | • | • | • | • | | • | | | | | • | | | • |
| Metrosideros excelsa | | | | | • | | • | | | | • | | | | • | |
| Morus alba | | • | | | • | • | | • | | | | | | | • | • |
| Myoporum floribundum | | | | | | • | | | | | | | | | | |
| M. insulare | | • | | | | | | | | • | • | | • | | | |
| M. platycarpum | | • | | • | • | | | | | | | • | • | | | |
| Nothofagus cunninghamiana | | | • | | • | • | | | | | | | | | | |
| N. fusca | | | | | • | • | • | | | | | | | | | |

# USES

| SPECIES | Farm forests | Windbreaks Low | Windbreaks High | Shade | Parks | Road sides | Streets | Avenues | Control of Sand drift | Control of Gully erosion | H'dges | Fodder | Ornamental | Honey | Birds | Autumn foliage |
|---|---|---|---|---|---|---|---|---|---|---|---|---|---|---|---|---|
| N. obliqua | | | • | | • | • | | | | | | | • | | | |
| Nyssa sylvatica | | | • | • | | | | • | | | | | • | • | | • |
| Olea europaea | | • | | | | | • | | | | • | | • | | | |
| Oxydendrum arboreum | | • | | | • | • | • | | | | | | • | | | • |
| Parrotia persica | | | | | • | • | | | | | | | • | | | • |
| Paulownia tomentosa | | | • | • | • | • | | • | | | • | | • | | | • |
| Persoonia pinifolia | | | | | • | • | • | | | | | | | | | |
| Phebalium squameum | | • | | | • | • | | | | | • | | | | | |
| Photinia serrulata | | | | | • | • | • | • | | | • | | | | • | |
| Picea abies | • | | • | | • | | | • | | | | | | | | |
| P. engelmannii | • | | • | | • | | | • | | | | | | | | |
| P. omorika | • | | • | | • | | | • | | | | | | | | |
| P. pungens | • | | • | | • | | | | | | | | | | | |
| P. sitchensis | • | | • | • | • | | | | | | | | | | | |
| Pinus canariensis | • | | • | | • | • | | • | | | | | | | | |
| P. contorta | • | | • | | • | | | | | | | | | | | |
| P. halepensis | • | | • | | • | • | | | | | | | | | | |
| P. halepensis var. brutia | • | | • | | • | | | | | | | | | | | |
| P. nigra subsp. laricio | • | | • | | • | | | | | | | | | | | |
| P. patula | • | | • | | • | • | | • | | | | | • | | | |

## USES

| SPECIES | Farm forests | Windbreaks Low | Windbreaks High | Shade | Parks | Road sides | Streets | Avenues | Control of: Sand drift | Control of: Gully erosion | H'dges | Fodder | Ornamental | Honey | Birds | Autumn foliage |
|---|---|---|---|---|---|---|---|---|---|---|---|---|---|---|---|---|
| P. pinaster | • | | • | | • | • | | | | | | | | | | |
| P. pinea | • | | • | | • | | | • | | | | | | | | |
| P. ponderosa | • | | • | | • | • | | • | | | | | | | | |
| P. radiata | • | | • | | • | | | | | | | | | | | |
| P. strobus | • | | • | | • | | | | | | | | | | | |
| Pistacia chinensis | | | | | • | • | • | • | | | | | • | | | • |
| Pittosporum bicolor | | • | | | • | • | • | | | | • | | | | | |
| P. crassifolium | | • | | | • | • | • | | | | • | | | | | |
| P. eugenioides | | • | | | • | • | • | | | | • | | • | | | |
| P. phillyreoides | | • | | • | • | • | • | | | | • | • | • | | • | |
| P. tenuifolium | | • | | | • | • | | | | | • | | | | | |
| P. undulatum | | • | | • | • | • | | | | | • | | | | | |
| Platanus x acerifolia | | | • | • | • | • | | • | | | | | | | | • |
| P. orientalis | | | • | • | • | • | | • | | | | | | | | • |
| Podocarpus elatus | | • | | | | | • | | | | • | | | | | |
| P. lawrencei | | • | | | | | | | | | • | | | | | |
| P. totara | | • | | | | | | • | | | • | | | | • | |
| Populus alba | • | | • | • | • | • | | | | • | | | • | | | • |
| P. alba var. pyramidalis | • | | • | • | • | • | | | | • | | | • | | | • |
| P. candicans | • | | • | • | • | • | | | | • | | | • | | | • |

# USES

| SPECIES | Farm forests | Windbreaks | | Shade | Parks | Road sides | Streets | Avenues | Control of | | H'dges | Fodder | Ornamental | Honey | Birds | Autumn foliage |
|---|---|---|---|---|---|---|---|---|---|---|---|---|---|---|---|---|
| | | Low | High | | | | | | Sand drift | Gully erosion | | | | | | |
| P. canescens | ● | | ● | | | | | | | ● | | | | | | ● |
| P. deltoides *var.* missouriensis | ● | | ● | ● | ● | ● | | | | ● | | | ● | | | ● |
| P. deltoides *var.* angulata | ● | | ● | ● | ● | ● | | ● | | ● | | | ● | | | ● |
| P. x generosa | ● | | ● | ● | ● | ● | | | | ● | | | ● | | | ● |
| P. nigra *var.* italica | ● | | ● | ● | ● | ● | | ● | | ● | | | ● | | | ● |
| P. serotina *var.* aurea | ● | | ● | ● | ● | ● | | | | ● | | | ● | | | ● |
| P. simonii *c.v.* 'fastigiata' | ● | | ● | ● | ● | ● | | ● | | ● | | | ● | | | ● |
| P. tremula | ● | | ● | ● | ● | ● | | | | ● | | | ● | | | ● |
| P. tremuloides | ● | | ● | ● | ● | ● | | | | ● | | | ● | | | ● |
| P. trichocarpa | ● | | ● | ● | ● | ● | | ● | | ● | | | ● | | | ● |
| P. yunnanensis | ● | | ● | ● | ● | ● | | ● | | ● | | | ● | | | ● |
| Prostanthera lasianthos | | ● | | | | ● | ● | | | | | | ● | | | |
| P. nivea | | ● | | | | ● | | | | | ● | | ● | | | |
| P. rotundifolia | | ● | | | ● | ● | ● | ● | | | ● | | ● | | | |
| Prunus cerasifera *var.* nigra | | ● | | | ● | ● | ● | | | | | | ● | | | |
| P. laurocerasus | | ● | | | | | | | | | ● | | ● | | | ● |
| P. lusitanica | | ● | | | | | | | | | ● | | | | | |
| P. serrulata | | | | | ● | ● | ● | | | | | | ● | | | ● |
| Pseudotsuga menziesii | ● | | ● | | ● | ● | | ● | | | | | | | | |
| P. menziesii *var.* glauca | ● | | ● | | ● | ● | | ● | | | | | | | | |

343

## USES

| SPECIES | Farm forests | Windbreaks Low | Windbreaks High | Shade | Parks | Road sides | Streets | Avenues | Control of — Sand drift | Control of — Gully erosion | H'dges | Fodder | Ornamental | Honey | Birds | Autumn foliage |
|---|---|---|---|---|---|---|---|---|---|---|---|---|---|---|---|---|
| Pyracantha crenulata | | • | | | | | | | | | • | | | | • | |
| Pyrus calleryana | | | | | • | • | | | | | | | • | | | • |
| P. ussuriensis | | | • | • | • | • | | | | | | | • | | | • |
| Quercus aegilops | | | • | • | • | • | | | | | | | • | | | • |
| Q. alba | | | • | • | • | • | | • | | | | | • | | | • |
| Q. borealis | | | • | • | • | • | | • | | | | | • | | | • |
| Q. canariensis | | | • | • | • | | | | | | | | | | | • |
| Q. castaneifolia | | | • | • | • | • | | | | | | | • | | | • |
| Q. cerris | | | • | • | • | | | | | | | | | | | • |
| Q. coccinea | | | • | • | • | • | | | | | | | • | | | • |
| Q. faginea | | | • | • | • | • | | • | | | | | | | | • |
| Q. ilex | | | • | • | • | • | | | | | | | | | | |
| Q. lyrata | | | • | • | • | | | | | | | | | | | • |
| Q. palustris | | | • | • | • | • | | • | | | | | • | | | • |
| Q. palustris 'Macedon' | | | • | • | • | | | • | | | | | • | | | • |
| Q. robur | | | • | • | • | | • | | | | | | | | | • |
| Q. suber | | | • | • | • | | | | | | | | | | | |
| Q. virginiana | | | | • | • | | | | | | | | | | | |
| Rhus succedanea | | | | | • | • | • | | | | | | • | | | • |
| Robinia pseudoacacia | • | | • | • | • | • | | | | • | | | | | | • |

344

# USES

| SPECIES | Farm forests | Windbreaks | | Shade | Parks | Road sides | Streets | Av-enues | Control of | | H'dges | Fodder | Orna-mental | Honey | Birds | Autumn foliage |
|---|---|---|---|---|---|---|---|---|---|---|---|---|---|---|---|---|
| | | Low | High | | | | | | Sand drift | Gully erosion | | | | | | |
| *Salix alba* var. *coerulea* | | | • | • | • | | | | | • | | | | | | • |
| *S. alba* subsp. *vitellina* | | • | | • | • | | | | | • | | | • | | | • |
| *S. alba* subsp. *vitellina* var. *pendula* | | • | | • | • | • | | | | • | | | • | | | • |
| *S. babylonica* | | • | | • | • | | | • | | • | | • | • | | | • |
| *S. discolor* | | • | | • | • | | | | | • | | | • | | | • |
| '*S. fragilis*' | | • | | • | • | | | | | • | | | | | | • |
| *S. humboltiana* | | • | | • | • | • | | • | | • | | | | | | • |
| *S. matsudana* c.v. 'tortuosa' | | • | | • | • | | | • | | • | | | • | | | • |
| *Sambucus nigra* | | • | | | • | | | | | | • | | | | | • |
| *Schinus molle* | | • | | • | • | • | | | | | | | | | | • |
| *S. terebinthifolius* | | • | | • | • | • | | | | | | | | | | |
| *Sequoia sempervirens* | | | • | | • | • | • | • | | | | | | | | |
| *Sequoiadendron giganteum* | | | • | | • | • | • | • | | | | | | | | |
| *Sophora japonica* | | | | | • | • | • | • | | | | | • | | | • |
| *S. microphylla* | | | | | • | • | • | | | | | | • | | | |
| *S. tetraptera* | | | | | • | • | | | | | | | • | | | |
| *Sorbus aucuparia* | | • | | • | • | • | | | | | | | • | | • | • |
| *S. decora* | | | | | • | • | | | | | | | • | | | • |
| *Stenocarpus sinuatus* | | | | • | | • | | • | | | | | • | | | |
| *Stewartia pseudocamellia* | | | | | • | | | | | | | | • | | | • |

## USES

| SPECIES | Farm forests | Windbreaks Low | Windbreaks High | Shade | Parks | Road sides | Streets | Avenues | Sand drift | Gully erosion | H'dges | Fodder | Ornamental | Honey | Birds | Autumn foliage |
|---|---|---|---|---|---|---|---|---|---|---|---|---|---|---|---|---|
| Styrax japonica | | | | | • | • | • | • | | | | | • | | | • |
| Syncarpia glomulifera | • | | • | • | | | | | | | | | | | | |
| Tamarix aphylla | | • | | | • | • | | | | • | • | | | | | |
| T. juniperina | | • | | | • | • | | | | • | • | | | | | |
| T. parviflora | | • | | | • | • | | | | • | • | | | | | |
| T. pentandra | | • | | | • | • | | | | • | • | | | | | |
| T. tetrandra | | • | | | • | • | | | | • | • | | | | | |
| Taxodium distichum | • | | • | | • | | | • | | | | | • | | | • |
| Taxus baccata | | • | | | • | | • | | | | • | | • | | | |
| Telopea speciosissima | | | | | • | | | | | | | | • | | | |
| Templetonia retusa | | • | | | • | • | | | | | • | | | | | |
| Thryptomene calycina | | | | | • | | | | | | | | | | | |
| Thuja plicata | • | | | | • | • | | • | | | | | • | | | |
| Thujopsis dolobrata | | | • | • | • | • | | | | | | | • | | | |
| Tieghemopanax sambucifolius | | | | | • | • | | | | | | | • | | | |
| Tilia x vulgaris | | | • | | • | • | | • | | | | | | • | | • |
| Tristania conferta | | • | | | • | • | • | | | | | | | | | |
| T. laurina | | | | • | • | • | | | | | | | | | | |
| Tsuga heterophylla | • | | • | | • | | | • | | | | | | | | |
| Ulmus glabra | | | • | • | • | • | | • | | | | | | | | • |

# USES

| SPECIES | Farm forests | Windbreaks Low | Windbreaks High | Shade | Parks | Road sides | Streets | Avenues | Control of: Sand drift | Control of: Gully erosion | H'dges | Fodder | Ornamental | Honey | Birds | Autumn foliage |
|---|---|---|---|---|---|---|---|---|---|---|---|---|---|---|---|---|
| U. x hollandica | | | • | • | • | • | | • | | | • | | | | | • |
| U. parvifolia | | | • | • | • | • | | • | | | | | • | | | • |
| U. procera | | | • | • | • | • | | • | | • | | | | | | • |
| U. procera *var.* van houttei | | | • | • | • | • | | • | | | | | | | | • |
| U. pumila | | | • | • | • | • | | | | | | | | | | • |
| Viminaria juncea | | | | | • | • | | | | | | | • | | | |
| Virgilia divaricata | | | | | • | • | | | | | | | • | | | |
| V. oroboides | | | | | • | • | | | | | | | • | | | |
| Zelkova carpinifolia | | | • | • | • | • | | | | | | | | | | • |
| Z. serrata | | | • | • | • | • | | | | | | | | | | • |

347

# REFERENCES

Anderson, R. H. *The Trees of New South Wales*. Government Printer, Sydney, 1947. (pp. 453)

Andrews, H. V. *Ancient Plants and the World They Lived in*. Comstock Publishing Association, New York, 1964. (pp. 279)

Bailey, L. H. *The Standard Cyclopedia of Horticulture*. 3 vols. The Macmillan Company, New York, 1947. (pp. 3639)

Baker, F. S. *Principles of Silviculture*. McGraw-Hill Book Company Inc., New York and London, 1952. (pp. 414)

Baver, L. D. *Soil Physics* (2nd ed.). John Wiley & Sons, New York. Chapman and Hall Ltd, London, 1948. (pp. 398)

Beard, J. S. *Descriptive Catalogue of West Australian Plants*. Society for Growing Australian Plants and Surrey Beatty & Sons, Chipping Norton, New South Wales. (pp. 122)

Black, J. M. *Flora of South Australia*. 3 vols. 1963. (pp. 683). Supplement by H. Eichler, 1965. (pp. 385). Government Printer, Adelaide.

Blackall, W. F. *How To Know Western Australian Wildflowers*. University of Western Australia Press, Perth, 1954. (pp. 370)

Blakeley, W. F. *A Key to the Eucalypts*. Forests and Timber Bureau, Canberra, 1965. (pp. 359)

Blunt, N. *The Compleat Naturalist and a Life of Linnaeus*. William Collins, Sons & Co. Ltd, London, 1971. (pp. 256)

Bond, H. C. *The Plant Kingdom* (3rd ed.). Foundations of Modern Biology Series, Prentice-Hall Inc., 1970. (pp. 190)

Boomsma, C. D. *Native Trees of South Australia*. Woods and Forests Department, Adelaide, 1972. (pp. 224)

Brooks, A. E. *Australian Native Plants for Home Gardens*. Lothian Publishing Co. Pty Ltd, Melbourne and Sydney, 1967. (pp. 149)

Brown, A. and Hall, N. *Growing Trees on Australian Farms*. Commonwealth Government Printer, Canberra, 1968. (pp. 397)

Caborn, J. M. *Shelterbelts and Windbreaks*. Faber and Faber Ltd, London, 1965. (pp. 288)

Cain, S. A. *Foundations of Plant Geography*. Harper & Brothers, New York and London, 1944. (pp. 556)

Cayley, N. W. *What Bird Is That?* Angus and Robertson Ltd, Sydney and London, 1943. (pp. 319)

Chippendale, G. M. (Ed.). *Eucalyptus Birds and Fruits*. Forests and Timber Bureau, Canberra, 1968. (pp. 96)

Chippendale, G. M. *Eucalypts of the Western Australian Goldfields, (and the adjacent wheatbelt)*. For. and Timber Bur.; Aust. Govt. Pub. Serv. 1973. (pp. 216)

Cochrane, G. R., Fuhrer, B. A., Rotherham, E. R. and Willis, J. H. *Flowers and Plants of Victoria*. A. H. & A. W. Reed, Wellington, Auckland and Sydney, 1971. (pp. 216)

Cook, D. I. and van Haverbeke, D. F. 'Trees and shrubs for noise abatement', *Research Bulletin 2461*, Forests Service, US Department of Agriculture, 1971. (pp. 77)

Cox, W. J. 'Lead contamination in soils, plants and animals', in *Soil News*, **33**, Sept. 1972. p. 27

Crocker, R. L. 'Post Miocene climatic and geological history and its significance in relation to the genesis of the major soil types of South Australia', *Bulletin 193*, CSIRO, 1946. (pp. 66)

Curtis, O. F. and Clark, D. G. *An Introduction to Plant Physiology*. McGraw-Hill Book Company Inc., New York and London, 1950. (pp. 752)

Curtis, W. M. *The Students' Flora of Tasmania—Part 1*. Government Printer, Hobart, 1956. (pp. 240)

Dakin, W. J. *Australian Seashores*. Angus and Robertson Ltd, Sydney and London, 1953. (pp. 372)

Dallimore, W. and Jackson, A. B. *A Handbook of Coniferae*. Edward Arnold & Co., London, 1966.

Dasmann, R. F. *Environmental Conservation*. John Wiley & Sons, New York. (pp. 375)

Esau, K. *Anatomy of Seed Plants*. John Wiley & Sons, New York, 1960. (pp. 376)

Esau, K. *Plant Anatomy*. John Wiley & Sons, New York, 1965. (pp. 767)

Ewart, A. J. *Handbook of Forest Trees for Victorian Foresters*. Government Printer, Melbourne, 1925. (pp. 523)

Ewart, A. J. *Flora of Victoria*. Government Printer, Melbourne, 1930. (pp. 1257)

Falvey, D. A. 'Plate tectonics: or dynamic approach to modern geological theory', in *Australian Natural History*, **17**, 8, Dec. 1972. pp. 258-264

FAO. *Forest Influences*. Forestry and Forest Products, FAO, Rome, 1962. (pp. 307)

Flemer, W. *Shade and Ornamental Trees in Colour*. Grosset and Dunlap Inc., New York, 1965. (pp. 144)

Fogg, G. E. *The Growth of Plants*. Penguin Books Pty Ltd, Middlesex, England, 1966. (pp. 288)

Francis, D. F. and Southcott, R. V. 'Plants harmful to Man in Australia', *Mis. Pub. No. 1*, Botanic Garden, Adelaide, 1967. (pp. 52)

Francis, W. D. *Australian Rain Forest Trees*. Angus and Robertson Ltd., Sydney and London, 1951. (pp. 469)

Gates, D. N. *Energy Exchange in the Biosphere*. Reprint Series in Plant Physiology. Harper & Row, New York and London, and John Weatherhill Inc., Tokyo, 1965. (pp. 151)

Geiger, R. *The Climate Near the Ground*. Harvard University Press, Cambridge, Massachusetts USA, 1959. (pp. 494)

Goor, A. Y. and Barney, C. W. *Forest Tree Planting in Arid Zones*. Ronald Press Co., New York, 1968. (pp. 409)

Griffiths, J. R. 'Australia and Gondwanaland', in *Australian Natural History*, **17**, 8 Dec. 1972. pp. 249-253

Hall, N., Johnston, R. D. and Chippendale, G. M. *Forest Trees of Australia*. Australian Government Publishing Service, Canberra, 1970. (pp. 334)

Harlow, W. M. and Harrar, E. S. *Textbook of Dendrology*. McGraw-Hill Book Company, New York, 1958. (pp. 561)

Harrison, R. E. *Handbook of Trees and Shrubs for the Southern Hemisphere*. A. H. & A. W. Reed, Wellington, Auckland and Sydney, 1967. (pp. 347)

Harrison, R. E. and C. R. *Trees and Shrubs*. A. H. & A. W. Reed, Wellington, Auckland and Sydney, 1967. (pp. 207)

Hazlewood, W. G. *A Handbook of Trees, Shrubs and Roses*. Angus and Robertson Ltd, Sydney, 1968. (pp. 271)

Hinds, H. V. and Reid, S. S. *Forest Trees and Timbers of New Zealand*. New Zealand Forest Service, Wellington, 1957. (pp. 207)

Holliday, I. and Hill, R. *A Field Guide to Australian Trees*. Rigby Ltd, Melbourne, 1969. (pp. 227)

Hosie, R. C. *Native Trees of Canada* (7th ed.). Canadian Forestry Service. Queen's Printer, Ottawa, Canada, 1969. (pp. 380)

Hunt, P. *Garden Shrubs*. Paul Hamlyn Publishing Group Ltd, London, 1969. (pp. 160)

Jacobs, M. R. *Growth Habits of the Eucalypts*. Commonwealth Government Printer, Canberra, 1955. (pp. 262)

Jacobson, J. S. and Clyde Hill, A. (Eds.). *Recognition of Air Pollution Injury to Vegetation: a Pictorial Atlas*. Air Pollution Control Association, 1970. (pp. 111)

Kelly, S. *Eucalypts*. Text by G. M. Chippendale and R. D. Johnston. Thomas Nelson (Australia) Ltd, Melbourne and Sydney, 1969. (pp. 82 plus 250 illustrations)

Kittredge, J. *Forest Influences*. McGraw-Hill Book Company, New York, 1948. (pp. 514)

Kozlowski, T. T. *Growth and Development of Trees*. Volumes I and II of Physiological Ecology— a series of monographs, texts, and treatises. Academic Press, New York, 1971. (pp. 514)

Laidlaw, W. B. R. *Guide to British Hardwoods*. Leonard Hill (Books) Ltd, London, 1960. (pp. 240)

Laseron, C. F. and Brunnschweiler, R. O. *Ancient Australia*. Angus and Robertson Ltd, Sydney, 1969. (pp. 253)

Leach, J. A. *An Australian Bird Book*. Whitcombe and Tombs Pty Ltd, Melbourne, Sydney and Perth, 1945. (pp. 200)

Leeper, G. W. *Introduction to Soil Science*. Melbourne University Press, Melbourne, 1964. (pp. 352)

Leeper, G. W. (Ed.). *The Australian Environment*. CSIRO, Australia and Melbourne University Press, 1970. (pp. 163)

Lord, E. E. *Shrubs and Trees for Australian Gardens*. Lothian Publishing Co., Pty Ltd, Melbourne and Sydney, 1967. (pp. 462)

Lothian, T. R. M. *The Practical Home Gardener*. Lothian Publishing Co., Melbourne, 1963. (pp. 390)

Lowry, W. P. *Weather and Life: an Introduction to Biometeorology*. Academic Press, New York and London, 1970. (pp. 305)

Macoboy, S. *What Flower Is That?* Paul Hamlyn Pty Ltd, Sydney, 1969. (pp. 317)

Makin, F. K. *The Identification of Trees and Shrubs*. J. M. Dent & Sons, Ltd, London, 1936. (pp. 326)

Meyer, B. S., Anderson, D. B. and Bohning, R. W. *Introduction to Plant Physiology*. Van Nostrand & Co., London, 1968. (pp. 541)

Miller, P. R., McCutchan, M. H. and Milligan, H. P. 'Oxidant air pollution in the Central Valley, Sierra Nevada foothills, and Mineral King Valley of California', in *Atmospheric Environment*, **6**, pp. 623-633. Pergamon Press, 1972.

Mitchell, A. J. *Conifers in the British Isles*. HMSO, London, 1972 (pp. 322).

Newbey, K. *West Australian Plants for Horticulture*. Society for Growing Australian Plants and Surrey Beatty & Sons, Chipping Norton, New South Wales, 1968. (pp. 128)

Northcote, K. H. and Skene, J. K. M. 'Australian soils with saline and solodic properties *Soil Publication No. 27*, CSIRO, 1972. (pp. 62)

Ovington, J. D. 'Some factors affecting nutrient distribution within biosystems', in *Ref. Natural Resources Series*, **5**, pp. 95-105. Pear, Copenhagen, 1968.

Parsons, W. T. 'Pesticides in the control of vermin and noxious weeds', in *Victoria's Resources*, **14**, 4, Dec. 1972. pp. 13-18

Parsons, W. T. *Noxious Weeds of Victoria*. Inkata Press Pty Ltd, Melbourne, 1973. (pp. 300)

Perry, F. and Greenwood, H. *Flowers of the World*. Paul Hamlyn Publishing Group Ltd, London, 1971. (pp. 320)

Pescott, R. T. M. *Gardening in Australia*. Thomas Nelson (Australia) Ltd, Melbourne and Sydney, 1971. (pp. 205)

Pirone, P. P. *Tree Maintenance*. Oxford University Press, New York, 1959. (pp. 483)

Pitts, J. N. and Metcalf, R. L. *Advances in Environmental Sciences—Vol. I*. Wiley Interscience, New York, 1969. (pp. 356)

Pokorny, J. *Trees of Parks and Gardens*. Spring Books, London, 1967. (pp. 134)

Porter, C. L. *Taxonomy of Flowering Plants*. Freeman & Co., San Francisco, 1967. (pp. 472)

Raven, P. H. 'An introduction to continental drift', in *Australian Natural History*, **17**, 8, Dec. 1972. pp. 245-248

Redher, A. *Manual of Cultivated Trees and Shrubs Hardy in North America*. The Macmillan Company, New York, 1947. (pp. 996)

Rodway, L. *The Tasmanian Flora*. Government Printer, Hobart, 1903. (pp. 320)

Rogers, F. J. C. *Growing Australian Native Plants*. Thomas Nelson (Australia) Ltd, Melbourne and Sydney, 1971. (pp. 82)

Russell, E. J. *Soil Conditions and Plant Growth*. Longmans, Green & Co., London, 1973. (pp. 849)

Salisbury, F. B. and Ross, C. *Plant Physiology*. Wadsworth Publishing Co., Belmont, California, 1969. (pp. 763)

Sargeant, H. *Garden Trees and Shrubs in Australasia*. Specialty Press Ltd, Melbourne, 1952. (pp. 256)

Sargent, C. S. *Trees of North America—Vol. I and Vol. II*. Dover Publications Inc., New York, 1922. Reprinted 1972 (?). (pp. 910)

Seagel, R. C. *et alia*. *Plant Diversity—an Evolutionary Approach*. Wadsworth Botany Series. Wadsworth Publishing Co., Belmont, California, 1969. (pp. 460)

Simpfendorfer, K. J. *Site Index of* P. radiata D. Don *in Relation to Some Physical Factors of Site in Ovens Plantations*. Thesis M.Sc.F., University of Melbourne, 1963. (pp. 205)

Simpson, A. E. 'Park standards in Australia', *Australian Parks*, Part I. **6**, 1, Aug. 1969. pp. 7-17; Part II. **6**, 2, Nov. 1969. pp. 5-12

Sinnott, E. W. and Wilson, K. S. *Botany—Principles and Problems*. McGraw-Hill Book Company, New York and London, 1963. (pp. 515)

Smith, D. M. *The Practice of Silviculture*. John Wiley & Sons, New York and London, 1962. (pp. 578)

Sporne, K. R. *The Morphology of the Gymnosperms*. Hutchison University Library, London, 1971. (pp. 216)

Streets, R. J. *Exotic Forest Trees in the British Commonwealth*. Clarendon Press, Oxford, 1962. (pp. 765)

Thimann, K. V. (Ed.). *The Physiology of Forest Trees*, papers presented at the Symposium, Harvard Forest, 1957. Ronald Press Co., New York, 1957. (pp. 678)

Thomas, M. *Plant Physiology*. J. & A. Churchill Ltd, London, 1949. (pp. 504)

Tribe, I. *The Plant Kingdom*. Paul Hamlyn Publishing Group Ltd, London, 1970. (pp. 160)

Troughton, E. *Furred Animals of Australia*. Angus and Robertson Ltd, Sydney and London, 1946. (pp. 376)

Weaver, J. E. and Clements, F. E. *Plant Ecology*. McGraw-Hill Book Company Inc., New York and London, 1948. (pp. 601)

Wilkins, M. B. *The Physiology of Plant Growth and Development*. McGraw-Hill, London, 1970. (pp. 695)

Williams, C. H. 'Cadmium, soil contaminant?' in *Soil News*, **33**, Sept. 1972. p. 27

Willis, J. M. *A Handbook to Plants in Victoria—Vol. I* (pp. 481) and *Vol. II* (pp. 832). Melbourne

University Press, 1972.

Woodward, M. *Leaves from Gerard's Herball*. Thorsons Publishers Ltd, London, 1972. (pp. 305)

Wulff, E. V. *An Introduction to Historical Plant Geography*. Botanica, Waltham, Mass., USA, 1943. (pp. 223)

——*Woody Plant Seed Manual*. Forests Service, US Department of Agriculture, US Government Printing Office, Washington, D.C., 1948. (pp. 416)

——*Brunnings Australian Gardener*. Revised by E. E. Lord. Robertson and Mullens Ltd, Melbourne, 1958. (pp. 505)

——*Manual of Meteorology for Fire Control Officers*. Bureau of Meteorology, Commonwealth of Australia, 1964.

——*New Zealand Forestry*. Government Printer, Wellington, 1964. (pp. 204)

——*Restoring the Quality of Our Environment*. Report of Environmental Pollution Panel, President's Science Advisory Committee, The White House, Washington, 1965. (pp. 317)

——*Climatic Averages, Australia*. Bureau of Meteorology, Commonwealth of Australia, Melbourne, 1969. (pp. 107)

——*Dormancy and Survival, Symposia of the Society for Experimental Biology No. XXII*. University Press, Cambridge, 1969. (pp. 598)

——*Research into Environmental Pollution*. Report of five WHO Scientific Groups. WHO, Geneva, 1969. (pp. 83)

——*Selected Flowering Eucalypts of Western Australia*. Forestry Department of Western Australia, 1972. (pp. 46)

# APPENDICES

The following information is reproduced by kind permission of the Forests Commission, Victoria

APPENDIX 1

# SEED EXTRACTION

Seed of trees, particularly Australian native species is commonly found in a woody or leathery fruit, e.g. 'gum nuts', wattle 'pods', pine cones, etc. Normally seed is dispersed naturally but if some is required for sowing collection of the fruit needs to be done before the seed is shed.

Ripe fruit of most species is a brown colour and must be collected at this stage. If green it contains unripened seed of low or nil viability, while if grey or weathered it is the remnant of a previous crop and contains little, if any, seed. Fruits must be collected before opening on the tree; for example eucalypt and melaleuca fruits are closed by easily seen valves, in conifers the cone scales would not have separated while in banksias and hakeas the cone valves would be tightly closed.

Once collected seed can be extracted by gentle warmth. This can be done by spreading the fruit on a piece of canvas out in the sun, or in a suitable container in a sunroom, glasshouse, etc. Very low warmth in an oven is also acceptable. Temperature should not exceed 60°C otherwise the seed may be injured. In mid summer do not dry seed in a metal container or on such metal items as a sheet of tin since temperatures under these conditions can exceed 60°C.

A more convenient way is to make a simple seed extractor which uses sunlight. This consists of:

1. A bottom tray about 8–10 cm deep.
2. An upper tray to fit on top of (1) with sides about 10–15 cm high and a wire mesh bottom of a size to stop the fruit falling through, and
3. A glass top (sheet of glass, old window, etc.) to give protection from weather, birds, mice, etc.

In use, line the lower tray with paper or plastic film to prevent loss of seed through any cracks. Cooking foil should not be used because of the risk of excessive heat being generated. Put the fruit in the upper section, assemble the whole unit and place in the sun for a few days. The warmth of the sun will open the fruit and the seed will drop through on to the lower tray; an occasional 'stirring' of the fruit (once a day) will dislodge more seeds and aid extraction. Provided the unit is reasonably waterproof it need not be covered during rain.

Once extracted seed can be cleaned by sieving out the larger and smaller particles while gentle winnowing on a day with a slight breeze or an electric fan will remove the lighter fragments. Store in a container with a tight lid such as a glass jar, to prevent ants, mice, etc., gaining access to the seed, and keep in a cool place. Keeping the seed in a hot shed or where the sun can shine on it through a window may cause a sufficient rise in the seed temperature to induce dormancy or even damage.

APPENDIX 2

# RAISING CONTAINER STOCK FROM SEED

This method of propagation applies specifically to the raising of eucalyp seed-lings, but the general principles are applicable to other species normally raised in containers (pots, tubes, etc.).

Seed should be sown in late winter, that is just before the beginning of the growing season. It is sown in boxes about 10 cm in depth and of any convenient size; a box 30–50 cm square will hold several hundred seedlings. Holes about 5–10 mm diameter should be bored through the bottom of the box and a 1 cm layer of charcoal, screenings, coarse gravel, or similar material placed on the bottom to give a good drainage. Fill the box with soil using a good open loam or sandy loam mixture which drains freely. Level the soil surface, and firm it by pressing lightly with a flat board. Scatter the seed evenly over it and just cover with fine sieved soil, again pressing lightly with a flat board. Mixing the seed with 2–3 times its volume of fine dry sand will give easier and more even distri-bution. Seed should be covered to a depth of no more than about twice its diameter; one of the most frequent causes of failure particularly with eucalypts and most other native species is covering the seed too deeply. Keep moist, water with a fine rose spray, cover the box with a sheet of glass and place in a warm shaded area.

The soil should be kept moist, but not wet. When the seedlings appear, remove the glass and place the box under shade to protect the plants from sun scorch. As the seedlings develop the shade cover should be removed gradually, for an increasing period each day, to allow the plants to harden off.

When the seedlings are 3–5 cm high, that is large enough to handle by hand, they need to be transplanted into containers to enable them to develop a bushy root system. Tubes 4–5 cm diameter, and about 15 cm long are preferred but similar containers such as tins, drink cans, etc., can be used. Tubes can be rolled from wood veneer, plastic film, tin plate, tarred paper, or similar water-proof material and tied with rubber bands, staples, hayband, etc., as con-venient. The advantages of tubes over pots are that they permit greater down-ward development of the root system, require less soil, and take up less space under shade frames and during transport.

Use a good friable soil mixture for tubing; one of the best all-round soils consists of a composted mixture of black or red loam, coarse river sand and well decayed animal manure in the proportions 3 : 2 : 1. The plants should be lifted very carefully from the seed box (using a flat trowel) to avoid damaging the rootlets. Place some soil in the bottom of the tube to about a quarter of its depth, firm by tamping, then hold a seedling in the centre with the roots hang-ing free, and fill in soil around the roots. When the tube is full, firm the soil with the fingers; the soil surface should then be about 1 cm from the top of the container. Stand the plants in a shady place such as under trees, in a shade house, or under a hessian cover, and keep well-watered. As the plants grow, gradually increase the exposure. Towards the end of the growing season (summer) hardening-off needs to be initiated; this is done by reducing watering and cover but not to the point where the plants wilt or suffer other damage.

When the plants are 25–30 cm high they are ready for planting in their permanent positions. Normally they should reach this size in 3–5 months after tubing. If the plants cannot be planted out in the first year they may be trans-ferred to larger containers and held for another year. Remove the tube but retaining the cylinder of soil around the roots, transplant into a container of about the same proportions, but with 3–4 times the soil's previous volume.

Plants are frequently seen in small containers about 5–8 cm long. While such small tubes are very good for raising cuttings, growing-on stock etc, they are far from ideal for plants which are intended for planting-out under field conditions. For field plantings, a container about 15 cm length is preferred. This encourages a good downward development of roots giving more below the main grass root zone. While grass roots may extend to some depth, most of the main competitive ones are in the top 8–10 cm so the roots of plants grown in short containers would be in direct competition with the main zone of grass roots. If ground and/or growing conditions in the first season after planting are not good, heavier losses will occur in plants raised in short containers.

APPENDIX 3

# RAISING OPEN-ROOTED PLANTS FROM SEED

These notes apply to the growing of *Pinus radiata* in open beds, but with slight modification can be used for other conifers and deciduous trees which are also usually raised as open-rooted stock.

Well drained soil in a sunny but sheltered position is necessary. Prepare the soil to a fine tilth. Seed may be sown either in autumn (February–March) or spring (September–October). Prior to sowing soaking the seed in water over-night will improve germination while coating it after soaking with a fungicide (Thiram, Captan, etc.) or red lead will reduce losses from fungi, insects and birds.

Seed is sown in shallow drills about 15–20 cm apart; sow 50–60 seeds per metre of drill and cover to a depth of about 1 cm. If conditions are dry, watering immediately following sowing and again when the seedlings break through the surface is desirable. Germination and emergence of the seedlings takes 4–6 weeks after which weeding and cultivation should be undertaken regularly.

About 6–8 weeks before planting out, undercut the seedlings by pushing a spade into the ground midway between the rows and at an angle so as to cut the seedling roots at about 15 cm depth. This stimulates the formation of rootlets and root hairs prior to planting out and improves the prospects of survival.

Small quantities of seed can be sown in seed boxes. Boxes should be about 15 cm deep and filled with an open, loamy to sandy soil. Good drainage is essential. Seeds are sown at about 1000 to the square metre i.e., about 3 cm between seeds and may be sown in rows or broadcast. Treatment otherwise is the same as for open-grown plants but of course being confined to the soil in the box, undercutting is not necessary.

Seeds sown in spring should be ready for planting out the following winter, while autumn sown seeds in the winter about 15 months after sowing. Optimum size for planting out is 30–40 cm; if the plants are much taller than this they can still be planted but losses in the field are likely to be higher. It is preferable that they be trimmed to 30–40 cm by cutting off the tops with shears, mowers, or when lifted, trimming bundles of plants to about this height with a hatchet.

Some people prefer larger and sturdier plants in which case it is necessary to retain the young trees for another year in the nursery. The aim is to produce stocky, bushy plants with a strong, fibrous root system, so it is necessary to transplant the seedlings in order to give them sufficient room to develop. Lift the seedlings carefully and transplant into lines 15–20 cm apart with 8–10 cm between plants. Make sure the roots are hanging freely in all directions and are not turned up or confined to one plane. Weed and cultivate regularly to keep the area weed-free and to prevent the soil surface crusting. The young trees are retained in the transplant bed until the following winter, i.e., they will be in the nursery for 21–27 months from seed.

Transplants, usually regarded as 'two year' old plants, are larger, sturdier and withstand vermin attack more readily, but are much costlier to produce. For most purposes, particularly where the planting site can be cultivated, the 'one year' old seedling is quite adequate; the Forests Commission in its own planting programmes has used only 'one year' plants for many years.

APPENDIX 4

# TRANSPLANTING LARGE TREES

As trees get older and larger they sometimes become unsuitable for the site in which they have been planted but could be suitable for another location. Technically, many trees of reasonable age can be successfully moved provided sufficient care is taken. The limiting factors are usually the availability of adequate equipment and finance.

Generally deciduous trees transplant fairly readily, conifers require much more care, while other evergreens, including most Australian species, are often difficult. Eucalypts are very difficult and it is usually impractical to transplant large specimens successfully.

SEASON

Trees should be moved in the winter months when they are comparatively dormant. This minimizes planting shock and allows the longest possible time for roots to re-establish themselves before the summer.

PREPARATION

Before transplanting, the development of roots close to the tree, that is within the ball of soil to be removed with the tree, needs to be encouraged. During the early autumn before transplanting, trench around the tree to about 0.5 m depth. The trench should be some distance from the tree trunk; a rough rule of thumb is about 10 cm radius for each centimetre of trunk diameter at about 20 cm above ground. Do not cut the roots under the tree at this stage. On completing the trench fill it with water and allow this to soak away. Rain-water collecting in the trench need not be drained off, but if a prolonged dry spell occurs in autumn a further watering by filling the trench is desirable.

A few days before moving is due, make sure the soil within the area enclosed by the trench is damp; if not, give it a good watering by soaking the entire surface area surrounded by the trench.

At the new site dig a hole sufficiently large to give about 20 cm clearance around the ball of earth which will be enclosing the tree roots.

EQUIPMENT

As a cubic metre of damp soil weighs more than 1 tonne some fairly heavy equipment may be required. In addition to a suitable crane or lifting gear you will need:

1. Hessian or material for wrapping to prevent the soil ball breaking up. For smaller trees this may be used to lift the ball of soil and tree.
2. For larger trees some form of sling which by a series of 'straps' holds the soil ball intact, and a spreader bar to which the 'straps' can be attached.
3. Padding to protect the trunk of the tree from injury by chains, ropes and the spreader bar.

Tie back branches of the tree as far as practicable to avoid damage and to make working conditions more convenient. Attach the crane or lifting gear to the butt of the tree—suitably padded to prevent damage to the bark—and by alternately lifting and tilting the tree, cut all the roots beneath the ball of earth. During this operation work the hessian underneath the tree to the opposite side, and then wrap it around the ball of earth to prevent soil breaking away from the roots. Attach the lifting straps in suitable positions and hook up to the sling.

TRANSPORT

The tree can be transported by the crane and slings or laid on its side in a truck, depending on the distance to be moved.

### REPLANTING

Good soil is placed in the bottom of the hole to bring the tree to the original level. Place the tree upright in the hole, loosen the straps and remove the hessian by tilting the tree back and forth as previously. Support the tree by at least three stays with suitable padding inserted where they contact the tree.

The soil may now be replaced, pressing well by trampling or gentle working with a rammer. Give a thorough watering to ensure the soil is soaked to the bottom of the hole.

Some pruning of the branches is advisable (up to 30%) to reduce transpiration, but this must be done with care to preserve the shape of the tree.

### IMMEDIATE REMOVAL

The above technique is recommended as being the most reliable but it is possible to move a tree in one operation in mid-winter. However, the risk of loss is greater. Select a cool day with dull moist weather, prepare the hole at the new site, then trench around and under the tree and remove it immediately to its new position. Otherwise take the same precautions in handling as outlined previously.

### AFTER-CARE

The most frequent cause of loss is failure of the root system to extend fast enough to supply the tree's water requirements in the warmer weather. The 'ball and trench' area should therefore be kept moist, but not wet, during the next summer period.

Leave the supporting stays firmly in position until the ground has consolidated further and the tree is more wind-firm. They should be left at least until early summer, that is after the period of spring growth and winds.

APPENDIX 5

# PLANTING NEAR DRAINS AND SEWERS

Plants listed below are ones which according to the South Australian Sewerage Act 1929–1966 may be planted in the locations shown but in some cases do not conform to current usage as used in this book. This list is offered as a guide only, and while Forests Commission experience has shown it to be generally satisfactory, conditions in individual cases can vary widely so that no recommendation, implied or otherwise, can be given.

## 1. PLANTED NOT CLOSER THAN 2 M TO ANY SEWER MAIN OR CONNECTION

| BOTANICAL NAME | COMMON NAME | BOTANICAL NAME | COMMON NAME |
|---|---|---|---|
| *Acacia cultriformis* | Knife-leaf Wattle | *Eucalyptus forrestiana* | Fuchsia Gum |
| *Acacia cyclops* | Western Coastal Wattle | *Eucalyptus leptophylla* | Slender-leaf Mallee |
| *Acacia howittii* | Sticky Wattle | *Eucalyptus orbifolia* | Round-leaf Mallee |
| *Acacia iteaphylla* | Gawler Range Wattle | *Eucalyptus preissiana* | Bell-fruit Mallee |
| *Acacia longifolia* | Sallow Wattle | *Eucalyptus pyriformis* | Large-fruit Mallee |
| *Acacia microbotrya* | Manna Wattle | *Eucalyptus spathulata* | Swamp Mallet |
| *Acacia retinodes* | Wirilda | *Eucalyptus stoatei* | Pear Gum |
| *Acacia sophorae* | Coast Wattle | *Eucalyptus websteriana* | Webster Mallee |
| *Acacia sowdenii* | Western Myall | *Euonymus japonicus* | Evergreen Spindle-Tree |
| *Acacia trineura* | Three-nerved Wattle | *Feijoa sellowiana* and forms | Pineapple Guava |
| *Acacia victoriae* | Bramble Wattle | *Geijera parviflora* | Wilga |
| *Acmena (Eugenia) smithii* | Lilly Pilly | *Hakea elliptica* | Oval-leaf Hakea |
| *Acmena paniculata* | Bush Cherry | *Hakea laurina* | Pincushion Hakea |
| *Acmena coolminiana* and related species | Blue Lilly Pilly | *Hakea petiolaris* | Sea-urchin Hakea |
| *Actinostrobus pyramidalis* | Swamp Cypress-Pine | *Hakea saligna* | Willow-leaf Hakea |
| *Bauhinia variegata* and forms | Ebony-Wood | *Hakea undulata* | |
| *Betula pendula* | Silver Birch | *Hakea sulcata* | |
| *Callistemon citrinus* | Crimson Bottlebrush | *Koelreuteria paniculata* | Golden Rain Tree |
| *Callistemon* 'Gawler Hybrid' | 'Gawler Hybrid' | *Jacaranda* spp. | Jacaranda |
| *Callistemon lilacinus* | Lilac Bottlebrush | *Lagerstroemia indica* | Pin Crepe-Myrtle |
| *Callistemon macropunctatus* | Scarlet Bottlebrush | *Leptospermum laevigatum* | Coast Tea-Tree |
| *Callistemon phoeniceus* | Fiery Bottlebrush | *Malus* spp. | Crab-apple |
| *Callistemon rigidus* | Stiff Bottlebrush | *Melaleuca elliptica* | Granite Honeymyrtle |
| *Callistemon salignus* | Willow Bottlebrush | *Melaleuca fulgens* | Scarlet Honeymyrtle |
| *Callistemon viminalis* and similar species | Weeping Bottlebrush | *Melaleuca hypericifolia* | Grey Honey Myrtle |
| *Cercis siliquastrum* | Judas Tree | *Melaleuca incana* | Red Honeymyrtle |
| *Citharexylum* spp. | Fiddlewood | *Melaleuca lateritia* | Robin Redbreast Bush |
| *Cotoneaster frigida* | Cotoneaster | *Melaleuca nesophila* | Showy Honeymyrtle |
| *Crataegus lavallei* | French Hawthorn | *Melaleuca pentagona* | Oval-leaf Honeymyrtle |
| *Crataegus oxyacantha* and other forms of Crataegus | English Hawthorn | *Melaleuca radula* | Graceful Honeymyrtle |
| *Duranta repens* | Sky Flower | *Nerium oleander* | Oleander |
| *Eucalyptus caesia* | Gungurru | *Photinia serrulata* | Chinese Hawthorn |
| *Eucalyptus calycogona* | Red Mallee | *Pittosporum crassifolium* and variegated form | Karo |
| *Eucalyptus cosmophylla* | Cup Gum | *Pittosporum phylliraeoides* | Weeping Pittosporum |
| *Eucalyptus crucis* | Silver Mallee | *Pittosporum tenuifolium* | Kohuhu |
| *Eucalyptus dielsii* | Diel Mallee | *Prunus* spp. | Flowering fruit trees |
| *Eucalyptus eremophila* | Tall Sand Mallee | *Pyracantha coccinea lalandi* | Fire-thorn |
| *Eucalyptus erythrocorys* | Red-cap Gum | *Pyracantha crenulata* | Nepal Firethorn |
| *Eucalyptus erythronema* | Lindsay Gum | *Pyracantha rodgersiana* | Yellow-Berry Firethorn |
| | | *Sophora tetraptera* | Yellow Kowhai |
| | | *Spartium junceum* | Spanish Broom |
| | | *Stenolobium stans velutina* | Florida Yellow-Trumpet |

| BOTANICAL NAME | COMMON NAME |
|---|---|
| *Stenolobium alatum* | Winged Yellow-Trumpet |
| *Viburnum tinus* | Laurustinus |
| *Vitex agnus-castus* | Lilac Chaste-tree |

## 2. PLANTED NOT CLOSER THAN 4 M TO ANY SEWER MAIN OR CONNECTION

| BOTANICAL NAME | COMMON NAME |
|---|---|
| *Acacia acuminata* | Raspberry-jam Wattle |
| *Acacia cyanophylla* | Orange Wattle |
| *Acacia pendula* | Weeping Myall or Boree |
| *Acacia salicina* | Willow Wattle |
| *Acacia saligna* | Golden Wreath Wattle |
| *Acer negundo* | Box-Elder Maple |
| *Agonis flexuosa* | Willow-Myrtle |
| *Albizzia julibrissin* | Pink Silk Tree |
| *Angophora cordata* | Dwarf Apple Myrtle |
| *Angophora costata* | Smooth Barked Apple Myrtle |
| *Arbutus unedo* | Strawberry Tree |
| *Bauhinia carroni* and related species | Queensland Bean or Ebony |
| *Brachychiton acerfolius* | Flame Tree |
| *Brachychiton discolor* | White Kurrajong |
| *Callitris columellaris* | Murray Pine |
| *Brachichiton populneus* | Kurrajong |
| *Casuarina cristata* | Belah |
| *Casuarina stricta* | Drooping She-oak |
| *Casuarina torulosa* | Rose She-oak |
| *Celtis australis* | Nettle-Tree |
| *Celtis occidentalis* and related species | Hackberry or Sugarberry |
| *Cotoneaster serotina* | Cotoneaster |
| *Erythrina indica hybrida* | Indian Coral-tree |
| *Eucalyptus cneorifolia* | Kangaroo Island Narrow-leaf Mallee |
| *Eucalyptus conglobata* | Port Lincoln Mallee |
| *Eucalyptus ficifolia* | Red-flowered Gum |
| *Eucalyptus gardneri* | Blue Mallet |
| *Eucalyptus incrassata* | Yellow Mallee |
| *Eucalyptus intertexta* | Gum-barked Coolabah |
| *Eucalyptus lansdowneana* | Crimson Mallee Box |
| *Eucalyptus lehmannii* | Bushy Yate |
| *Eucalyptus leucoxylon rosea* | Red-flowered Yellow Gum |
| *Eucalyptus megacornuta* | Warty Yate |
| *Eucalyptus nutans* | Red-flowered Moort |
| *Eucalyptus pileata* | Ravensthorpe Mallee |
| *Eucalyptus platypus* | Round-leaf Moort |
| *Eucalyptus sargentii* | Sargents Mallet |
| *Eucalyptus sideroxylon* | Red Ironbark |
| *Eucalyptus steedmanii* | Steedmans Gum |
| *Eucalyptus stricklandii* | Stricklands Gum |
| *Eucalyptus torquata* | Coral Gum |
| *Eucalyptus viridis* | Green Mallee |
| *Ficus rubiginosa variegata* | Variegated Rusty-Fig |
| *Fraxinus excelsior* var. *aurea* | Golden Ash |
| *Fraxinus ornus* | Manna Ash |
| *Hakea suaveolens* | Sweet Hakea |
| *Harpephyllum caffrum* | Kaffir Plum |

| BOTANICAL NAME | COMMON NAME |
|---|---|
| *Hymenosporum flavum* | Native Frangipani |
| *Ligustrum japonicum* and variegated forms | Golden Hedge Privet |
| *Ligustrum lucidum* and variegated forms | Glossy Tree Privet |
| *Liquidambar styraciflua* | Liquidamber |
| *Melaleuca alternifolia* | |
| *Melaleuca armillaris* | Bracelet Honeymyrtle |
| *Melaleuca halmaturorum* | Kangaroo Paperbark |
| *Melaleuca huegelii* | Chenille Honeymyrtle |
| *Melaleuca lanceolata* | Moonah |
| *Melaleuca linariifolia* var. *trichostachya* | Flax-leaf Paperbark |
| *Metrosideros excelsa* | N.Z. Christmas Tree |
| *Myoporum insulare* | Boobialla |
| *Myoporum montanum* | Water Bush |
| *Parkinsonia aculeata* | Jerusalem Thorn |
| *Pittosporum rhombifolium* | Queensland Pittosporum |
| *Pittosporum undulatum* | Sweet Pittosporum |
| *Pittisporum* 'Variegata' | Variegated Pittosporum |
| *Sophora japonica* | Pagoda Tree |
| *Sorbus aucuparia* | Rowan Ash |
| *Tamarix juniperina* | Flowering Tamarisk |
| *Tristania conferta* | Brush box |

## 3. SHOULD NOT BE PLANTED IN ANY STREET OR ROAD IN A DRAINAGE AREA

| BOTANICAL NAME | COMMON NAME |
|---|---|
| *Araucaria heterophylla* | Norfolk Island Pine |
| *Casuarina cunninghamiana* | River She-Oak |
| *Casuarina glauca* | Grey Buloke |
| *Ficus*—all species | Fig |
| *Eucalyptus bridgesiana* | But But |
| *Eucalyptus camaldulensis* | River Red Gum |
| *Eucalyptus citriodora* | Lemon Scented Gum |
| *Eucalyptus cladocalys* | Sugar Gum |
| *Eucalyptus cornuta* | Yate |
| *Eucalyptus maculata* | Spotted Gum |
| *Eucalyptus occidentalis* | Swamp or Flat-topped Yate |
| *Eucalyptus salmonophloia* | Salmon Gum |
| *Fraxinus oxycarpa* | Desert Ash |
| *Fraxinus raywoodii* (unless grafted or budded onto certified *Fraxinus ornus* (Manna Ash) root stock) | Claret Ash |
| *Lagunaria patersoni* | Pyramid Tree |
| *Platanus*—all species | Plane Tree |
| *Populus nigra* and related species | Poplar |
| *Robinia pseudoacacia* | False Acacia |
| *Salix babylonica* and related species | Weeping Willow |
| *Schinus molle* | Pepper Tree |
| *Tamarix aphylla* | Athel Tree |
| *Ulmus procera* and related species | English Elm |

APPENDIX 6

# COMMERCIAL POPLAR CULTIVATION

For successful commercial plantations, poplars require warm, moist, well-drained soils with a good water supply. For optimum growth a high evenly distributed annual rainfall of about 1 300 mm is desirable, or the equivalent in irrigation water should be available. Sandy loam soils are preferred; on heavy loam or clay loam types growth can be restricted. The best sites then are also suitable for traditional high value farming but, as in other countries, poplar growing can be used as a means of diversifying farm income.

For the profitable production of commercial timber it is important that the right type or strain is planted. Many varieties of poplars are quite suitable for amenity planting as shelter, shade and beauty but growth rate would not be fast enough for profitable timber production nor do they grow well under plantation conditions. *Populus deltoides* var. *angulata* (South Africa strain) and strain 1488 are tolerant of a wide range of geographical conditions and do well under irrigation in northern Victoria. Southern Victoria is a little too cold for growth and at present it is suggested that commercial poplar growing should be limited to suitable irrigated areas of the Murray Valley using the South African strain of *P. deltoides* var. *angulata*, 1488, or 1488 Cobram semi-evergreen. The main commercial plantation in Victoria is operated by Brymay Forests Pty. Ltd. between Cobram and Yarrawonga on the Murray River.

During the summer of 1971–72 the fungal disease Golden Glow Rust (*Melampsora medusae* and *M. laricii-populina*) was first identified in eastern Australia. These diseases can cause serious defoliation and although up to May 1975 no grave problems have resulted from them in Victorian plantations, efforts are being made to develop resistant strains which will grow sufficiently fast to make commercial plantations profitable. Until these have been evolved the above strains are still the best but some caution is needed in considering new commercial ventures.

Poplars require intensive care for most of their growing period of 15 to 20 years. The planting site needs to be deeply cultivated before planting. Poplars are planted out as cuttings without roots or as sets (rooted cuttings). Rooted cuttings are grown in nursery beds for 9–12 months before transplanting and give a better strike. Normal spacing is 4 m by 4 m to 5.5 m by 5.5 m (600 and 400 trees per hectare respectively). Competition from weed growth should be eliminated for at least two years while young plants need to be protected from stock, rabbits and hares for the first few years. Grazing may be permitted at a later stage.

Pruning, that is removal of the lower branches flush with the trunk, needs to be undertaken regularly and frequently. Since branches produce knots the object of pruning is to ensure that future wood growth is free of knots, as poplar wood, to command a good price, must be free of knots. Normally pruning is commenced at about two years. At least annual pruning is necessary, sometimes 2–3 times a year if the growth rate is fast, but it should not at any stage exceed half the height of the tree otherwise growth will be reduced. As the height increases ladders will need to be used. Epicormic shoots which may develop soon after each pruning must be rubbed off within a few weeks of their appearance.

Thinning is required to keep the tree growing quickly. On good sites thinning to about one third the original number of trees should be undertaken at four years and thereafter at about every 3–4 years, removing about a third of the trees remaining at each thinning. Final clear felling is normally at 15 years.

Poplar wood if clear of defects such as knots and shakes is used for veneers and plywood, or if some defects are present for joinery and cabinet timbers. Reject material or small logs can be used for pulpwood, wood wool, 'aspen pads' for air conditioners, and similar uses. In most cases because of the high value of the land required and high maintenance costs it will be necessary to manage the plantation for the highest yield of defect-free timber. While at this time the market is irregular, there is no doubt that good quality poplar timber will sell readily when available in a reasonable volume.

# GLOSSARY OF TECHNICAL TERMS

Many terms are defined in the text where indexed.

alien.   plant not a native of the country where found.

alternate.   borne singly at each node.

anther.   the part of the stamen which contains pollen

axil.   the angle between a leaf and a stem.

biological control.   the use of living organisms to control pests, animals or plants.

bract.   a modified leaf subtending a flower or flower cluster.

bole.   stem or trunk.

C.   conifer.

calyx.   the whorl of sepals surrounding the petals of a flower.

catkin.   a deciduous spike of unisexual flowers.

coppice.   (verb) to put out copious new shoots from cut off trunks or major branches; to sucker; such shoots are called suckers.

corolla.   the whorl of petals in a flower.

corymb.   a flat-topped racemose inflorescence.

D.   deciduous; losing leaves in autumn.

downy.   densely covered with short soft hairs.

E.   evergreen; not losing leaves in autumn.

entire.   unbroken; the leaf edge is a smooth line, contrast 'serrate'.

escape.   (noun), a cultivated plant gone wild or naturalised.

family.   a group of allied genera of plants with similar characteristics.

flora.   the plants of a particular region or epoch.

funicle.   thread attaching seed to pod.

genus. (plural, genera).   group of plants with common structural characters distinct from those of any other group (the genus ranks next under the family and above the species.)

glabrous.   hairless.

glaucous.   of a bluish colour.

H.   hardwood.

hydrolysis.   a double decomposition with water as one of the reactants.

inflorescence.   arrangement of flowers of a plant in relation to axis and to each other; collective flower as contrast solitary flower.

juvenile regrowth.   as promoted by pollarding; in many eucalypts the leaves of such vigorous young growths differ markedly in shape from the adult leaves.

L.C.   long cultivated.

metabolism.   process by which nutritive material is built up into living matter or broken down into simpler substances.

morphology.   study of the form of plants and animals.

mutant.   a new species or form produced by accidental or artificially induced change within a cell nucleus.

native.   in Australian usage means a plant naturally occurring in the country or district before introductions by white men.

naturalised.   alien plant now perpetuating spontaneously in the country or district.

node.   joint on a stem where a leaf or branch arises.

nitrify.   botanical meaning is to combine atmospheric nitrogen in or in the vicinity of the plant's roots into a form assimilable by the plant.

operculum.   cap covering the stamens in a bud of the genus *Eucalyptus*.

order.   a group of plant families with similar characteristics.

ovary.   lower part of the pistil consisting of one or more carpels; seed vessel.

ovule.   female germ cell contained in carpel.

palmate.   shaped as a hand with fingers.

panicle.   a branched raceme or corymb.

petal.   one of the parts of the corolla, often coloured.

pinnate.   a compound leaf with leaflets on opposite sides of a common stalk. (as a feather).

pistil.   female organ of flower comprising stigma style and ovary.

pollarded.   tree polled so as to produce close rounded head of young branches.

pollen.   fine powdery grains containing male germ cell discharged from anther.

pollinate.   shed pollen upon.

propagate.   multiply specimens of plants from parent stock by seeds or asexually by cuttings, layering, grafting.

provenances.   forms originating from a particular geographical region.

raceme.   stalked flowers arranged on a prolonged axis, the lower flowers opening first.

rhizome.   underground stem which produces leafy shoots and true roots.

363

secretary tissues.   plant parts adapted as a gland to produce and discharge special products.

sepals.   the parts of the calyx.

serrate.   as a saw with toothed edge.

shrub.   woody plant of less size than a tree and usually divided into separate stems from near the ground.

solitary flowers.   single on a stem as contrast those in an inflorescence.

spine.   sharp pointed stiff woody structure.

stamen.   the male organ of the flower.

species.   a division of a genus, members differing only in minor details.

stigma.   receptive surface of female organ of flower where pollen germinates.

succulent.   fleshy and soft.

sucker.   see coppice.

synonym.   botanical usage includes unscientific or superseded but well known equivalent names.

viability.   (of a sample of seeds), is usually expressed as the percentage which germinates in a standard test.

whorl.   ring of leaves or other organs around stem & etc. of plant.

windfirm.   resistant to wind damage.

# INDEX

AN INTRODUCTION TO TREES FOR SOUTH EASTERN AUSTRALIA

TEMPLETONIA 257, 311, 346
  *T. retusa* 257
Templetonia
  Red 257
texture of soil 62
thallophyta
  algae 21
  fungi 21, 22
  lichens 22
  male and female gametes 23
Thorn
  Cockspur 173
  Washington 173
THRYPTOMENE 257, 311, 346
  *T. calycina* 257
Thryptomene
  Grampian 257
Thuja
  Mock 257
THUJA 257, 311, 346
  *T. plicata* 257
THUJOPSIS 257, 311, 346
  *T. dolobrata* 257
TIEGHEMOPANAX 257, 311, 346
  *T. sambucifolius* 257
TILIA 258, 311, 346
  *T. x vulgaris* 258
Totara 241
trace elements 52, 53, 54, 81
  in fertilizer 83
transpiration 41
transplanting trees 32
Tree Caper 162
tree groups
  conifers—hardwoods 274
  deciduous—evergreen 275
  native introduced 274
Tree Hakea 217
Tree Lucerne 170
Tree of Heaven 148
tree planting 85
  diagrams 86, 87
tree structure, Fig. 5 30
  annual rings 34
  crown 36
  fibro vascular system 32
  hardwoods, Fig. 7 35
  softwoods, Fig. 6 34
  stem (trunk) 32–34
  wood structure 34
    see also cell structure
    and growth
trees, lists of
  common names 267–273
  characteristics 278–312
  synonyms 265–266
  uses 313–347
TRISTANIA 258, 311, 346
  *T. conferta* 258
  *T. laurina* 258
trunk 32, 33

see tree structure
TSUGA 258, 311, 346
  *T. heterophylla* 258
Tuart 193
Tulip Tree 225
Tupelo 232
Turpentine 255
ULMUS 259, 311–12, 346–7
  *U. glabra* 259
  *U. x hollandica* 259
  *U. parvifolia* 259
  *U. procera* 259
  *U. procera* var. *van houttei* 260
  *U. pumila* 260
Umbrella Cedar 230
uses of trees
  see tables 313–347
variety 26
vegetative propagation 122
VIMINARIA 260, 312, 347
  *V. juncea* 260
VIRGILIA 260, 312, 347
  *V. divaricata* 260
  *V. oroboides* 260
Virgilia 260
Virgilia
  Cape 260
  Pink 260
viruses 20
Wait-a-while 133
Wallowa 132
Wallace's Line 15
Walnut
  Black 220
Wandoo 210
Waratah
  N.S.W. 256
Washington Thorn 173
Water Bush 215
water in function of
  carrying nutrients 42
  evaporation and stomata,
    Fig. 9 44
  osmosis 39
  photosynthesis 47–50
  respiration 50
  transpiration 41
    see also growth
water in soil 63–65
water table, perched 81
water use
  by deciduous trees 43
  by evergreen trees 43
Wattle
  Awl leaf 143
  Black 138
  Black, Early 134
  Boomerang 131
  Box Leaf 132
  Bramble 144
  Cape 149
  Cedar 143

Coastal 138
Coastal, W.A. 134
Cootamundra 131
Crowded Leaf 133
Deane 134
Drummond 135
Flax 137
Flinders 139
Fringed 135
Frosty 140
Gawler Range 137
Gladstone 145
Gland 136
Golden 141
Golden Glory 131
Golden Rain 140
Golden Wreath 142
Graceful 134
Hairy 140, 144
Hairy Pod 136
Hakea 136
Hickory 140
Hickory, Mountain 139
Hindmarsh 143
Hop 142
Jamwood 131
Jones 137
Knife Edge 133
Leper 137
Mallee 139
Manna 139
Mealy 135
Mount Morgan 140
Mudgee 142
Myrtle 139
Orange 133
Perfume 135
Red 142
Red Stem 141
Rice 141
River 133
Silver 134
Silver, Queensland 140
Snowy River 132
Spur Wing 143
Sticky 136
Sunshine 132
Sweet 142
Sword Leaf 136
Umbrella 139
Varnish 144
Wallangarra 131
Wavy Leaf 144
Wyalong 133
Waxflower, Geraldton 170
weedicides 85, 87
Weeping Myall 140
White Cedar 230
windbreaks 100–103
  see Fig. 15 102
Wilga 214
Willow